About Island Press

Island Press is the only nonprofit organization in the United States whose principal purpose is the publication of books on environmental issues and natural resource management. We provide solutions-oriented information to professionals, public officials, business and community leaders, and concerned citizens who are shaping responses to environmental problems.

Since 1984, Island Press has been the leading provider of timely and practical books that take a multidisciplinary approach to critical environmental concerns. Our growing list of titles reflects our commitment to bringing the best of an expanding body of literature to the environmental community throughout North America and the world.

Support for Island Press is provided by the Agua Fund, The Geraldine R. Dodge Foundation, Doris Duke Charitable Foundation, The Ford Foundation, The William and Flora Hewlett Foundation, The Joyce Foundation, Kendeda Sustainability Fund of the Tides Foundation, The Forrest & Frances Lattner Foundation, The Henry Luce Foundation, The John D. and Catherine T. MacArthur Foundation, The Marisla Foundation, The Andrew W. Mellon Foundation, Gordon and Betty Moore Foundation, The Curtis and Edith Munson Foundation, Oak Foundation, The Overbrook Foundation, The David and Lucile Packard Foundation, Wallace Global Fund, The Winslow Foundation, and other generous donors.

The opinions expressed in this book are those of the author(s) and do not necessarily reflect the views of these foundations.

About SCOPE

The Scientific Committee on Problems of the Environment (SCOPE) was established by the International Council for Science (ICSU) in 1969. It brings together natural and social scientists to identify emerging or potential environmental issues to address jointly the nature and solution of environmental problems on a global basis. Operating at an interface between the science and decision-making sectors, SCOPE's interdisciplinary and critical focus on available knowledge provides analytical and practical tools to promote further research and more sustainable management of the earth's resources. SCOPE's members, thirty-eight national science academies and research councils and twenty-two international scientific unions, committees, and societies, guide and develop its scientific program.

SCOPE 67

Sustainability Indicators

SCOPE Series

SCOPE 1 – 59 in the series were published by John Wiley & Sons, Ltd., U.K. Island Press is the publisher for SCOPE 60 as well as subsequent titles in the series.

SCOPE 60: *Resilience and the Behavior of Large-Scale Systems,* edited by Lance H. Gunderson and Lowell Pritchard Jr.

SCOPE 61: *Interactions of the Major Biogeochemical Cycles: Global Change and Human Impacts,* edited by Jerry M. Melillo, Christopher B. Field, and Bedřich Moldan

SCOPE 62: *The Global Carbon Cycle: Integrating Humans, Climate, and the Natural World,* edited by Christopher B. Field and Michael R. Raupach

SCOPE 63: *Invasive Alien Species: A New Synthesis,* edited by Harold A. Mooney, Richard N. Mack, Jeffrey A. McNeely, Laurie E. Neville, Peter Johan Schei, and Jeffrey K. Waage

SCOPE 64: *Sustaining Biodiversity and Ecosystem Services in Soils and Sediments,* edited by Diana H. Wall

SCOPE 65: *Agriculture and the Nitrogen Cycle: Assessing the Impacts of Fertilizer Use on Food Production and the Environment,* edited by Arvin R. Mosier, J. Keith Syers, and John R. Freney

SCOPE 66: *The Silicon Cycle: Human Perturbations and Impacts on Aquatic Systems,* edited by Venugopalan Ittekkot, Daniela Unger, Christoph Humborg, and Nguyen Tac An

SCOPE 67: *Sustainability Indicators: A Scientific Assessment,* edited by Tomáš Hák Bedřich Moldan, and Arthur Lyon Dahl

SCOPE 67

Sustainability Indicators

A Scientific Assessment

Edited by
Tomáš Hák
Bedřich Moldan
Arthur Lyon Dahl

A project of SCOPE, the Scientific Committee on
Problems of the Environment, of the
International Council for Science

UNEP

ICSU
International Council for Science

European Environment Agency

IHDP

ISLANDPRESS
Washington • Covelo • London

Library of Congress Cataloging-in-Publication Data

Measuring progress towards sustainability : assessment of indicators : a project of SCOPE, the Scientific Committee on Problems of the Environment, of the International Council for Science / edited by Tomáš Hák, Bedřich Moldan, and Arthur L. Dahl.
    p. cm.
Includes bibliographical references and index.
ISBN-13: 978-1-59726-130-2 (acid-free paper)
ISBN-10: 1-59726-130-0 (acid-free paper)
ISBN-13: 978-1-59726-131-9 (pbk. : acid-free paper)
ISBN-10: 1-59726-131-9 (pbk. : acid-free paper)
1. Environmental indicators. 2. Sustainable development—Evaluation.
3. Environmental monitoring—Evaluation. I. Hák, Tomáš. II. Moldan, Bedřich.
III. Dahl, Arthur L. IV. International Council for Science. Scientific Committee on Problems of the Environment.
GE140.M44 2007
333.7—dc22
                        2006035641

British Cataloguing-in-Publication data available.

Printed on recycled, acid-free paper

Manufactured in the United States of America

10 9 8 7 6 5 4 3 2 1

Contents

List of Figures, Tables, Boxes and Appendices

Figures

Tables

Boxes

Appendices

Foreword: Finding the Right Indicators for Policymaking

Jacqueline McGlade

Degradation and extreme alterations to the natural environment pose some of the deepest challenges to modern society (Vitousek et al. 1997). The effects of humans on the planet can be found everywhere, from the interstices of the polar ice caps to the depths of the oceans. Although many governments and institutions have accepted that action must be taken to tackle the most urgent problems, increasing levels of consumerism and the inexorable drive to improve the living conditions of people in the developing world mean that society is being pushed up against a wide range of environmental limits. This is the challenge facing sustainable development.

The sheer scale of the flow of materials from nature to society and back is remarkable: Even in the most modern and efficient industrial economies, the average per capita consumption is 60,000 kg of natural resources per year, the weekly equivalent of 300 shopping bags filled with materials, or the weight of a luxury car. Given population growth, resource use will have to become much more efficient by 2030 just to keep environmental degradation at present levels (Daly 1997; McGlade 2002).

What can also be observed in many parts of the world is that through our ability to manipulate and alter the fundamental relationships underpinning the planet's ecosystems, we have begun to expose ourselves to a variety of gradual and unexpected ecological changes leading to the loss, severe decline, and shifts in the ecosystem services on which we rely (Gewin 2002; Ayensu et al. 1999; Millennium Ecosystem Assessment 2003; EEA 2004).

There is a growing body of evidence from many bioregions that the accumulation of small, seemingly insignificant changes can lead to greenlash: flips or dramatic shifts in the structure and dynamic behavior of ecosystems (Rand et al. 1994; McGlade 1999). Greenlash can happen without warning; sometimes this is because we inadvertently lose a set of functional relationships or keystone species, which hold together networks of feeding relationships, because they are geographically distant or hidden within trophic dynamics (Hogg et al. 1989). It has also been observed that the removal or loss of keystone species can cause irreversible changes. This was noted in Paine's early work (1974) and in many other areas. For example, in lakes (Carpenter et al. 1985) and the Sea of Azov, large-scale hydrographic changes caused by increased use of freshwater from rivers for domestic, industrial, and agricultural purposes led to significant increases in salinity, which caused the loss of the key planktonic food items for the major fish species and the collapse of many fisheries (Mee 2001).

Many changes can occur without early warning signals. Changes in climate, levels of toxic chemicals, habitat fragmentation, and loss of biodiversity often appear to occur gradually, but ecosystem responses can be striking and sudden. Predicting which types of change will occur and over what time and space scales is fundamental to protecting our environment. Sentinel indicators—those that capture the dynamics of change—are essential in this context and may not coincide with any keystone species.

Detecting Changes on Different Time and Space Scales

Ecosystems have different levels of resilience, resistance, and hysteresis. Long-term data series can help predict which responses are most likely to occur, but often it is phenomena at the margins or on local scales that give us the clues. Changes in local conditions can result in local extinctions for certain species, but in most instances their loss is rapidly made up by surrogates. However, in some instances declines in species affected by long-range processes may not always be so easy to detect (Keeling et al. 1997; McGlade 1999).

Environmental degradation and changes such as global warming, the depletion of the ozone layer, and the presence of toxic polychlorinated biphenyls in Antarctica have arisen because of activities within national boundaries, often thousands of miles away, and a misdirected sense of concreteness in overall policy thinking.

Unfortunately, in many of today's environmental institutions there is still a belief that models coupled with management intervention can lead to predictable outcomes. But well-structured theories are conspicuous by their absence in environmental management, and many of the models used include only a limited number of possible future states. So it is extremely important to understand which indicators can best provide early and maturing signals of change.

The other downside of our false sense of security in interventions is that they very often go wrong: The introduction of rabbits and cane toads and the inadvertent transport of alien species around the world are constant reminders. Unfortunately, such experiences seem to have taught us nothing. It seems that the road to ecological disaster is littered with good intentions.

In the past, environmental decision making was made on an ad hoc basis, solving each particular problem in isolation from others. Now, however, a more profound thinking is needed about production and consumption patterns and how we can support different societies without engendering significant unintended shifts in the biosphere. Laws and institutions, no matter how efficient or well arranged, must be reformed or abolished if they are unresponsive. Overexploitation and misuse of resources must be curtailed or prohibited if they cause fundamental harm to environmental processes, but we need indicators of change to guide us along the way (McGlade 2001).

Sustainability Indicators

It is increasingly important for sustainability policies to be supported by information flows from heterogeneous sources. Whether these relate to economic, social, or environmental processes, they will need to be monitored in a transparent way, through electronic transactions across a wide range of communication media.

Indicators represent, at root, an approach designed to meet this challenge. The majority of sustainability indicators are derived from separate analyses of economic, social, and natural processes. In some instances, however, the indicators are integrated across more than one domain. Those charged with delivering sustainability are seeking this connectivity across varying levels of complexity and scale.

A characteristic of indicators is that they allow an expanding set of sentinel observations to be drawn into policymaking. As new knowledge becomes available or the focus of decision making shifts, underpinning data flows can be augmented or replaced. Indicators can be descriptive, related to performance, efficiency, policy effectiveness, or overall welfare, but in the context of sustainability it is their integration across different policy arenas that is most critical.

Perhaps the biggest bottleneck facing us today is our ability to choose the right signal or indicator to make a decision at the right time. What is needed is an indicator framework in which to successfully monitor, learn, decide, and act, to be able to obtain a clear view of where current and proposed policies are taking society.

Sustainability Indicator Framework

The main purpose of any sustainability indicator framework is to provide a comprehensive and highly scalable information-driven architecture that is policy relevant and understandable to members of society and will help people decide what to do.

It must contain sentinel indicators, ones that directly reflect changes across significant areas of interest to society and can be communicated easily. In this way we will be able to learn about changes and interpret the various forms of information as clear views of progress to date and possible future directions. In this way we will be able to achieve balance in our actions.

The framework must cover an end-to-end process—monitor, assess and learn, decide, and act—and be transparent throughout. The sustainability indicator framework calls for a modular approach, allowing new modules to be introduced, taking advantage of core infrastructures, reducing costs, identifying risks, and integrating different processes into the cycle. A modular approach allows continual refinement and improvement even for issues that do not yet warrant a dedicated monitoring solution or have not been anticipated. Throughout each step, the identification of sentinel information, its verification, and the links to policies and other indicators of economic, social, and environmental health must be accessible. Peer review and societal acceptance are key elements in building confidence in the use of sustainability indicators, which in the end must give us a clear view of where policies are taking society.

Summary

As is widely recognized, sustainable development must be central to its vision and practice. The objectives of economic prosperity, social well-being, and environmental recovery and protection must be better integrated into our practices and policies.

The enlargement and review of the European Union sustainable development strategy provide a unique opportunity to reinforce sustainable development. At the moment, however, many national sustainable strategies obscure a number of important challenges: how to turn ambitions into actions, how to ensure effective policy coherence, and how to best provide a focus and set priorities. Most strategies have been led by environment ministers and remain silent on how sustainable development priorities are to be integrated into the budgetary process. The strategies often are unclear as to how the costs and benefits of policies, including inaction, across different sectors can be systematically assessed to allow informed decisions to be made. The majority of strategies are too all-embracing and run the risk of poor implementation. A key consideration is how to harness the existing momentum at national and international levels so that they can become mutually reinforcing. The different strategies should define a common vision, encourage the creation and regular updating of information on sustainable development, reinforce progress using relevant indicators, promote leadership for sustainability, and create a wider public understanding.

Literature Cited

Ayensu, E., D. R. Claasen, M. Collins, A. Dearing, L. Fresco, M. Gadgil, H. Gitay, G. Glaser, C. Juma, J. Krebs, R. Lenton, J. Lubchenco, J. A. McNeely, H. A. Mooney, P. Pinstrup-Andersen, M. Ramos, P. Raven, W. V. Reid, C. Samper, J. Sarukhán, P. Schei, J. G. Tundisi, R. T. Watson, and A. H. Zakri. 1999. International ecosystem assessment. *Science* 286:685–686.

Carpenter, S. R., J. F. Kitchell, and J. R. Hodgson. 1985. Cascading interactions and lake productivity. *BioScience* 35:634–639.

Daly, G. C. (ed.). 1997. *Nature's services: Societal dependence on natural systems.* Washington, DC: Island Press.

EEA (European Environment Agency). 2004. Impacts of Europe's changing climate. An indicator-based assessment. *EEA Report* 2.

Gewin, V. 2002. Ecosystem health: The state of the planet. *Nature* 417:112–113.

Hogg, T., B. A. Huberman, and J. M. McGlade. 1989. The stability of ecosystems. *Proceedings of the Royal Society of London* B 237:43–51.

Keeling, M. J., I. Mezic, R. J. Hendry, J. M. McGlade, and D. A. Rand. 1997. Characteristic length scales of spatial models in ecology via fluctuation analysis. *Philosophical Transactions of the Royal Society of London* B 352:1589–1601.

McGlade, J. M. (ed.). 1999. *Advanced ecological theory*. Oxford: Blackwell Science.

McGlade, J. M. 2001. Governance and sustainable fisheries. Pp. 307–326 in *Science and integrated coastal management*, edited by B. von Bodungen and R. K. Turner. Berlin: Dalhem University Press.

McGlade, J. M. 2002. Primacy of nature: Earth democracy. *Resurgence* 214:40–41.

Mee, L. 2001. Eutrophication in the Black Sea and a basin-wide approach to its control. Pp. 71–92 in *Science and integrated coastal management*, edited by B. von Bodungen and R. K. Turner. Berlin: Dalhem University Press.

Millennium Ecosystem Assessment. 2003. *Ecosystem studies: Ecosystem science and management*. Washington, DC: Island Press.

Paine, R. T. 1974. Intertidal community structure: Experimental studies on the relationship between a dominant competitor and its principal predator. *Oecologia* 15:93–120.

Rand, D. A., H. Wilson, and J. M. McGlade. 1994. Dynamics and evolution: Evolutionarily stable attractors, invasion exponents and phenotype dynamics. *Philosophical Transactions of the Royal Society of London* B 343:261–283.

Vitousek, P. M., H. A. Mooney, J. Lubchenco, and J. M. Melillo. 1997. Human domination of Earth's ecosystems. *Science* 277:494–499.

Preface

A number of intergovernmental organizations and national governments, but also regional and local authorities, local communities, business organizations, other economic actors, academic institutions, and nongovernment organizations of many kinds, are developing and using sustainability indicators. At present, hundreds of different indicators and indices have been suggested and are used in many varied contexts, by different users, for diverse purposes. Specific indicators exist for all pillars of sustainable development. Some of them link selected phenomena to specific targets. So-called headline indicators seek to address the most important social, economic, or environmental issues. Aggregated indicators and indices try to capture a complex reality and propose a single and simple picture of it.

Indicators of sustainable development have figured prominently in research and policy agendas for many years. Agenda 21, adopted at the UN Conference on Environment and Development in Rio in 1992, expressed the need to formulate sets of indicators in order to better monitor and foster sustainable development. Many delegates reiterated this need at the first session of the UN Commission on Sustainable Development (CSD-1) in New York (1993). However, when concrete proposals for the development of such indicators were tabled during CSD-2 (1994), political will for their adoption was lacking. The CSD then commissioned the Scientific Committee on Problems of the Environment (SCOPE) and the United Nations Environment Programme (UNEP) to step in and undertake a joint project that was launched in 1994. SCOPE, established by the International Council of Science (ICSU) in 1969, acts at the interface between science and decision makers, providing advisors, policy planners, and decision makers with analytical tools to promote sound management and policy practices. SCOPE has the mandate to assemble and assess the information available on human-made environmental changes and the effects of these changes on people and to assess and evaluate the methods used to measure environmental parameters.

The SCOPE/UNEP project was designed

To bring together government delegates from all parts of the world, representatives of intergovernmental agencies, and scientific experts to discuss indicators of sustainable development in a nonpartisan context

To review existing sets of sustainability indicators developed (at that time) by various
national and international agencies

To provide the science base that subsequently helped initiate the political process that
finally resulted in the adoption of the CSD Work Programme on Indicators (1995)

The synthesis volume that resulted from this project, SCOPE 58, *Sustainability Indicators*, was distributed to all delegations at the UN General Assembly Special Session in 1997 and reached a wide public through commercial distribution channels.

During the 2001 CSD-9 session it was stated that indicators of sustainable development are now widely accepted and used and are recognized as an essential component of the process leading to a sustainable path of development. Subsequent international forums have affirmed the importance of indicators of sustainable development. In 2002 the Johannesburg World Summit on Sustainable Development encouraged further work on indicators for sustainable development by countries at the national level (including integration of gender aspects) on a voluntary basis, in line with national conditions and priorities. In December 2005 the CSD reaffirmed the importance of indicators for sustainable development

There has been useful progress since the Rio Earth Summit launched an international indicator development process. Many—perhaps too many—indicators, indicator sets, and indices have been assembled. Although sustainability indicators are used ever more extensively and intensively by a wide range of users and in many different contexts, it does not necessarily follow that they are scientifically sound or used appropriately. There has been no consensus on a common set of scientific and management criteria for evaluating indicators from several points of view (e.g., reliability of supporting data, scientific rigor of definitions of indicators, validity of underlying assumptions and concepts, relevance of positive or negative trends for sustainable development). At the time of the first SCOPE/UNEP project, research on indicators was in the early development stage, research questions were still being refined, and the data simply were not there or were insufficient.

This volume emerged as an outcome of the Assessment of Sustainability Indicators project, again implemented jointly by SCOPE and UNEP, together with the International Human Dimensions Programme on Global Environmental Change (IHDP) and the European Environment Agency (EEA), under the sponsorship of ICSU. A workshop held in Prague, Czech Republic, in May 2004 brought together thirty-five experts from seventeen countries. Three working groups—conceptual challenges, methodological frontiers, and policy relevance—built on background chapters that were written and circulated to participants before the workshop to assess selected indicators, providing cross-cutting perspectives in order to formulate a forward-looking framework for the assessment of sustainability indicators. The volume incorporates an overview and three cross-cutting chapters with others that are broadly concerned with sustainable development, including its economic, social, and environmental dimensions and other relevant perspectives, used at the international, national, regional, and local levels. It also reviews the specific features of indicators, both in general terms and in terms of some of the more widely known or innovative indicator sets, frameworks, and individual indicators or indices.

This review of the state of the art in sustainability indicators can draw a series of conclusions. There has been useful progress since the Rio Earth Summit in 1992 launched an international indicator development process. Many indicator sets have been assembled, countries have started their own indicator programs at the national level, and many cities and communities have developed and used indicators to measure their own progress. Methods are gradually becoming standardized, and policy decisions increasingly provide clear directions and targets. However, as individual chapters demonstrate, major conceptual challenges remain, methods warrant further development, and more must be learned about the most effective ways to influence policy. Progress has been sufficient to apply indicators at the national level and for international comparisons in support of sustainability goals and targets. What is needed is not a fixed approach to be applied everywhere but a process of adaptive implementation, with indicators evolving as the science of integrated indicators, frameworks, and models advances. We need to learn by doing. Each country or institution should select indicators and approaches suited to its needs, priorities, and means and use them to guide policy and action toward sustainable development.

The concept of sustainable development and the ability to measure progress toward its goals have become immensely important for many professions—researchers, educators, planners, nongovernment organizations, experts, policy analysts, and policymakers—and the wider public. In fact, the community that generates and makes use of indicators is vast and exists at all levels (e.g., sectoral, national, local), from experts to users, principally professionals interested in assessment, planning, and development. We hope that the considerations raised in this volume will assist in their effort to ensure a more sustainable society for future generations.

Bedřich Moldan
On behalf of the editors

John W. B. Stewart
Editor-in-Chief, SCOPE Publications

Véronique Plocq-Fichelet
Executive Director, SCOPE

Acknowledgments

The contributions of experts and financial support received from the United Nations Environment Programme (UNEP), the International Council for Science (ICSU), the European Environment Agency (EEA), and the International Human Dimensions Programme on Global Environmental Change (IHDP) were critical in the development of this volume and are gratefully acknowledged, as are the European Commission within grant EVG3-CT-2002-8007 INFOSDEV and the Ministry of the Environment of the Czech Republic. Charles University Environment Center hosted the workshop that was held at the "Karolinum," the historical complex of Charles University in Prague, and provided invaluable institutional and technical support.

1

Challenges to Sustainability Indicators

Bedřich Moldan and Arthur Lyon Dahl

The most difficult challenge facing policymakers is deciding the future directions of society and the economy in the face of often conflicting requirements for short-term political success, economic growth, social progress, and environmental sustainability. The wrong decisions can carry heavy consequences, increase human suffering, and even precipitate crises. Improving the basis for sound decision making, integrating many complex issues while providing simple signals that a busy decision maker can understand, is a high priority. At a time when modern information technologies increase the flow of information but not our ability to absorb it, we need information tools that condense and digest information for rapid assimilation while making it possible to explore issues further as needed. This is the goal of indicators.

Indicators are symbolic representations (e.g., numbers, symbols, graphics, colors) designed to communicate a property or trend in a complex system or entity. Traditionally, most indicators for decision makers have been numbers calculated by statistical services, including complex indices such as the gross national product (GNP) or percentages such as the unemployment rate.

Chapter 40 of Agenda 21 acknowledges that "commonly used indicators such as GNP and measurement of individual source or pollution flows do not provide adequate indications of sustainability" and states that "indicators of sustainable development need to be developed to provide solid bases for decision-making at all levels and to contribute to a self-regulating sustainability of integrated environment and development systems" (UNCED 1992:paragraph 40.4). Although 12 years later much progress has been made in refining the concept of sustainable development, the challenge of Agenda 21 has not been met in a satisfactory way. Many indicator sets have been assembled, but none has been widely implemented, and their integration to support self-regulating sustainability is still a major challenge. The development of

1

indicators is still seen as one of the major topics within sustainable development projects and programs (OECD 2004).

This book focuses primarily on the assessment of existing indicators in order to assist those who need to apply indicators now and to shed some light on the way ahead. The successes and failures are analyzed, gaps in knowledge exposed, and research needs identified. In addition, some new approaches are proposed. Progress in sustainability indicator development is reviewed in this book in the following three domains: conceptual challenges, methodological frontiers, and policy relevance. The review focuses on indicators broadly concerned with sustainable development, including its economic, social, and environmental dimensions and other relevant perspectives, largely as used at the national and international levels. Although many of the issues raised apply equally to indicators used at the local level, there are too many interesting initiatives at that level to include them in this review.

Probably the only generalization one can make about indicators or indices used or proposed is that there is no ideal indicator that fully encompasses all the desired qualities. There are always trade-offs, and the goal is not to eliminate the trade-offs completely but to make them transparent and to identify and avoid major constraints. Thus there is no one recommended indicator set but different approaches that may be appropriate for particular uses.

Conceptual Challenges of Sustainability

Assuming that indicators are intended to report on sustainability, the most important and difficult definition is that of sustainability itself. Sustainability is the capacity of any system or process to maintain itself indefinitely. Sustainable development thus is the development of a human, social, and economic system able to maintain itself indefinitely in harmony with the biophysical systems of the planet. Sustainable development is perhaps the most challenging policy concept ever developed. Its core objective—a kind of ethical imperative—is to provide to everybody everywhere and at any time the opportunity to lead a dignified life in his or her respective society. It is essentially an anthropocentric concept of sustained intergenerational and intragenerational justice (Grunwald et al. 2001), claiming for humans the right to a dignified life (Littig 2001). This demand for a high quality of life is assumed to include a decent standard of living, social cohesion, full participation, and a healthy environment (WCED 1987).

Sustainable development, as elaborated in Agenda 21 and confirmed at the World Summit on Sustainable Development, has three explicit dimensions, domains, or pillars: social, economic, and environmental. A fourth pillar, institutional, was included in the system of sustainability indicators adopted by the UN Commission on Sustainable Development (CSD), although very few institutional indicators were identified. More often, institutions are seen as providing the underlying enabling framework for action and change. Other approaches may recognize only human or social and natural or environmental dimensions or may subsume all compartmentalization into a more integrated and dynamic framework. Wherever such subdivisions are used for conceptual convenience or to make the concept more accessible and policy relevant, they immediately raise the challenge of integration across the subdivisions.

Most indicator sets for sustainable development have assembled indicators for each of the pillars while neglecting the links between them despite their key relevance for policy and planning. Development of interlinkage indicators is thus a particular challenge. The so-called decoupling indicators are highly relevant in this respect (OECD 2002), and other cross-cutting indices have been proposed.

Indicators of sustainability should measure characteristics or processes of the human–environmental system that ensure its continuity and functionality far into the future. Specifying the characteristics of the system or entity to be maintained can be very subjective and specific, and political, philosophical, and cultural differences may prevent any wide consensus. More effort is needed to apply the techniques of system science to this issue, developing more alternative models that reflect the diversity and complexity of human systems and cultures. These will help to explain the behavior of such complex nonlinear systems and their sensitivity, resilience, and capacity to switch between alternative steady states. The resulting understanding can contribute to more adaptive management, with indicators serving as monitoring and signaling mechanisms. The optimal sustainability indicators are those that capture the essential characteristics of the system and show a scientifically verifiable trajectory of maintenance or improvement in system functions. Science cannot always validate the goals set for the system, but it can validate the ability of the indicators chosen to measure the trajectory toward those goals or the reduction in damaging factors threatening the system's sustainability.

It is probably not possible or even desirable to arrive at one standard definition of sustainability. Such a dynamic concept must evolve and be refined as our experience and understanding develop. Rather than trying to resolve this issue, this review examines some of the basic underlying concepts and their relationship to indicators. The idea of carrying capacity, for instance, with the exception of the ultimate limits of life conditions on the planet, depends on political choices about an acceptable standard of living and thus is a very subjective and normative concept that cannot easily be captured in indicators. Resilience is a more useful aspect of sustainability that can be defined in terms of vulnerability, adaptability, and responsiveness or sensitivity. This allows a better understanding of the behavior of this system property integrating and clarifying synergies between various human economic, social, and cultural characteristics. The goal is to identify irreversible changes beyond which recovery is not possible. However, this entails defining how critical the loss is to overall system functions, whether substitutions are possible, what compensation may be needed, and the level of uncertainty that may necessitate the application of the precautionary approach.

The notions of weak and strong sustainable development have been debated in the recent literature, and a number of indicators or frameworks have been proposed to capture them. For weak sustainability, efforts have focused on whether the well-known macroeconomic indicators gross national product (GNP) and gross domestic product

(GDP) can be transformed to produce an indicator of sustainable development. For strong sustainability, the concept of critical natural capital (CNC) was introduced for the stocks of capital that cannot be substituted by other stocks of environmental or other capital to perform the same functions (Ekins et al. 2003).

The packaging of data into indicators is a way of simplifying complex and detailed information. Decision makers and the public lose interest rapidly if presented with more than a few indicators. It is therefore highly desirable to keep the number of indicators to a minimum while still representing the issues of sustainable development. The ultimate test of any indicator effort is its suitability for a specific use and the impact the indicators have on policies and public awareness. The issue of aggregation is very important in this respect because it can both generate useful information and facilitate its communication. Assessments that do not combine their indicators into a small set of indices are extremely hard to interpret, whereas those that do communicate their main findings instantly. When indicators are combined into indices, they provide a clear picture of the entire system, reveal key relationships between subsystems and between major components, and facilitate analysis of critical strengths and weaknesses. No information is lost if the constituent indicators and underlying data are available to be queried. However, there is a problem with the selection of indicators to be aggregated, which can intentionally or unintentionally introduce arbitrary weightings or other distortions.

Policy Relevance

Sustainability indicators generally are intended to target ongoing political processes, yet they often are developed with surprising political naïveté. Because such indicators are at the interface of science and politics, framing the issues in a policy-relevant way is particularly important and generally entails a participatory process. To be effective, indicators must be credible (scientifically valid), legitimate in the eyes of users and stakeholders, and salient or relevant to decision makers.

Policy has a life cycle, from the realization that there is a need for a policy instrument to tackle a certain issue, to the design of the policy and its implementation, evaluation, and adaptation, and finally to its phasing out or integration into another policy instrument. Indicators must meet different information needs at various stages of the policy life cycle. One function would be early warning, raising awareness of an unfavorable trend that may be evidence of a new and emerging issue or signaling a policy gap for an existing issue. Other indicators are used in impact assessments or outlooks, when new policy proposals are being developed, and still others contribute to the mid-term to long-term monitoring of policy implementation.

The policy life cycle for the design of indicator sets necessitates flexibility. Some indicators designed for monitoring will remain policy relevant for a long time, whereas others in the indicator set may need to change for maximum policy relevance. However, flexibility must be offset against the risk of losing familiarity and continuity in the indicators, which are both key elements in their adoption and use.

Indicators often are distinguished from raw data and statistics in that they contain reference values such as benchmarks, thresholds, baselines, and targets. Such values have various functions, but the most important is to transform meaningless data into information. A reference might be a target (distance to target), a baseline (distance to a certain meaningful state), a threshold value (distance to a collapse), a reference year (change in time), or a benchmark (difference with another country). All these reference points lend meaning and political weight to data and are used mostly in the interpretation of indicators. There are still many indicators for which reference values or baselines have yet to be established.

Distance-to-target indicators measure performance in reaching policy goals. Three distinct types of targets can be identified: political or hard targets; soft targets, such as those for sustainability reference values, minimum viable populations, and thresholds; and benchmarks.

Hard targets are set through political processes and can be very useful in producing effective indicators. Where hard targets are vague or qualitative and need clarification or definition, the indicator community can highlight this. Indicator producers can also draw attention to the lack of targets.

Indicators for soft targets, despite their uncertainty, can use ranges transparently, especially when such scientific targets highlight the inadequacy of political targets derived through a consensus process.

Benchmarking adds context to indicators, for instance by ranking countries. The comparison should be acceptable and relevant. However, if all countries are doing badly on a measure, benchmarking against the average will not encourage sustainability. Generally, peer pressure is a good thing among partners in any community, be they scientists, enterprises, or countries. National-level indicators therefore should be developed to allow intercountry comparison. The comparability should be as direct, simple, and evident as possible.

Capturing Diversity

In a world of great differences between countries, it is not easy to select indicators that provide a completely objective assessment of sustainable development. Those living within the Western economic paradigm naturally will choose indicators that reflect their conception of development. Data also are more available in industrialized countries. Developing countries may have different perspectives and priorities. For example, traditional economic indicators may overlook the informal economy and subsistence sector. Designers of indicators for use at the global level must accept a plurality of legitimate perspectives reflecting economic, cultural, environmental, and other differences, with particular attention to a better balance of indicators relevant to different stages of development. Any biases or value assumptions should be acknowledged.

Cultural differences may be expressed in the choice of indicators or in the levels or targets that are seen as sustainable. Because most indicator work has been done in the

North, one of the present challenges is to look again at the various indicator sets dealing with sustainable development, such as the Millennium Development Goals (MDGs), the United Nations Environment Programme (UNEP) Global Environment Outlook, and the indicators of the CSD and others, from different cultural perspectives and see whether significant cultural biases can be identified and, if possible, repaired. Cultures probably differ most in social needs such as freedom, acceptance, respect, equity, participation, and gender issues. Nevertheless, a common foundation of values is expressed in UN declarations and conventions on human rights, child labor, and others that identify specific issues (the indicator) and minimum levels. There is a need to distinguish the objective and normative components of indicators and to develop indicators for dimensions such as equity and participation.

Ideally, indicators should be chosen that are not too influenced by such diversity. Features of a robust indicator include a simple and unified method, commonly agreed issues and targets of wide applicability, transparency in the process, and agreement between partners on the process.

Use and Users

Indicators are by definition communication tools. Failure to communicate makes the indicator worthless. However, because sustainable development is a multistakeholder process, indicators must communicate to a variety of different actors. It is the capacity of the indicator to reach its target audience that determines its success as an indicator of sustainable development. Some users need simple, structured information (voters, the nonspecialist media, and decision makers), whereas others prefer an intermediate level of detail (local government, policy implementers, nongovernment organizations, funding bodies, and industries), and policymakers and academics may need more technical information.

In targeting governments, it is useful to distinguish between ministers and parliamentarians who make decisions, policy implementers and enforcers such as regulatory bodies and environmental protection agencies, and policymakers who are mostly civil servants, scientists, economists, and social scientists who design policy portfolios, evaluate policy alternatives, construct and evaluate indicators of sustainability, and brief ministers.

Most present indicators have been developed by governments and intergovernmental bodies in response to their needs. This ensures policy relevance, but it often fails to capture what is going on at the grass roots of society. Other indicators have been created by nongovernment organizations or academics to draw attention to policy issues. There are few indicators by and for the real agents of change: businesses and individuals operating at a decentralized level in all societies. Because the most effective indicators and feedback loops are those created and managed directly by users for their own purposes, only broad processes of education for sustainable development can equip individuals, local institutions, and small businesses with knowledge and indicators they can use to make their own behavior more sustainable. The issue of how to reconcile the centralized approaches needed to produce standard comparable indicators and the decentralized nature of most decision making affecting sustainability has not yet been explored.

The business community is an essential actor for sustainability that is not captured well in indicators. Indicators of sustainable business behavior would complement indicators at the government level. Although many corporate reports now include information on environmental and social performance that could be used for indicators, it is still difficult to get businesses to share the information they collect. Some information is seen as confidential because it provides a commercial advantage, and businesses are not motivated to share negative information that might damage their reputation or profitability. Yet much of the effort to move toward sustainability involves identifying and reducing problems such as pollution. This is an important gap that must be filled, particularly for small and medium enterprises that are responsible for the bulk of business activity.

User involvement is important to indicator design and acceptance. Stakeholders may have local knowledge that can contribute to more effective indicators. Participation also ensures relevance to the decision-making process, political commitment, and ownership of the results. Participatory processes can reveal conflicting social interests, values, and preferences that must be taken into account. The quality of the process is important.

Acceptance and use of indicators are a continuing challenge. Indicators that reflect badly on politicians, corporate executives, and senior officials will be rejected or suppressed, and most indicators of sustainable development show negative trends. Careful indicator development processes, outside pressure, and objectivity will be necessary to overcome this obstacle.

Democracy and Equity

Sustainable development includes an important ethical component, expressed as the right of every human being to a fair share of the benefits this planet offers. Democratic processes help to ensure access to all the dimensions of development. In particular, with the transformation to a knowledge-based service economy, access rights become essential for societal well-being and a critical element of a dignified life. These rights include the following:

- *Biophysical environment:* This includes access to land and natural resources, safe drinking water and sanitation, and housing and energy, both from the environment and through adequate infrastructure, including the technologies of a modern information society (computer, telephone, Internet).
- *Economic dimension:* On the individual level this includes a secure minimum income to guarantee active participation in society, including access to the sociocultural system. In the context of the national economy, this should include fair access to the benefits of the economy and the ability to contribute to wealth generation, in both the market economy (salaries and employment) and the nonmarket economy (unpaid

work, caring work, voluntary community work, and the resulting services). This entails access to markets for all potential producers, with no entrance barriers, and access to finance (i.e., nondiscriminatory credit conditions). On the international level this entails the removal of obstacles to participation in the global economy, such as old (and long written-off) debt and trade barriers erected by the affluent societies.

- *Social dimension:* This includes access to knowledge, information, and experience, such as nondiscriminatory education, the opportunity to work and participate in social processes, access to information technology, and the ability to select and transform information into relevant knowledge.
- *Institutional dimension:* This includes access to information (newspapers, the Internet, oral communication, and expertise), information exchange (the right to free speech and the right to provide content), and decision making. Components include the legal right to participation, equal access (e.g., racial, ethnic, and gender equity), a participatory political system, nondiscriminatory social security systems, access to justice, and legal provisions for access to economic, social, and environmental resources.

The distribution of access is a measure of the intergenerational justice within a society. There is increasing recognition of the risk, in choosing the conceptual framework and governance of sustainable development, that certain ideas become embedded as authoritative whereas others are marginalized. Democratic processes therefore are particularly important in defining ends, means, and indicators of sustainable development in order to ensure access to and inclusion of the diverse perspectives in a society. In the end, the institutional structures should reflect the aggregated preferences expressed by this process. Democratic representation helps these institutions to be transparent, accessible, and accountable.

One of the critically important enabling conditions for democratic participation is capacity building. This applies broadly to all indicator processes, not only to indicators of sustainable development, and is needed in several areas. First, there is a need to improve the capacity of decision makers to understand and use indicators, especially in relation to setting and monitoring targets. Capacity also must be built, particularly in developing countries, to increase public participation in the processes of defining indicators and setting targets for sustainability. This requires that the public understand the role and use of indicators. The introduction of indicators as a topic in school curricula or other awareness or educational initiatives should be explored, as in the UN Decade of Education for Sustainable Development (2005–2014) and the European Consumer Citizenship Network.

Scales and Frameworks

The economic, social, and environmental dimensions or pillars of sustainable development have different characteristic time scales, ranging from a long-term view of sustainability in general to the short-term perspective of policy and economic measures. Environmental systems evolve slowly and have longer time lags between cause and effect than economic systems. A parallel mismatch in time scales occurs between the methods in the disciplines that study the different pillars. This makes it hard to present sustainability to policymakers, who tend to act on experience rather than insight and therefore take action only when a problem is observable, not when a problem is predicted, especially if the prediction is uncertain. A key challenge for sustainability indicators therefore is to reflect time lags, the trade-offs between the short and long term, and the distinction between weak and strong sustainability.

There are similar challenges in relating indicators at different spatial scales, where the same indicator may have different meanings in different contexts or when applied at different scales. Unsustainable states, trends, and drivers may be apparent only when indicated at the appropriate scale. A local community can appear sustainable if it exports its unsustainable consumption or waste disposal. Similarly, indicators may show a high per capita income at the national scale, for example, while hiding significant inequities between subregions and societal groups. To compensate for this we need indicator sets in a nested hierarchical structure covering different geographic scales or units. One significant gap is in indicators appropriate for measuring global sustainability and planetary limits.

Policy is implemented on the basis of political boundaries. In indicator design, reporting by political units usually is needed for policy relevance. In the environmental field, this generally means remapping ecological boundaries onto political boundaries. In deciding on reporting units, a key factor is that averaging for large political units or regions may not always capture important issues of sustainability, especially if there are large disparities within regions.

Sustainability indicators may be easier to understand and interpret when assembled in some conceptual framework, perhaps with a hierarchical arrangement of subdomains. The three pillars (economic, social, and environmental) are one such framework, but many others are possible. Such frameworks may reflect different values and weightings, which should be transparent. Frameworks may help to interrelate indicators from the natural and social sciences, to position both stock and rate indicators, and to identify interlinkages.

Measurability

Indicators necessarily limit themselves to the sphere of the measurable. Like models, indicators can reflect reality only imperfectly. However, even within the measurable, the quality of indicators is determined largely by the way reality is translated into measures and data, be they quantitative or qualitative. Although present scientific knowledge is inadequate to understand many aspects of human–environment interactions, and some feedback loops between human and environmental systems are irreducibly complex, many issues are sufficiently well understood to necessitate scientifically accurate indicators. The quality of indicators inevitably depends on the underlying data that are used to compose them. The prevailing data gaps in monitoring of human–environment interactions and the poor quality of many databases (especially on the global and local levels) are potential threats to the quality of the related indicators.

The quality of an indicator can be judged on five methodological dimensions: purpose and appropriateness in scale and accuracy, measurability, representation of the phenomenon concerned, reliability and feasibility, and communicability to the target audience. There is seldom a perfect indicator, so the design generally involves some methodological trade-offs between technical feasibility, societal usability, and systemic consistency.

Although it may always seem desirable to enhance data quality and to develop new data sets on a number of issues and scales, indicators can grow into costly enterprises. It should not be forgotten that indicators are merely assessment tools, for which the cost of improvements should not limit the capacity to implement policy. The two must be matched in cost-effective ways.

Data Availability

Many indicator projects are constrained by the availability of relevant and reliable data. Data availability often is a selection criterion to ensure a rigorous quantitative underpinning for the indicators. As a result, most indicators have been constructed on the basis of existing data. This often rules out the inclusion of "ideal" indicators because of the paucity of appropriate information, and new indicators must be derived from existing data. It can take 5–10 years to develop new data flows. This excludes relevant indicators for newly emergent issues, based on more recent scientific insights and political priorities. To furnish these flows often means breaking away from old data sets and collecting new data. If indicators can be selected only when there are existing data, a vicious circle ensues, blocking the desired evolution.

The root cause of this problem lies not just in changing data demands, inefficient data collection processes, and resource constraints but also in a lack of clarity on data needs in the first place. This has meant that demanding data collection processes have been set in train that limits future flexibility. Therefore, it is very important to get the conceptual understanding right about the phenomena to monitor and the indicators to use before invoking the data availability criterion.

Types of Indicators

Most existing sustainability indicators are quantitative. They are based on quantitative measurements of variables from which indicators and indices are derived. Some definitions of indicators actually include quantification as a defining characteristic of indicators alongside simplification and communication. This limitation of quantitative indicators can exclude significant factors or otherwise bias indicators of sustainability. Some relevant issues can be assessed only through qualitative measurement (e.g., social cohesion, happiness, or sense of place). The social sciences are generating qualitative indicators, such as through surveys that can be answered on scales ranging from "not happy–noncompliant–disagree" to "totally happy–compliant–agree." Integrating these data with quantitative data remains a critical methodological issue. What is important is that the feasibility and reliability criteria for indicators impose strict scientific quality standards regardless of the quantitative or qualitative nature of their underlying measurements.

Indicator presentation may also use qualitative representations such as smilies, barometers, and color coding to strengthen communication of the results. These qualitative outputs give an indicator value or direction and signal whether it is good or bad.

Where direct indicators do not exist, perhaps because of missing data or insufficient knowledge of interactions, proxy or substitute indicators are widely used. Proxy indicators usually are representations of complex systems and can be useful for communicating complex issues. Examples are greenhouse gas emissions instead of climate change, bird presence as a proxy for biodiversity, and GDP as a proxy for economic welfare. Although most proxy indicators do not comprehensively represent the issue, they will change with that issue and thus signal general trends. However, they are less suitable for identifying the precise dynamics of change and possible intervention points.

A few basic types of indicators or indices may be distinguished by their methods of construction and level of aggregation:

- *Indicator:* This includes results from the processing (to various extents) and interpretation of primary data. Examples include SO_2 emissions for a particular country per year and employment rates.
- *Aggregated indicator:* This combines, usually by an additive aggregation method, a number of components (data or subindicators) defined in the same units (e.g., tons, monetary units). Examples include material flow aggregates as domestic material consumption, the Living Planet Index, or the GDP.
- *Composite indicator :* This combines various aspects of a given phenomenon, based on a sometimes complex concept, into a single number with a common unit (e.g., years, hypothetical hectares). Examples include life expectancy and the Ecological Footprint.
- *Index:* This generally takes the form of a single dimensionless number. Indices mostly require the transformation of data measured in different units to produce a single number. Examples include the Human Development Index and Air Quality Index.

Assessment of Specific Indicators

The SCOPE review examined the specific features of indicators both in general terms and in terms of some of the more widely known or innovative indicator sets, frameworks, and individual indicators or indices. Special attention was devoted to the ways in which the following indicator approaches illustrated certain issues or challenges or represented the present state of the art in indicator development:

Commission on Sustainable Development indicator set (UNCSD 2001)
Millennium Development Goals (MDGs) indicators
UNEP Global Environment Outlook indicators
Structural indicators (European Commission)
Human Development Index (HDI)
United Kingdom Headline indicators
Material Flow Analysis–based indicators
Energy Flow Analysis–based indicators
Ecological Footprint
Living Planet Index
Environmental Sustainability Index (ESI)
Environmental Vulnerability Index (EVI)
Well-being of Nations
Biodiversity indicators
Driving force–Pressure–State–Impact–Response framework
Three-pillar versus four-pillar frameworks
Corruption Perception Index, Freedom Index
Well-being Index

Conclusions

What conclusions can be drawn from this review of the state of the art in indicator development for sustainability? There has been useful progress since the Rio Earth Summit in 1992 adopted Agenda 21 and launched an international indicator process. Many indicator sets have been assembled; countries have started their own indicator programs at the national level, as called for by the Commission on Sustainable Development; and many aspects of sustainability have been given a more precise definition or measure through indicators. Methods are gradually becoming standardized, and policy decisions increasingly provide clear directions and targets, as exemplified by the MDGs and their indicators. However, as the following chapters demonstrate, major conceptual challenges remain, methods need further development, and more must be learned about the most effective ways to influence policy. We are still far from fully integrated sets of indicators or indices to support self-regulating sustainability.

There is also at present no international strategy or clear future direction for indicators of sustainability, including its environmental, economic, and social dimensions, and no mechanism is providing international leadership in this area. Since the Rio Earth Summit launched the CSD indicator process, there has been a healthy multiplication of international initiatives, some intergovernmental but regional (Organization for Economic Co-operation and Development [OECD], EU, South Pacific Applied Geoscience Commission [SOPAC], EVI), but most nongovernment and academic (ESI, Ecological Footprint, Wealth of Nations, Living Planet Index). Some of these have been one-off efforts, and others have continued without really becoming operational. The quality, often poor at first, has been slowly but steadily improving, with some of the latest versions such as the EVI and ESI launched in January 2005 being sufficiently good to earn some intergovernmental credibility and political momentum. The demand for indicators as status and performance measures relevant to international policy goals and targets will only increase. At some point, one or more appropriate indices will need to become institutionalized in some intergovernmental process to provide sufficient stability and credibility for widespread use by governments, but it is not clear how this will happen.

There are two options for the future of indicators of sustainability: letting the present anarchy continue until survival of the fittest prevails or implementing more strategic intervention and guidance of the process. The former might lead to the survival of the financially and politically strongest rather than the scientifically most appropriate, with a bias toward the wealthiest countries. It would be in the interest of the international community to try to make the process more balanced and objective by giving it some direction or leadership.

Governments need to buy into this process, but this is politically perilous because a good index makes some countries look bad, and no leader likes that. Country rankings do attract the attention of political leaders and can motivate governments to do something about the problems uncovered. Such indicators bring to the public sector a little of the competitive spirit that helps make the business sector more efficient. However, only UN bodies have the strength and universality to ensure that scientific credibility and objectivity win out over political expediency, as the United Nations Development Program (UNDP) has demonstrated with the HDI. The process still needs to be designed with care, perhaps with an ongoing dialogue with government experts to breed familiarity before the process becomes too political.

There is a vacuum in this area at the moment. Indicators are not a visible part of the CSD extended work program. There has been some acceptance in the UN of the work on economic vulnerability indices, but more could be done on long-term economic sustainability measures. UNEP might logically lead on the environmental side, but the principal environmental sustainability and vulnerability indices are outside UNEP and gaining their own momentum and acceptance. There is still a big gap on the social indicator side, with no adequate measures of social resilience or sustainability despite some work in Latin America. The work on indicators should also be linked more closely to the global observing systems and the Global Earth Observation System of Systems process, which could generate new global data sets. Some combination of indices, appropriately harmonized for complementarity while responding to relevant policy mandates, may be most appropriate. At a minimum, sustainability indicator activities should be the subject of regular informal consultations among the UN partners most concerned.

The most important message from this assessment probably is that progress is sufficient to apply indicators now at the national level and make international comparisons in support of sustainability goals and targets. What is needed is not a fixed approach to be applied everywhere but a process of adaptive implementation, with indicators evolving as the science of integrated indicators, frameworks, and models advances. We need to learn by doing. Each country or institution should select indicators and approaches suited to its needs, priorities, and means and use them to guide policy and action toward sustainable development. This is the only way to ensure a more equitable and sustainable society for future generations.

Appendix 1.1. Comments on selected indicators, indices, and indicator sets.

During the deliberations of the working groups, several categories of concrete indicators were assessed. Some of them were discussed by all three working groups, some by only one or two of them. There was no attempt to reach a final authoritative opinion on any indicator or indicator set, but many useful comments were made. Here we summarize the results of the groups' discussions. We present opinions that did not encounter any serious objections. References and contacts for the presented indicators can be found in the Annex.

TITLE, TYPE, AND SOURCE OF THE INDICATOR OR SET	DESCRIPTION	COMMENTS
Corruption Perceptions Index (CPI) *An index developed by Transparency International (TI)*	The TI CPI ranks countries based on experts' perception of corruption (CPI sources are surveys). It measures the overall extent of corruption (frequency and amount of corruption) in the public and political sectors. CPI ranks a record 146 countries.	• Focuses on corruption in the public and political sector (corruption means the abuse of public office for private gain). • Sources do not distinguish between administrative and political corruption or between petty and grand corruption.
Dashboard of Sustainability (DS) *An indicator software tool developed by the European Joint Research Center in Ispra*	The DS presents complex relationships between economic, social, and environmental issues in a highly communicative format aimed at decision makers and citizens interested in sustainable development. It contains various indicator sets, including the UNCSD and MDG sets.	• A tool, not an indicator itself. • Allows various aggregation mechanisms. • User friendly. • Includes statistical tools for testing simple hypotheses. • Attractive design. • Allows good links between pillars.

Appendix 1.1. Comments on selected indicators, indices, and indicator sets (*continued*).

TITLE, TYPE, AND SOURCE OF THE INDICATOR OR SET	DESCRIPTION	COMMENTS
Ecological Footprint (EF) *A composite indicator introduced by Redefining Progress*	The EF is the corresponding area of productive land and aquatic ecosystems needed to produce the resources used and assimilate the wastes produced by a defined population at a specified material standard of living, wherever on Earth that land may be located. Thus EF is a measure of the load imposed by a given population on nature.	• Deals with important aspects of sustainability (carrying capacity, overconsumption, and biocapacity) but covers only the environmental pillar. • High level of aggregation (underlying information is less accessible). • The methods are not unified. • Data quality varies across indicators and countries. • Used mainly at the national level but can be applied at any scale, including individuals. *Global targets included.* • Used by the scientific community and the media. • Strongly communicative on public and policy levels (raises public awareness efficiently). • Low global and supranational policy relevance but stronger at local level. • Uses ranking. • Communicates the urgency of environmental sustainability, emphasizing effects of exported impacts.

Appendix 1.1. Comments on selected indicators, indices, and indicator sets (*continued*).

Title, Type, and Source of the Indicator or Set	Description	Comments
Economy-wide material flow indicators *A framework of aggregated pressure indicators standardized by Eurostat*	The material flow indicators are based on economy-wide material flow analysis, which quantifies physical exchange between the national economy, the environment, and foreign economies on the basis of total material mass flowing across the boundaries of the national economy	• Focuses on the environmental pillar (use of natural resources). • Used for different scales (supranational, national, local). • Methods discussed by scientific community. • Highly aggregated indicators. • Data quality differs between indicators. • No ranking, but possible to find clusters of countries. • Likely to be long-lived.
European Environment Agency (EEA) core set indicators *A set of environmental indicators developed by the EEA*	In early 2004 the EEA proposed its core indicator set. These indicators aim to cover the entire environmental pillar of sustainable development. Indicators are sorted into 10 subgroups: climate change, fisheries, water, agriculture, energy, transport, biodiversity, waste, air pollution, and terrestrial.	• EEA core set will be internationally evaluated (based on 11 criteria such as data availability, timeliness, and representativeness). • A standard set of environmental indicators. • Some important issues are not covered: forests (both healthy and cut), soil (erosion, desertification), and material flows.

Appendix 1.1. Comments on selected indicators, indices, and indicator sets (*continued*).

TITLE, TYPE, AND SOURCE OF THE INDICATOR OR SET	DESCRIPTION	COMMENTS
Environmental Sustainability Index (ESI) *An index developed by the Yale Center for Environmental Law and Policy and the Center for International Earth Science Information Network (CIESIN) at Columbia University*	The ESI is an aggregated index capturing the environmental dimension of sustainability. It is based on a set of 21 core indicators, each of which combines 3–6 variables for a total of 76 underlying variables.	• High aggregation. • Covers environmental domain of sustainable development. • Arbitrarily selected variables. • Mix of variables and components from different parts of causal chain. • Good communication tool, has media attention. • Reliability is lessened by flaws in international databases. • No attempt to capture linkages.
Eurostat sustainable development indicators *A set of sustainability indicators developed by Eurostat*	The set contains 63 indicators, of which 22 are mainly social, 21 are mainly economic, and 16 mainly environmental. This list is structured along a more policy-oriented classification than the previous one, according to the relevant sustainability dimensions (4), themes (15), and subthemes (38).	• The set draws on and extends the UNCSD list of 58 core sustainable development indicators (more than 66% of the indicators are comparable with those in the UNCSD core list). • Driven by data available at European level. • Based on the existing work on pressures and sectoral indicators. • No major policy impact so far (structural indicators receive more attention); work is under way.

Appendix 1.1. Comments on selected indicators, indices, and indicator sets (*continued*).

TITLE, TYPE, AND SOURCE OF THE INDICATOR OR SET	DESCRIPTION	COMMENTS
Freedom in the World *An index developed by Freedom House*	Freedom in the World is the annual comparative assessment of global political rights and civil liberties. The survey includes both analytical reports and numerical ratings for 192 countries and 14 territories.	• Measures freedom according to two broad categories: political rights and civil liberties. • The survey method establishes universal standards that are derived in large measure from the Universal Declaration of Human Rights. • An element of subjectivity is inherent in the survey findings. • Assists policymakers, the media, and international organizations in monitoring trends in democracy and tracking increases and decreases in freedom worldwide. • No use of targets and ranks.
Global Environmental Outlook (GEO) indicators *A set of indicators highlighting key global and regional environmental issues, developed by UNEP*	A set of 18 indicators first published in the *GEO Year Book* in 2003. They highlight some of the key global and regional environmental issues and trends identified in GEO reports. Indicators are structured along the following themes: • Atmosphere • Natural disasters • Forests • Biodiversity • Coastal and marine areas • Freshwater	• Provides an annual overview of major environmental changes. • Availability of reliable, up-to-date global data sets still limits the choice. • The indicators are not very well balanced: too many indicators on climate change relative to other priority areas, and some important issues are missing; no indicators on urban issues; no land indicator besides forest cover; no

Appendix 1.1. Comments on selected indicators, indices, and indicator sets (*continued*).

TITLE, TYPE, AND SOURCE OF THE INDICATOR OR SET	DESCRIPTION	COMMENTS
	• Global environmental issues	direct climate measures; water quality is not measured; no indicator on use of chemicals (an index of land use combining forest, cropland, and urban land use might usefully be included). • Two or three indicators per issue would be desirable (or just one indicator per issue; this would entail using aggregated indicators). • No framework, therefore not very useful for policy guidance.
Human Development Index (HDI) *An index developed by UNDP*	The HDI is a composite index that measures a country's average achievements in 3 basic aspects of human development (quality of life): longevity, knowledge, and a decent standard of living, measured by income.	• Integrates a small number of variables to keep the indicator simple. • High data quality and reliability. • Used for national level. • No use of targets. • Neglects environmental issues. • One easy-to-communicate number. • Frequently used by developing country governments.
Indicators to measure decoupling of environmental pressure from economic growth *Set of sustainable development indicators developed by the OECD*	The set comprises 31 indicators covering a broad spectrum of environmental issues such as climate change, air pollution, water quality, waste disposal, material use,	• Shows linkages between environmental and economic pillars. • Plots environmental parameters against economic parameters (GDP).

Appendix 1.1. Comments on selected indicators, indices, and indicator sets (*continued*).

TITLE, TYPE, AND SOURCE OF THE INDICATOR OR SET	DESCRIPTION	COMMENTS
	and natural resources. They aim at measuring and plotting the decoupling of environmental pressure from economic growth.	• Used for different scales (national, sectoral). • Aggregated as well as headline indicators. • Data quality varies across indicators and countries. • Uses ranking. • Issue of decoupling highly policy relevant.
Living Planet Index (LPI) *An aggregated indicator promoted by the World Wildlife Fund*	The LPI is an indicator of the state of the world's biodiversity. It measures trends in populations of vertebrate species living in terrestrial, freshwater, and marine ecosystems around the world. The LPI is the average of 3 separate indices measuring changes in abundance of 555 terrestrial, 323 freshwater, and 267 marine species around the world.	• LPI includes national and global data on human pressures on natural ecosystems arising from the consumption of natural resources and the effects of pollution. • All three individual components are given an equal weighting. • Appropriate for respective countries. • Participatory process. • Simple, appealing, and measurable (measures changes in species abundance).
Millennium Development Goals (MDGs) *A set of goals commonly accepted as a framework for measuring development progress, developed by the UN*	MDGs grew out of the agreements and resolutions of world conferences organized by the UN in the past decade. The goal is to assist in achieving significant, measurable improvements in	• Under the goal "ensure environmental sustainability," 3 targets and concrete indicators (e.g., forest areas, protected areas, energy sources, CO^2 emissions) are defined.

Appendix 1.1. Comments on selected indicators, indices, and indicator sets (*continued*).

TITLE, TYPE, AND SOURCE OF THE INDICATOR OR SET	DESCRIPTION	COMMENTS
	people's lives in 8 areas: poverty, education, gender equality, child mortality, maternal health, HIV and AIDS, other diseases, environment, and global partnership. The first 7 goals are mutually reinforcing and are directed at reducing poverty in all its forms. The last goal, global partnerships for development, is about the means to achieve the first 7.	• Environmental target indicators are not directly related to sustainability (e.g., land covered by forest is a poor proxy indicator for degradation of the terrestrial environment). • MDG set uses existing data sets. • Not picked up by scientific community (no challenging methodological issues). • Media attention is growing. • Not appropriate for communicating sustainability.
OECD core environmental indicators (CEI) *One of the indicator sets developed by OECD*	This set of indicators helps track environmental performance and progress toward sustainable development in this domain. It is based on the Pressure-State-Response (P-S-R) framework and covers 15 major issues (e.g., biodiversity, climate). It contains about 50 indicators. CEI can be disaggregated at the sectoral level (SEI) or territorial level (TEI).	• Covers mostly environment. • Uses some less sensitive indicators. • Used for influencing environmental policy. • Helpful for state reporting and national comparison. • Good coverage of links. • Does not include developing countries. • Widely used by organizations and the scientific community. • Part of a greater framework.

Appendix 1.1. Comments on selected indicators, indices, and indicator sets (*continued*).

TITLE, TYPE, AND SOURCE OF THE INDICATOR OR SET	DESCRIPTION	COMMENTS
Structural indicators *A set of indicators developed by Eurostat for the European Council*	The agreed set is to support assessment of annual progress by the EU member states in the *Synthesis Report.* Originally 35, then 42 indicators are organized along 5 policy domains: employment, innovation, economic reform, social cohesion, and environment and economic background indicators. In 2004 the set was shortened to 14 indicators.	• The indicators are by definition macro-level and performance-oriented indicators, focused on short-term development. • Purpose: comparison between countries, primarily on regional, social, and economic development. • High focus on employment issues. • Only one true environmental indicator. • No attempt to capture linkages. • Uses national data. • Good at present state and trends over time. • Might be meaningful to the private sector.
UNCSD indicators *Set (theme indicator framework) of sustainable development indicators developed by UNCSD*	The set comprises 58 indicators organized according to sustainable development dimensions and themes (e.g., education, atmosphere, economic performance). The set aims at covering sustainable development as a whole, addressing all 4 dimensions of sustainability.	• Covers all important aspects of sustainability but fragmented, no integration or linkages. • Top-down (closed) process, weak public participation. • Aggregated and headline indicators. • Data quality varies across indicators and countries. • Used for national level. •No use of targets and ranks. • Weak policy impact globally. • Likely to be supplanted by MDGs.

Appendix 1.1. Comments on selected indicators, indices, and indicator sets (*continued*).

TITLE, TYPE, AND SOURCE OF THE INDICATOR OR SET	DESCRIPTION	COMMENTS
Well-being Index (barometer of sustainability) *An index introduced and published by IUCN*	It combines 36 indicators of health, population, wealth, education, communication, freedom, peace, crime, and equity into a human well-being index and 51 indicators of land health, protected areas, water quality, water supply, global atmosphere, air quality, species diversity, energy use, and resource pressures into an ecosystem well-being index. The two indices are then combined into a Well-being/Stress Index that measure how much human well-being each country obtains for the amount of stress it places on the environment.	• Applies a concept of equal treatment of people and ecosystems; very illustrative symbols of the egg with its yolk (human) and white (ecosystem). • Relative scaling of results is used in the framework of sustainability (bad, poor, medium, fair, good); scaling of each indicator was affected by international standards, targets, expert opinion). • Target value for each component is set but is biased by specific development concept. • Results are user-friendly; indices are presented as a barometer of sustainability; both barometer and maps are easy to read. • Majority of indicators are based on existing and regularly updated sources (some elements are just theoretical and data are missing, e.g., shelter, culture, seawater). • Top-down process, isolated methods.
Environmental Vulnerability Index (EVI) *An index developed by SOPAC*	The EVI combines 50 indicators, each related to sustainability thresholds, to produce country profiles of	• Measures vulnerability and resilience of environmental systems. • Includes reference values.

Appendix 1.1. Comments on selected indicators, indices, and indicator sets (*continued*).

TITLE, TYPE, AND SOURCE OF THE INDICATOR OR SET	DESCRIPTION	COMMENTS
	the resilience and vulnerability of environmental systems and resources. Indicators for weather and climate, geology, geography, resources and services, and human population are also used to generate subindices for climate change, natural disasters, biodiversity, desertification, water, agriculture and fisheries, and human health.	• Created by developing country organization with wide consultation and country participation. • Country profiles useful guide to priority setting and policy action. • Covers only environmental vulnerability; similar indices needed for social and economic vulnerability.

Literature Cited

Ekins, P., C. Folke, and R. De Groot. 2003. Identifying critical natural capital. *Ecological Economics* 44:159–163.

Grunwald, A., R. Coenen, J. Nitsch, A. Sydow, and P. Wiedemann (eds.). 2001. *Forschungswerkstatt Nachhaltigkeit: Wege zur Diagnose und Therapie von Nachhaltigkeitsdefiziten.* Berlin: Reihe Global zukunftsfähige Entwicklung, Perspektiven für Deutschland 2.

Littig, B. 2001. *Zur sozialen Dimension nachhaltiger Entwicklung.* Vienna: Strategy Group Sustainability.

OECD. 2002. *Indicators to measure decoupling of environmental pressures from economic growth.* SG/SD(2002)1. Paris: OECD.

OECD. 2004. *Measuring sustainable development: Integrated economic, environmental and social frameworks.* Paris: OECD.

UNCED. 1992. *Rio declaration on environment and development. Report of the United Nations Conference on Environment and Development.* Distr. General, August 12, 1992, A/CONF.151/26 (Vol. I).

WCED. 1987. *Our common future. World Commission on Environment and Development.* Oxford: Oxford University Press.

PART I
Cross-Cutting Issues

2

Meeting Conceptual Challenges

Sylvia Karlsson, Arthur Lyon Dahl, Reinette (Oonsie) Biggs, Ben J. E. ten Brink, Edgar Gutiérrez-Espeleta, Mohd Nordin Hj. Hasan, Gregor Laumann, Bedřich Moldan, Ashbindu Singh, Joachim Spangenberg, and David Stanners

The concept of sustainability and its measurement with indicators may seem intuitively simple, but it is difficult to implement in practice. Over the past two decades some problems have been resolved and much progress made on others, but major challenges remain. This chapter summarizes the present conceptual challenges, illustrated by selected indicator approaches. The challenges are grouped in two clusters: measuring sustainability and sustainable development, and developing indicators through processes that ensure their universal applicability. These various conceptual challenges suggest future research agendas and approaches to indicator use.

Measuring Sustainability and Sustainable Development

The usefulness of any indicator intended to measure how sustainable (or unsustainable) the world is, or the progress society is making toward sustainable development, naturally depends on how these terms are defined. Although this was discussed briefly in chapter 1, it still provides the major conceptual challenges for indicator development. First, it is necessary to go beyond a sectoral approach to a system approach. Does sustainable development fit a linear model with three or four pillars? Are there alternative, system-based approaches to understanding and measuring sustainability? Second, the entity to be measured must be defined in temporal and spatial scales and related to some model for sustainable development. What does sustainability mean for a village, a country, or the planet? Over what time span should the world, its ecosystems, and humanity sustain themselves and in what form?

From Pillars to Linkages to Systems

It has become common to consider sustainable development within a certain conceptual framework, and this also influences indicator development. In Agenda 21 and at the UN World Summit on Sustainable Development (WSSD), the international community refers to economic, social, and environmental dimensions or pillars of sustainable development (UNCED 1993; United Nations 2002). In some contexts a fourth institutional pillar is added, as in the framework for indicators adopted by the Commission on Sustainable Development (CSD).[1] Alternatively, institutions are seen as providing the underlying enabling mechanism for effecting action and change in any of the pillars. Most of the present approaches to indicator development compile indicators for these pillars. There has been much progress in developing indicators within each pillar, and many such indicators are being implemented.[2]

Others prefer to see sustainable development as the interaction of the environmental and human systems in a two-part coupled framework (e.g., Prescott-Allen 2001). Many national and international bodies arrange their indicators in a "pressure–state–response framework" (sometimes expanded with "driving forces" and "impacts": DPSIR). Although this framework implies causal relationships between indicators, research has only recently begun to develop the data and models necessary to interrelate indicators in such a framework.[3] Significant conceptual challenges remain even when we consider single indicators (and further challenges for aggregated indices). For example, there is limited understanding of how the complex properties of the social dimension reinforce or obstruct sustainable social development. Concepts such as social capital are emerging as aspects relevant to social cohesion but have not yet been captured in indicators, and more work is needed to evaluate their usefulness. There are similar challenges in the institutional pillar, as indicated by the very few institutional indicators included in the CSD list (UN Division for Sustainable Development 2001). Economic indicators are well established but do not include measures of long-term sustainability in the economic system.

All these approaches are limited in that they address isolated elements of sustainability. Sustainability and sustainable development are characteristics of integrated systems with multiple linkages, feedback loops, and interdependencies. Although political approaches to sustainable development often are narrowly sectoral, with little focus on integration in practice, decision makers are increasingly asking for indicators to help build mutually reinforcing links between pillars. The challenge of defining and quantifying links between the pillars has not been resolved, but some progress has been made in the last decade, and examples are available (Table 2.1).

The fundamental conceptual challenge is to go beyond a mere collection of parts and apply a more system science–oriented approach to consider the sustainability of whole systems composed of interacting subsystems with emergent system properties (Chapter 10, this volume). It is the underlying properties that determine the dynamics and behavior of these systems and ultimately how sustainable the systems are over long periods of time. Examples of such system properties are resilience, carrying capacity, energy and material flows, and intergenerational knowledge transfer.

Table 2.1. Interlinkage indicators in the four-pillar sustainable development framework.

Linkage	Indicators
Environmental–economic	Resource productivity (gross domestic product/total material input) (Eurostat 2001; OECD 2001). Transport intensity (Böge 1994; SDC 2004).
Socioeconomic	Labor productivity (production per capita; see any national labor statistic). Income distribution per decentile (see any national social statistic).
Socioenvironmental	Environmental health problems (no clear definition so far; work under way by the World Health Organization, European Environment Agency, and others). Access to common goods (to be specified regionally, available in Scandinavia under traditional law).
Economic–institutional	Corruption rate (Transparency International Index I). Share of taxes on labor, capital, and the environment in total tax revenues (not often calculated, but the basic data often are available from national statistical offices).
Socioinstitutional	Co-decision rights of workers (e.g., according to the European Works Council directive; in Europe, data are available from Eurostat labor market statistics, from the EU Commission, the trade unions, and others). Reliability of the health care and social security system (reliability is a subjective term and so far undefined).
Environmental–institutional	Nongovernment organizations' right to file suit (data for this are collected under the Aarhus Convention demanding such access). Freedom of information (in Europe, North America, and Central Asia regulated by an UN Economic Commission for Europe directive adopted in 1998 as a minimum standard).

Source: Spangenberg and Hinterberger (2002).

Only rudimentary efforts have been made to look at system properties and processes of integrated human, social, and economic systems. Many of the best-known indicators and indicator sets fail to include any such overall system properties, and they focus on more limited aspects of sustainable development (Chapter 11, this volume). For example, the CSD set assembles a large number of indicators but does not permit any judgment on the sustainable behavior of the system as a whole (Chapter 10, this volume). A more qualitative approach is to develop scenarios for alternative futures that span all the dimensions (e.g., Millennium Ecosystem Assessment 2005; UNEP 2002). These are sometimes generated by models (e.g., Meadows et al. 1992), in which case quantitative data and indicators can be both fed into and generated from the models. Modeling approaches provide tools to explore system behavior, identifying which factors are important and how sensitive the system is to variations in different indicators. Linking indicators with models will eventually provide more integrated perspectives on measuring progress toward sustainability.

Expanding the Temporal and Spatial Scales

Sustainability is a concept inherently related to time and space. What spatial unit should be sustained for how long? The approach to such questions differs between policymakers and scientists and depends on their focus on specific sectors or pillars of sustainable development.

With respect to the temporal dimension, each pillar is characterized predominantly by different time dynamics in, for example, lag times between cause and effect or the time horizon for policymaking. Sustainability in general is a long-term concept. Environmental issues have the longest range of temporal horizons, from floods or toxic emissions, to gradual changes over decades in the atmosphere, oceans, and climate caused by human action, to slow natural processes over millennia such as evolution and species formation, to the "death" of the sun. At the other extreme, economic issues involve very short-term decisions and impacts ranging from daily exchange figures to a few decades for infrastructure investments, with the future being discounted so that anything beyond that becomes irrelevant. Social issues generally fall between these two extremes, taking the length of a human life as an appropriate time frame, although negative effects on the social life of a generation, such as mass unemployment or poverty, can have impacts on the self-esteem and behavior of future generations (Arendt 1981).

The challenge for sustainability indicators is to anticipate such time lags and the trade-offs between the short and long terms. In developing highly aggregated indices, it may be worthwhile to consider weighting the time scales of the different pillars, giving higher weights to long-term or irreversible effects, for example, in order to improve their comparability.

For the spatial dimension, there are similar differences and specificities for each pillar. The relevant boundaries of a function or process may or may not correspond to the political boundaries of nation-states. Economic transactions increasingly span the globe, communities and cultures transcend national borders, and ecosystem boundaries range from puddles of water to biomes.[4] Yet indicators for all three pillars are generally remapped onto political (usually national) boundaries. Indicators for local communities are also common, but there are almost no indicators measuring the sustainability of the planetary biogeochemical life support system or of humanity as a species at the global scale.[5]

A consequence of this spatial fragmentation is that trends and drivers are easily hidden when analyzed at one particular scale. A nation or local community can appear sustainable if it does not consider its impact on the sustainability of close and distant neighbors. Similarly, indicators portraying good average values, for example in income, can hide significant inequities between subregions and societal groups at smaller spatial scales.

The challenge is to develop indicators that capture issues of sustainability at different spatial scales, in a nested hierarchical structure that links the scales with some scientific consistency while reflecting what can be managed at each level. The same indicator may have different meanings for sustainability in different contexts or when applied at different scales, so each use is context specific. Some approaches such as material flow analysis and energy flow analysis with proper data and modifications can be scaled up and down.[6]

Strong versus Weak Sustainability

One approach that discusses and even measures degrees of sustainability is the notion of weak and strong sustainability (Turner 1993). It was derived from the economic concept of capital, defined as a stock of resources with the capacity to give rise to flows of goods and services.[7] Ecological economists have expanded the concept to disaggregate the capital stock into four types (Ekins 1992) and linked them to the four dimensions of sustainability, including the institutional dimension (Spangenberg 2001):

- Manufactured capital: result of past material production (excess of output over immediate consumption)
- Human capital: people, skills, and knowledge
- Social and organizational capital: social networks and organization
- Natural capital: all features of nature providing resources for production and consumption, absorbing wastes, and furnishing amenities such as natural beauty

The issue is the substitutability of the different forms of capital in achieving sustainability of the whole system. Weak sustainability requires that the total capital stock (aggregated over the four types) does not decline. This presumes that all types of capital are substitutable in their capacity to generate human welfare and maintain system functioning. Strong sustainability requires that the stock of natural capital be maintained above critical levels. This assumes that substitutability regarding welfare generation is limited, or it applies sustainability criteria broader than welfare maximization. It entails the physical protection of certain absolute levels of natural capital, which cannot be substituted without provoking major and unpredictable system perturbations. The challenge for indicator development is not to give the final answer to the question of whether weak sustainability is sufficient but rather to map out where on the scale of weak to strong sustainability current drivers and policies are heading regarding the fraction of human well-being that can be expressed in monetary values. The subjectivity of assigning such values to all dimensions, describing them as different types of capital, is controversial, but it does enable the integration of social and natural aspects with the economic indicators that usually dominate the political agenda.[8] However, other elements of sustainability must be measured in other units, providing complementary but indispensable information.

A number of indicators and frameworks have been developed in the context of this discussion. For example, the "green" net national product has been proposed as a measure of the return on the aggregate of all capital types, but it requires reliable market values of all elements of the natural capital stock. On the other hand, the Critical Natural Capital (CRITINC) Framework recently introduced the concept of critical natural capital as those stocks of capital that cannot be substituted by other stocks of environmental or other capital to perform the same functions (Ekins et al. 2003). However, these indicators are very scale dependent, and evaluations of weak and strong sustainability must be considered at different temporal and spatial scales.

Finding the Planetary Limits

Carrying capacity is a familiar concept in ecology and refers to the population sizes of species that a particular ecosystem can sustain over time. In the context of measuring sustainability, it has been extended to the human species and refers to the numbers of people that can be maintained in the coupled human–environment system within planetary limits.

The difficulty in considering a species such as humans, who are able to raise the productivity of their own environment, is that carrying capacity, like sustainability itself, is a subjective and normative concept that depends on political choices of the spatial and temporal horizons considered and the preferred types of environmental, social, and economic systems. In environmental systems, for example, the carrying capacity depends on the limits set for the subsystem being analyzed and the acceptable level of degradation that can be tolerated in the system.

The only objective dimension to carrying capacity concerns the ultimate limits of maintaining conditions for life on the planet. Except for energy, the biosphere is essentially a closed system, and this imposes ultimate biophysical limits on growth in any material parameter at the planetary level. As resource limits are reached, further growth can come only from increases in efficiency.[9] Human technological development allows us to reach those limits and has even given us the military capacity to exterminate ourselves, and our ignorance of biosphere systems can give us the illusion that there is no need to be precautionary. The difficulty is that the inertia in planetary systems produces long time lags between our impacts and the resulting consequences (Meadows et al. 1992). This again justifies efforts to develop global-scale indicators (Chapter 10, this volume).

Although science should determine the ultimate biophysical carrying capacity of the planet as the outer limit for long-term sustainability, only subjective choices can decide the second crucial dimension of human carrying capacity: the acceptable standard of living for the people within the system. A finite set of resources can provide abundance for a few or bare subsistence for many. This gives sustainable development an important dimension of redistributive justice both in space (relative wealth and poverty) and in time (intergenerational equity).

The environmental space concept and Ecological Footprint index effectively communicate the concept of planetary carrying capacity in one dimension, the spatial one, by calculating how much space is needed to meet the needs of an individual, community, or nation, as related to the space available (EEA 1997).

Exploring Vulnerability and Resilience

Vulnerability and *resilience* are two terms that are increasingly used in scientific analysis of sustainable development from a system perspective. Whereas many concepts focus on the outer limits to sustainability, these terms apply to the inner limits to sustainability in a particular system. Although there is still a need to clarify and consolidate how these concepts are defined and applied, resilience is the capacity of the coupled human–environment system to cope with internal or external disturbance and its ability to adjust and adapt (Gutiérrez-Espeleta 1999). This applies to the social, economic, and environmental subsystems. Vulnerability is a characteristic of the lower end of the resilience spectrum. A system with high resilience has low vulnerability and vice versa. Systems need resilience not only to normal variations but also to extreme events, whether floods, droughts, or sudden drops in the stock market.

The increasing attention to these concepts has led to policy requests for indicators of resilience and vulnerability, such as in the UN Programme of Action for Small Island Developing States (United Nations 1994). Examples of responses are various economic vulnerability indices (Briguglio 1995) and the Environmental Vulnerability Index (EVI) (Pratt et al. 2004). The latter focuses on the environmental resources and ecosystems on which human society depends and profiles how vulnerable they are to further disturbance.

Indicating Irreversibility

When a system is not resilient enough to absorb disturbance and is degraded, the changes often are irreversible. It is the irreversible changes that are critical to sustainability. Irreversibility defines an absolute limit beyond which reestablishing the status quo is not possible. The concept of irreversibility is inherent in any analysis of coevolving systems. However, so far it is used primarily for environmental systems, describing events such as species extinctions or permanent loss of vital ecosystem functions, and is an essential part of identifying reference values. The approach is very different in the social and economic sphere. Social systems are characterized by permanent change, and irreversible changes would be those that have impacts lasting more than two generations, for the better or for the worse. In economics, whereas more recently emerged subdisciplines such as evolutionary and ecological economics analyze irreversibility and the resulting path dependency of system development as a characteristic of complex, nonlinear systems such as nature, society, and the economy, the neoclassical mainstream of economic thinking holds that everything in the economic system is reversible by definition, as reflected in the concept of weak sustainability.

The application of the concept of irreversibility in sustainability analysis depends on a number of factors:

• How critical the loss is to the overall system functions or productivity
• Whether substitutions for the loss are possible or desirable
• What compensations are needed to reduce the loss and the costs to the system
• What level of uncertainty is involved

The concept thus is not easily defined in scientific terms because it also depends on normative choices such as the social acceptance of compensation for degradation.

Defining irreversible limits to critical life support systems is a major conceptual challenge, as is the development of indicators providing early warning of the risk of irreversible damage. Setting such limits and predicting the risk for passing them are highly uncertain. Research has shown that the behavior of global systems such as biogeochemical cycles is characterized by thresholds and surprises (Steffen et al. 2002). Indeed, it was for situations with risk of irreversible damage that the precautionary principle was first formulated (EEA 2001). The Organisation for Economic Co-operation and Development (OECD) applies irreversibility only to ecosystems and the need to safeguard the natural processes capable of maintaining or restoring the integrity of ecosystems. Other areas of irreversibility may be difficult to define with the present state of knowledge and therefore are highly controversial.

Adding Meaning with Reference Values, Trends, and Targets

Indicators often are distinguished from raw data and statistics in that they are given meaning in relation to some type of reference value.[10] In the simplest case of two data points, the user interprets the trend indicated as positive or negative depending on the desired outcome. A reference value may be a baseline for which the indicator measures the distance to a meaningful state, such as a background value, standard, or norm. Or it can be a threshold value for irreversibility or instability, and the indicator measures the distance to a limit or point of no return. If a reference year has been set, the indicator measures changes over time related to that year, and a benchmark indicator measures differences between countries, companies, and so on. A reference value may become an explicit soft or hard target for policy, with distance-to-target indicators measuring the distance to the desired target or the limit to be avoided (see also Chapter 4, this volume). All types of reference values lend meaning and importance to data and therefore contribute to the function of indicators to communicate useful information.

Reference values are broadly accepted in such fields as health care, economics, environmental quality, climate change, and education. Physicians assess a patient's health by comparing measured values (e.g., blood pressure or blood sugar level) to baseline values corresponding to his or her sex, height, weight, and age. In the quality assessment of soil, water, and air, preindustrial background values play a prominent role. For biodiversity indicators, data on the number of species or the size of an animal population are meaningless without a baseline or reference value to which they can be compared, and there are a number of alternatives in this respect (Figure 2.1). A national species richness of 30,000 or a population of 1,000 dolphins is meaningful only when compared with a baseline value. The choice of that baseline is a normative and political challenge.[11]

A baseline in this context is not the targeted state. When policymakers have agreed on specific targets for an issue, they become another type of reference value to which indicators can be linked, such as the Millennium Development Goals (MDGs) (for a longer discussion of targets see Chapter 4, this volume). Most indices provide only relative rankings, as in country comparisons. Only the Environmental Vulnerability Index (Pratt et al. 2004) systematically proposes indicators referenced to specific parameters of environmental sustainability.

Avoiding "Data Drivenness"

Many indicator projects are driven by the availability of relevant and reliable data because indicators are useful only when there are sufficient data to give meaningful

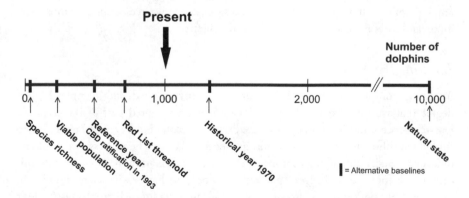

Figure 2.1. Six different baselines for one indicator value (1,000 dolphins present). A current population of 1,000 dolphins means different things when compared with historical data (1,300 in 1970), viability of the population (250), threat status (750) or the natural state (10,000) (Brink 2000).

results. In assembling indicator sets for sustainable development, data availability usually is a selection criterion to ensure rigorous quantitative underpinning. Even so, the limited quantity and quality of data underlying indicators of sustainability leave them open to criticism. Data collection is expensive, and countries are already under great pressure to supply data to international organizations from which they often receive multiple, overlapping, and uncoordinated requests for information. They are reluctant to accept indicators that imply new data collection.

This creates another conceptual challenge by producing biased and incomplete indicator sets that fall far short of measuring sustainability. We are forced to use indicators that were created for other purposes and describe only limited parts of the human–environment system. There are still extensive gaps in our knowledge, often reflecting inadequate supporting data. The result is both spatial and temporal bias. Scientific research and statistical data collection are strongest in industrialized countries, whose concerns and priorities dominate existing indicators.[12] Temporal biases come from the lack of long-term data sets and the concentration of most research on a very narrow time frame linked to the present.

It takes a long time to initiate new data collection processes, often 5–10 years, even in wealthy countries. Thus indicators being implemented now reflect issues identified at least 5 years ago. Finding the best indicators of sustainability entails breaking away from data availability constraints and determining the appropriate phenomena to monitor and the indicators needed. This implies switching from a deductive to an inductive research process. Data gaps can initially be filled with pilot collections and sampling, the use of remote sensing data, or the use of proxies (see Chapter 3, this volume).

Once development processes and sustainability issues are better understood and modeled with suitable indicators, it should be possible make data collection simpler and more flexible, for example with optimal spatial and temporal sampling, as a guide to institutionalizing long-term monitoring. However, changing data collection practices requires political and legal authorization and significant resources, calling for more flexibility and careful consideration of costs and benefits.

Limiting the Numbers

Analyzing complex systems and their properties involves reducing complexity to a degree that we can understand. Simplification is an accepted part of the scientific research process and is naturally associated with difficult choices about how much to simplify and how to do it without misrepresenting reality. The process of developing indicators entails simplifying complex and detailed information to provide communication tools for larger audiences. Specialists may be quite happy with a large number of indicators, but policymakers often request a single number for each problem to be dealt with. The latter may help policymakers attract attention but has limited usefulness in determining management action. The purpose and target audience determine the effective number of indicators, ideally the minimum necessary. At the same time, indicators of sustainable development should represent the large number of relevant issues. Selecting indicator sets and aggregating them into fewer indices are two of the most challenging aspects of indicator development.

Although few believe it is practical to develop just one aggregated index for sustainable development, incorporating all three pillars, there have been efforts to develop one index for each pillar. This still ignores the need for more systemic indicators linking the pillars that can provide higher-level integration. Any such cluster of highly aggregated indices has to be conceptually sound, with a supporting second layer of data that is easy to disaggregate, as demonstrated by the Dashboard of Sustainability, for example.[13]

Challenges in Process and Universality

The previous sections have amply demonstrated how many aspects of indicators involve normative choices or are biased by data availability. This raises questions about how indicators are developed, who is involved, and whose normative choices they reflect. Are indicators developed in one environmental, socioeconomic, and cultural setting valid in other settings? Can at least some indicators be universally applicable? The conceptual challenges raised by these types of questions require diligent efforts to make implicit value assumptions explicit and hidden biases visible.

Approaching the North–South Continuum

Countries are often compared by levels of development along a continuum from industrialized, developed countries (largely in the North) to the least developed countries (in the South). The size of this North–South divide and its increase over time are repeatedly confirmed by socioeconomic indicators in reports from various intergovernmental organizations. However, the conceptual challenge is to reflect the significant diversity of industrialized and developing countries in the design and selection of sustainable development indicators and determine how it affects their universal validity.

A number of indicators are clearly biased toward industrialized countries and their stage of socioeconomic development. In the economic domain, indicators such as GDP and income per capita are usually based on data for money flows generated through wage labor in the formal economy, partly because of the difficulty of defining and measuring nonmarket and subsistence activities. GDP has been repeatedly criticized for failing to incorporate the value of the informal economy, which constitutes a significant proportion of productive activity: 55 percent (Fukami 1999; Statistisches Bundesamt 1995, 2003), even in developed countries. In developing countries, the informal economy can reach up to 90 percent. The informal economy meets many development needs, such as child care, education, subsistence food production, water supply, fuel, housekeeping, handcrafts, and other domestic products. Where environmental conditions are good and traditional rural social and subsistence systems intact, the quality of life can be quite high, even in countries classed statistically as least developed. It would be valuable to determine how paid and unpaid work differ in their contributions to social, economic, and environmental sustainability.[14]

Another bias linked to the level of socioeconomic development lies in data availability and the effectiveness of statistical services. Most indicators originally were developed by industrialized countries, according to their own priorities, and reflect what they do best. Countries that are less developed economically may be more advanced in areas such as social cohesion or solidarity that are not captured in standard indicators. For example, an indicator of strong family relationships within and between generations would highlight the social and economic benefits of extended families, village communities, clans, and tribes, and the costs of the high divorce rates would be more important in countries without public health care and other welfare systems where people rely on these informal security systems.

The interpretation of indicators is often biased toward developed countries as well. For example, the number of cars in an industrialized country is usually used as an indicator of air pollution and consequent impact on human health. In a developing country, on the other hand, it may indicate improved access to markets and education. Different contexts may call for different indicators that have the same meaning. For example, coronary heart disease may be a more relevant health indicator in a developed country, whereas infant mortality may be more appropriate in a developing country.[15]

Recognition of this problem is leading to the revision of some indices. For example, the Growth Competitiveness Index of the World Economic Forum (WEF 2004), which has previously emphasized technological development, is being redesigned to reflect competitiveness at various stages of economic development.

Comparing Countries

Generally, peer pressure is considered a good thing among partners in any community, whether they be scientists, enterprises, or countries, if it leads to a healthy effort to strive for improvement and excellence. Comparison also helps to show what does and does not work and why. Many well-established national-level social and economic indicators and indices not only measure the development performance of a country over time but are also used to rank countries. For sustainable development indicators, there have been obstacles, both scientific and political, to the use of national-level indicators for worldwide country comparisons. The CSD indicators were explicitly endorsed by governments solely to measure each country's own sustainable development, by using a selection of indicators according to its own national priorities and circumstances, out of concern that intercountry rankings would be misused to impose conditions on development assistance.

Despite these concerns, good reasons remain for developing national-level indicators that allow country comparisons. All nations need to contribute to global sustainability.

The conceptual challenge is how to design indicators so that the comparison is legitimate and useful.

The approach to developing indicators suitable for intercountry comparison, whether at global or regional scale, should strive to

• Develop a sound, simple, and unified method for the selected indicators.
• Select indicators that reflect common agreed aspects of sustainable development or commonly agreed targets for action.
• Avoid indicators that are highly influenced by diversity in natural, socioeconomic, and cultural circumstances.
• Have full transparency of the whole process (development of indicators, methods, data collection, and presentation).
• Obtain agreement among the partners involved on the process, including public availability of results.

It is important to stress that indicators for intercountry comparison are only a complement to other indicator sets developed according to local, national, or regional priorities.

Reflecting Cultural Diversity

Different cultures usually have different views on what constitutes sustainable development. Such differences can be small variations in what types of economic or political policies should be adopted to promote sustainability, or they can represent significant divergences from the underlying development paradigm. This will influence both what a society would like to measure with indicators and which reference levels are seen as desirable or sustainable. The indicator sets most in use today are biased toward the dominant values of a Western-style market economy. For example, the reliance on GDP and related indicators reflects an economic development paradigm with a strong emphasis on the individual rather than the community and on material rather than social or spiritual dimensions of society, which may not be shared across all cultures.

This narrow focus also means that many aspects of society that are crucial to sustainability but are not part of the dominant political paradigm are absent in indicators. For example, there is a body of research on the importance of community values, such as trust and cooperation, for fostering collective action to manage common resources that could provide a basis for useful indicators (see Ostrom 1992).

The first challenge confronting indicator developers is to look again at the various indicator sets, particularly those used for intercountry comparison on sustainable development, to see whether significant cultural biases can be identified and made transparent, if not reduced. The second challenge is to develop indicators for a broader range of sustainable development issues identified within cultures that are largely underrepresented in the scientific and political debate on indicators.

Despite such cultural differences, there are still many values common to all human beings that should be reflected at the core of any indicator set. Everybody, regardless of culture, needs a minimum amount of food, clean water and air, shelter, space, health care, security, self-respect, social relations, respect for other living beings, and time, access, and opportunity to develop one's abilities.

The need to preserve the ecological balance of the world is also universal. The ability of diverse cultures and countries to agree on common values and priorities and to reflect them in indicators is exemplified in the Convention of Biological Diversity, where indicators to address different aspects of biodiversity at the ecosystem, species, and genetic levels were agreed on in 2004 (Conference of the Parties 2004). Although a common target is set for these indicators to achieve a significant reduction in the loss of biodiversity by 2010, countries are free to choose more ambitious targets. Another example is the MDGs and their derived targets and indicators for areas such as food, water, and health (United Nations General Assembly 2000).

Closing In on Equity

Global sustainability is a concept with solid physical limits, but sharing responsibility below that level is largely about how much is fair and for whom. Equity and justice are implicit in the sustainable development concept, both temporally in intergenerational equity, respecting the development needs of future generations, and spatially in intragenerational equity, stressing poverty eradication today (Chapter 19, this volume).

Most of the focus on equity and its measurement is at the lower end, at the extremes of poverty, focusing on the ability of people to meet basic needs. Less attention has been paid to the upper end of the equity continuum, the extremes of wealth and related overconsumption. For example, there are limited data on wealth at the national level, even for a proxy such as the number of millionaires, and few indicators of overconsumption (UNDP 1998). Because measurement often leads to management, there are strong incentives to ensure a lack of political attention on the issue of wealth.

National-level indicators that aggregate data into averages can hide significant inequity. National economic statistics are not easily disaggregated to measure equity along the gradients between rich and poor, urban and rural, men and women, and children and adults, or between racial or ethnic groups. The Gini coefficient captures income inequity within countries, and recent editions of the United Nations Development Program Human Development Report have highlighted aspects of social inequity (UNDP 2004). However, the conceptual challenge is to develop a range of indicators that capture the equity dimension of sustainable development.

Closing In on Democracy

The concept of participation and majority decision making expressed in the term *democracy* is related to equity. Although democracy may be interpreted differently in various intercultural contexts, there is a claim for democracy as a universal principle for institutionalizing sustainable development. This can include access to and participation in processes of generating knowledge, developing indicators, and using them to guide action. There is a risk that certain ideas become embedded as authoritative in the conceptual framework and governance of sustainable development, whereas others are marginalized. Given the normative dimensions of indicators and the biases they contain, democratic processes are particularly necessary to ensure access to and inclusion of different types and sources of knowledge in indicator development (Berkhout et al. 2003:25). This entails engaging scientists and users from a much broader spectrum of countries (particularly developing countries), cultures, and disciplines (see Chapter 4, this volume).

Multistakeholder processes of dialogue, decision making, and implementation are increasingly institutionalized across governance levels, as in local Agenda 21 roundtables, the practices adopted by the CSD, and the emphasis on partnerships at the WSSD (see UNCED 1993; United Nations 2002; CSD 2004). Principle 10 of the Rio Declaration outlines the right of access to information, participation, and justice embodied in the Aarhus Convention.[16] Although there are indicators designed to account for the degree of implementation of democratic principles, most developed by nongovernment organizations such as the Corruption Perceptions Index (Lambsdorff 2003), the International Standards of Elections (OSCE 1990), the Worldwide Press Freedom Index (Reporters without Borders 2004), and key indicators for the violation of human rights (Amnesty International 2004), the challenge remains to develop indicators for democratic practice concerning sustainable development.[17]

Winning Acceptance

The need for indicator users, particularly decision makers in political processes, to agree to and take ownership of sustainable development indicators creates its own conceptual challenge. Politicians, corporate executives, and many other senior officials can retain their positions only if they are seen to do well. Any indicator that reflects well on their performance will be supported, but indicators that show they are doing badly will meet strong opposition or rejection. In the current state of the world, most indicators of sustainability show how unsustainable present trends are. Given the proverbial tendency to shoot the messenger bearing bad news, it is very difficult to win acceptance for indicators that reflect negatively on the performance of decision makers. Only a careful indicator development process, and often peer pressure from others who support the process or from a demanding electorate, can win reluctant consensus to adopt and use sustainability indicators. The process itself must be seen as transparent, inclusive, fair, and legitimate, and strive to be independent of pressures from narrow organizational or personal self-interest, if it is to succeed.

Conclusions

This review of the conceptual challenges in developing indicators of sustainable development shows both progress made and work remaining. Table 2.2 assesses the stage of development of existing indicator efforts to meet each conceptual challenge discussed in this chapter.

Some indicators are ready for use, such as for linkage between economy and environment, economic equity, and certain efforts to consider the spatial aspects of sustainability. Most challenges still need further research and development, including the following:

• Developing indicators for global sustainability and planetary limits
• Completing indicator sets for each pillar with reference values
• Continuing to develop linkage indicators between pillars
• Exploring alternative models for understanding the sustainability of systems and indicators of system characteristics
• Interrelating different temporal and spatial scales, especially the short-term economic and long-term environmental perspectives
• Correcting the balance of indicators relevant to countries' stages of development and cultural diversity
• Distinguishing the objective and normative components of indicators and developing indicators for dimensions such as equity and participation

The significant progress over the last decade has made it possible to focus more precisely on the remaining conceptual challenges and to define some ways forward. Many indicators relevant to sustainable development have been assembled, and although many gaps remain, it is already possible to start addressing the key issue of integration. Economic, social, and environmental subsystems interact so fundamentally that all must be considered together in an exploration of feasible pathways toward sustainable development. In particular, successful governance for sustainable development depends on an appropriate analysis of these links.

A system perspective alerts us to the complex and often nonlinear character of these subsystems, to the existence of thresholds for irreversible switching from one stable system state to another, and to other surprises that are difficult to unravel and even more difficult to predict. Thus, a system perspective implies a more humble approach to governance, recognizing the limits of our ability to fully understand and control the impact of policies and actions. It encourages us to use indicators in a more responsive learning mode, acknowledging the need for wide participation in adaptive management to achieve the dynamic state that is sustainability.

Table 2.2. Stage of development in indicators to meet conceptual challenges.

Sustainability			Process and Universality		
Challenge	Stage of Development		Challenge	Stage of Development	
Measuring sustainability and sustainable development	In general	I	North–South issues	For strengths of countries	I
	Some indicators capturing some components of sustainability	II		For Northern development bias	II
Frameworks	Biodiversity indicators	III	Country comparability	For indicator definitions	III
Linkages between pillars	Economy–environment linkage	III	Cultural diversity	International	I
	Social–environment linkage	I		National (some countries)	II
Temporal and spatial scales	In general terms	I	Equity	Environmental equity	I
	In individual cases	III		Social and institutional equity	II
Strong–weak sustainability	Where applicable	I		Economic equity	III
Planetary limits		I	Democracy		II–III
Vulnerability–resilience	Environmental systems	II	Acceptability		II
	Social and economic systems	I			
Irreversibility		I			

Note: The stages of indicator development are classified as follows: I, research stage; II, in development, some progress made; III, ready for implementation.

Notes

1. See Spangenberg (Chapter 7, this volume) for a detailed discussion of the institutional dimension of sustainable development and related indicator developments. See also Spangenberg et al. (2002) and Spangenberg (2002).

2. For indicators in use, see Rosenström and Palosaari (2000), UCR and MINAE (2002), and the list of indicator Web sites in the Annex.

3. A prominent example of indicators developed within this framework is the decoupling indicators, focusing on the links between the driving force and pressure component (Chapter 13, this volume). This can refer to pressures on the environment from material and energy flows and land requirements, as in the Geobiosphere Load Index (Chapter 14, this volume). Haberl et al. (Chapter 17, this volume) argue for pressure indicators for biodiversity loss and describe one example of a comprehensive pressure indicator to meet this need. For health, a modified framework of driving force, pressures, state, exposures, health effects, and actions (DPSEEA) has been applied (Chapter 15, this volume).

4. These scales are expanding with globalization. Jesinghaus (Chapter 5, this volume) discusses indicator approaches to measure various aspects and impacts (good or bad) of globalization.

5. Eisenmenger et al. (Chapter 12, this volume) discuss how global domestic extraction of raw materials can be related to specific scarcities such as global net primary production of biomass.

6. See Eisenmenger et al. (Chapter 12, this volume) for a detailed description of the approaches to make material flow analysis take into account transnational material flows and Moldan et al. (Chapter 14, this volume) for a similar introduction to energy flow analysis.

7. It is also closely associated with the evolution of the term *natural capital* (Victor 1991). Knippenberg et al. (Chapter 19, this volume) explore the concept of capital in discussions on sustainable development and show the normative implications of its use.

8. Zylicz (Chapter 6, this volume) discusses the value of greening GDP as a way to improve social welfare measures without having to assign relative weights to various aspects of well-being, which is often done in sustainable development indices.

9. Domestic extraction of raw materials (DE), when measured in DE per unit GDP at the global level, expresses the overall material intensity of the total human economy (Chapter 12, this volume).

10. The EEA database on Sustainability Targets and Reference Values contains definitions and links (star.eea.eu.int/default.asp).

11. Biggs et al. (Chapter 16, this volume) outline the baselines that have been proposed for biodiversity indicators by the Convention on Biodiversity.

12. Indeed, there are significant gaps in data collection for indicators in many OECD countries, such as for decoupling indicators (Chapter 13, this volume). For a discussion of the divide between the North and the South in scientific capacity in general and environmental knowledge production in particular, see Karlsson (2002).

13. See esl.jrc.it/envind/dashbrds.htm and Jesinghaus (Chapter 5, this volume) for

information on the Dashboard of Sustainability. Bauler et al. (Chapter 3, this volume) discuss the methodological challenges of aggregation in more detail.

14. This difference has clear gender dimensions, for example in how society values reproductive and caring work (Chapter 7, this volume).

15. von Schirnding (Chapter 15, this volume) discusses environmental health indicators in more detail.

16. The full name is the Convention on Access to Information, Public Participation in Decision-Making and Access to Justice in Environmental Matters (www.unece.org/env/pp).

17. Jesinghaus (Chapter 5, this volume) lists some democracy-related indicators in a cluster called "Culture and Governance."

Literature Cited

Amnesty International. 2004. *Amnesty International report 2004.* London: Amnesty International.

Arendt, H. 1981. (reprint). *Vita activa oder vom tätigen Leben.* Munich: R. Piper.

Berkhout, F., M. Leach, and I. Scoones. 2003. Shifting perspectives in environmental social science. Pp. 1–31 in *Negotiating environmental change. New perspectives from social science,* edited by F. Berkhout, M. Leach, and I. Scoones. Cheltenham, UK: Edward Elgar.

Böge, S. 1994. Die Transportaufwandanalyse. Ein Instrument zur Erfassung und Auswertung des betrieblichen Verkehrs. *Wuppertal Papers* 21:1–21.

Briguglio, L. 1995. Small island developing states and their economic vulnerabilities. *World Development* 23(9):1615–1632.

Brink, B. J. E. ten. 2000. *Biodiversity indicators for the OECD environmental outlook and strategy, a feasibility study.* RIVM report 402001014, in cooperation with WCMC. Bilthoven, Netherlands: National Institute for Public Health and the Environment.

Conference of the Parties. 2004. *Convention on Biological Diversity.* COP Decision VII/30. UNEP/CBD/COP/7/21, available at www.biodiv.org/decisions/default.asp?lg=0&m=cop-07.

CSD. 2004. *Commission on Sustainable Development 12th session, Review of thematic issues, Chair's Summary Part I and II.* Final unedited version. Commission on Sustainable Development. www.un.org/esa/sustdev/.

EEA. 1997. *The concept of environmental space: Implications for policies, environmental reporting and assessments.* Experts' Corner: 1997/2, edited by J. Hille. Copenhagen: European Environment Agency.

EEA. 2001. *Late lessons from early warnings: The precautionary principle 1896–2000.* Environmental issue report no. 22. Copenhagen: European Environment Agency.

Ekins, P. 1992. A four capital model for wealth creation. Pp. 147–153 in *Real-life economics: Understanding wealth creation,* edited by P. Ekins and M. Max-Neef. London: Routledge.

Ekins, P., S. Simon, L. Deutsch, C. Folke, and R. De Groot. 2003. A framework for the practical application of the concepts of critical natural capital and strong sustainability. *Ecological Economics* 44(2–3):165–185.

Eurostat European Statistical Office. 2001. *Material use indicators for the European Union, 1980–1997.* Eurostat Working Papers 2/2001/B/2. Luxembourg: Office for the Official Publications of the European Unions.

Fukami, M. 1999. *Monetary valuation of unpaid work in 1996 in Japan.* International Seminar on Time Use Studies, Ahmedabad, India.

Gutiérrez-Espeleta, E. 1999. *Environmental sustainability: A nonsense approach.* Paper presented at the Annual Conference of the Norwegian Association for Development Research, September 16–17, Oslo.

Karlsson, S. 2002. The North–South knowledge divide: Consequences for global environmental governance. Pp. 53–76 in *Global environmental governance: Options & opportunities,* edited by D. C. Esty and M. H. Ivanova. New Haven, CT: Yale School of Forestry and Environmental Studies.

Lambsdorff, J. G. 2003. *Background paper to the 2003 Corruption Perceptions Index.* Framework Document, available at www.transparency.org/cpi/2003/dnld/framework.pdf.

Meadows, D. H., D. L. Meadows, and J. Randers. 1992. *Beyond the limits: Confronting global collapse, envisioning a sustainable future.* Post Mills, VT: Chelsea Green.

Millennium Ecosystem Assessment. 2005. *Ecosystems and human well-being: Scenarios,* Vol. 2, *Findings of the Scenarios Working Group.* Washington, DC: Island Press.

OECD. 2001. *Environmental indicators to measure decoupling of environmental pressure from economic growth.* ENV/EPOC(2001)26. Paris: OECD.

OSCE. 1990. *International standards of elections.* Document of the Copenhagen Meeting of the Conference on the Human Dimension of the Council of the Conference for Security and Co-operation in Europe, Section 6–8, Copenhagen. Available at www.osce.org/documents/odihr/1990/06/1704_en.html.

Ostrom, E. 1992. The rudiments of a theory of the origins, survival, and performance of common-property institutions. Pp. 293–318 in *Making the commons work: Theory, practice and policy,* edited by D. W. Bromley. San Francisco: ICS Press.

Pratt, C., U. Kaly, J. Mitchell, and R. Howorth. 2004. *The Environmental Vulnerability Index (EVI): Update and final steps to completion.* SOPAC Technical Report 369. United Nations Environment Programme (UNEP), South Pacific Applied Geoscience Commission (SOPAC).

Prescott-Allen, R. 2001. *The Well-being of Nations: A country-by-country index of quality of life and the environment.* Washington, DC: Island Press.

Reporters without Borders. 2004. *Worldwide press freedom index.* Paris: Reporters without Borders. Available at www.rsf.org.

Rosenström, U., and M. Palosaari (eds.). 2000. *Signs of sustainability: Finland's indicators for sustainable development 2000.* Helsinki: Finnish Environment Institute.

SDC. 2004. *Shows promise. But must try harder.* London: Sustainable Development Commission.

Spangenberg, J. H. 2001. Investing in sustainable development. *International Journal of Sustainable Development* 4:184–201.

Spangenberg, J. H. 2002. Institutional sustainability indicators: An analysis of the institutions in Agenda 21 and a draft set of indicators for monitoring their effectivity. *Sustainable Development* 10:103–115.

Spangenberg, J. H., and F. Hinterberger. 2002. *Post Barcelona: Beyond Barcelona.* SERI discussion paper, Vienna, available at www.seri.de.

Spangenberg, J. H., S. Pfahl, and K. Deller. 2002. Towards indicators for institutional sustainability: Lessons from an analysis of Agenda 21. *Ecological Indicators* 2:61–77.

Statistisches Bundesamt. 1995. *Die Zeitverwendung der Bevölkerung. Familien und Haushalte 1991/1992* (Tabellenband III). Wiesbaden.

Statistisches Bundesamt. 2003. *Wo bleibt die Zeit? Die Zeitverwendung der Bevölkerung 2001/2002.* Berlin: Bundesministerium für Familie, Senioren, Frauen und Jugend.

Steffen, W., J. Jäger, D. Carson, and C. Bradshaw (eds.). 2002. *Challenges of a changing Earth.* Berlin: Springer Verlag.

Turner, R. K. 1993. Sustainability: Principles and practice. Pp. 3–36 in *Sustainable environmental economics and management: Principles and practice,* edited by R. K. Turner. London: Belhaven Press.

UCR and MINAE. 2002. *Indicadores del desarrollo sostenible de Costa Rica.* San José, Costa Rica: Universidad de Costa Rica (UCR) and Ministerio de Ambiente y Energía (MINAE).

UNCED. 1993. *Report of the United Nations Conference on Environment and Development.* Rio de Janeiro, June 3–14, 1992, A/CONF.151/26.Rev.1. New York: United Nations.

UN Division for Sustainable Development. 2001. *Indicators of sustainable development: Guidelines and methodologies,* 2nd ed. New York: United Nations.

UNDP. 1998. *Human development report 1998.* New York: Oxford University Press.

UNDP. 2004. *Human development report 2004: Cultural liberty in today's diverse world.* New York: United Nations Development Program.

UNEP. 2002. *Global environment outlook 3.* London: Earthscan.

United Nations. 1994. *Report of the Global Conference on the Sustainable Development of Small Island Developing States,* Bridgetown, Barbados, April 25–May 6. A/CONF.167/9, October 1994. New York: United Nations.

United Nations. 2002. Plan of implementation of the World Summit on Sustainable Development. In *Report of the World Summit on Sustainable Development,* Johannesburg, South Africa, August 26–September 4. A/CONF.199/20. New York: United Nations.

United Nations General Assembly. 2000. *Resolution adopted by the General Assembly: 55/2.* United Nations Millennium Declaration A/RES/55/2, September 18. New York: United Nations.

Victor, P. A. 1991. Indicators of sustainable development. Some lessons from capital theory. *Ecological Economics* 4:191–213.

WEF (World Economic Forum). 2004. *The global competitiveness report 2003–2004.* New York: Oxford University Press.

3

Identifying Methodological Challenges

Tom Bauler, Ian Douglas, Peter Daniels, Volodymyr Demkine, Nina Eisenmenger, Jasper Grosskurth, Tomáš Hák, Luuk Knippenberg, Jock Martin, Peter Mederly, Robert Prescott-Allen, Robert Scholes, and Jaap van Woerden

The methodological challenge in deriving indicators for sustainable development lies in constructing indicators that are accurate representations of environmental or societal states or trends but are easily understood by their target audiences. Methodological challenges thus involve two broad sets of questions: those concerned with the design and development of indicators and those concerned with the purpose and use of indicators. Basic concerns over data availability, data quality, and the adequacy of the algorithms used can be resolved largely through technical, scientific agreement. However, the central issue of adjusting methods to indicator relevance and use has to be addressed through trade-offs between form and function in specific societal and political settings.

Constructing a sustainable development indicator raises such methodological issues as the multidimensionality of domains, the complexity of the socioenvironmental system under scrutiny, and the presence of cross-scale (both temporal and spatial) effects and impacts. The translation of these issues into coherent procedural and substantive methods largely determines the formal quality of the assessment tool.

The use and purpose of the indicator are part of the process of developing awareness of sustainable development (i.e., contributing to the self-generation of sustainable development; see Chapter 1, this volume), essentially an iterative process wherein new indicators are developed, tested, reformulated, and improved as a result of interaction with and feedback from users in all walks of society. This chapter concentrates on the methodological challenges in the development of indicators. First we break sustainable development down into a hierarchical setting of subdomains necessary for the construction of indicators. Such constructed frameworks are necessary in order to link explicitly the different indicators stemming from different origins (economic, social, and environmental). Linked to the hierarchical level of the indicator is the definition of its degree of aggregation and the adequacy of the link between the type of the indicator and the particular use that is made of it. Indicator construction reflects a series of trade-offs between antagonistic, or at least noncomplementary, criteria. After having sketched five criteria, which can help us assess the methodological strength of an individual indicator, we apply them to a limited series of headline economic indicators. Finally, the last sections of the chapter link challenges of policy relevance and conceptual issues with methodological challenges as the three types of issues that occur during the construction of an indicator.

The complex issues related to the practical use of indicators and their acceptability to stakeholders in government, the private sector, and civil society are explored less in this chapter but are addressed in Chapter 4. Chapters 2, 3, and 4 should be read together.

Defining the Subdomains of Sustainable Development: Enhancing Methodological Transparency in Indicator Formulation

The concept of sustainable development recognizes that life depends on the earth's biophysical support systems. The state of the planet and its ecosystems is at least in part but often almost wholly a consequence of past and present human activity, determined by the interplay of social, economic, and political factors. Any overall indicator of the state or trends in sustainable use of the planet must reflect this interplay of Earth systems and ecological dynamics with economics, politics, and social dynamics. To evolve, good indicators need two dialogues: one between the scholars of natural science and social science, to achieve an academic consensus on indicators and the method for their construction, and the other among the scholars, the users, and society. Neither of these dialogues is an everyday occurrence. This chapter reflects the first kind of dialogue, but the second type of dialogue will have to assume an increasingly important role if indicators for sustainable development are to gain prominence in everyday thinking. A single indicator of sustainable development (the SD index) would be a valuable communication tool and could capture the public imagination. However, it might obscure the central issue of sustainable development: learning to recognize and evaluate the existence of trade-offs between the achievement of high levels of human well-being while restoring and maintaining ecosystem integrity. Such a two-domain framework facilitates both the equal treatment of people and the environment and the analysis and communication of interactions between people and ecosystems. Further subdivision of the human side (into social and economic domains, for instance) would reduce the weight given to the environment from half to one third or, if an institutional domain is added, to a quarter (Figure 3.1). Methodologically, increasing the number of subdivisions complicates the rep-

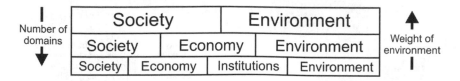

Figure 3.1. The weight of the environment decreases as the number of human domains increases.

resentation of causalities and feedback relationships between the domains and reduces our ability to develop accurate indicators of the interlinkages of domains. It also increases the difficulty of communicating the causes and directions of trends and hence of desirable pathways toward sustainable development.

In order to cope with the emphasis given to the role of the economy, still the main preoccupation of decision making, a division into three domains appears suitable. Such a threefold subdivision is really a methodological concession to political and societal adoption and use of indicators. Because indicators are usefully defined only in relation to their practical applicability for decision making, such concessions are inherent in their nature as decision-making instruments.

Adding a fourth, institutional domain would confer no advantage other than its current acceptance by the UN Commission on Sustainable Development (CSD). The more domains there are, the harder it is to portray human–ecosystem interactions clearly or to distinguish interactions that are real from those that are artifacts stemming from the design of the framework.

The contrasting economic, social, and environmental dimensions (domains) of sustainable development should not be oversimplified. They warrant a closer examination, best envisaged in terms of a hierarchy of indicators (Figure 3.2). The uppermost level of the hierarchy contains broad, widely held, but unavoidably normative concepts linked to the societal and cultural understanding of sustainable development (e.g., in terms of environmental, social, and economic capitals). Below these broad concepts, a limited number of headline issues that are widely, and beyond polity, agreed to be important may be understood as critical issues for human development for the coming decades. Each of these issues (e.g., climate change, biodiversity, or perception of well-being) is multidimensional (Atkinson et al. 2002) and synthesizes a complex range of processes and conditions, with an implicit value-laden and culturally influenced normative subdiscourse. Nevertheless, scientifically defensible aggregation frameworks for formulating indices or composite indicators at this level are possible. Within expert and policy-formulating communities, it should be possible to disaggregate the issues into their components, in a transparent and traceable fashion.

The individual component issues necessary to build the composite (headline)

	Sustainable development		
	Environmental	Social	Economic
Headline issues (about 10)	Biodiversity Air quality Water use and quality Land use Energy Resource use Climate change	Health and security Knowledge and education Perception of well being Institutional capacities	Material and energy flows and intensities Income distribution Economic growth Debt servicing
Component indicators (about 100)	Mean global temperature Tropospheric ozone N and S deposition River chemistry Area of forests Fish stock status etc.	Healthy life expectancy Disability losses Crime rate Participatory institutions School enrollment Literacy etc.	Gini coefficient Losses to natural disasters Waste intensities etc.
Base data	Land cover etc.	Population etc.	Economic structure etc.

Figure 3.2. Hierarchy and scales, from sustainable development to base data.

indicators will change over time, as will the importance assigned to them (e.g., when they are weighted in aggregated indicators), as particular issues change in significance in public and government perception and as scientific knowledge and information increase. Many component issues are selected because they are directly affected by policy (e.g., protected areas). Several uncontroversial but essential data sets of variables, such as population density, land cover, and economic structure, form the base of the indicator hierarchy.

The emphasis on links between the three domains provides the opportunity to design and produce integrating indicators carrying more powerful and detailed messages about the understanding of progress in various elements of sustainable development. In methodological terms, this opportunity poses further challenges in terms of designing frameworks and models that allow different data sets from diverse origins (e.g., combining biodiversity data sets with data on rural development) to be integrated in meaningful and transparent ways that can be readily communicated. With the increase in the number of domains, the methodological strength and communicative capacity of the frameworks (e.g., driving force, pressure, state, impact, and response [DPSIR] frameworks) developed to link the domains coherently and comprehensively will increase in importance. Although links between some issues built on adequate methodological and scientific foundations (e.g., energy–economy decoupling) are being widely used, many links, particularly between the environment and society, need much more work. Much research on these issues is needed to turn sustainable development indicators into decision-making tools that will help to identify alternative ways to promote sustainable development.

Adjusting the Level of Aggregation to Purpose

Aggregation is the combination of many components into one. One important role of aggregation is to extract information from data. The performance of an economy cannot be determined accurately from a few businesses, nor can the state of biodiversity be determined from the presence or absence of a single species. The aggregation of many components (transactions in an economy, species in an ecoregion) is needed to produce meaningful information.

Another role of aggregation is to produce information in a way that enhances communication. When indicators are combined into aggregates, they can provide a better picture of the entire system by concentrating on key relationships between subsystems and between major components and facilitate analysis of critical strengths and weaknesses. No information is lost if the constituent indicators, the underlying data, and the algorithms are there to be queried. However, the value of such queries depends on the technical capabilities of users. Without such capabilities, users might interpret aggregation as a loss in transparency.

Three main types of aggregates can be identified:

Aggregated indicators. These include summations of accounts constructed from raw data measured in the same unit, such as the System of National Accounts (money), material accounts (weight), and energy accounts (energy). The data are aggregated by simple addition, with no need for weighting. Examples are the gross domestic product (GDP), Total Material Requirement, and Total Energy Requirement. Reliability is affected by completeness of data coverage and the organizational consistency of the accounting framework.

Synthetic indicators. These are summations of data not derived from accounts. They combine the large number of measurements (or estimates) necessary to produce indicators of phenomena comprising many variables and rendered in a common unit, such as human health and longevity, species diversity, and freedom and security. Examples are health-adjusted life expectancy at birth (years of life minus years lost to disease and injury) and the Biodiversity Intactness Index (numbers of native species minus estimated numbers lost as a result of land use activities).

Indices. These are combinations of lower-level indicators. When indicators measure the same class of components and are in a common unit (e.g., a city's air quality index), aggregation is straightforward. It is more complex when many different components are measured in unlike units, as in the Human Development Index, the Well-being Index, the Environmental Sustainability Index, and the indices produced via the Dashboard of Sustainability and Compass of Sustainability. All of these indices convert indicator measurements to a performance score by applying standardized statistical normalization methods. They differ in how this is done and in the rigor of the procedures used to combine different components.

Aggregation requires measurements in the same unit. Transparency and reliability are affected by the method of converting base data to a common unit and by the procedure for combining (normalized) base data from different components. Indices are more prone to distortion because they combine unlike components. But aggregated and synthetic indicators are not immune either.

Simple base data may also aggregate information. In some cases this is desirable, as in measurement of water quality at the mouth (or downstream frontier point) of a river, which provides a summation of the water quality of the basin. In other cases it is undesirable, as when an average value masks major variations in performance within the spatial unit concerned.

Articulating between Types of Indicators

Most existing sustainability indicators are entirely quantitative. They are based on quantitative measurements of variables, from which indicators and indices are derived. Some definitions of indicators identify quantification as a defining part of indicators alongside simplification and communication (Adriaanse 1993). Reliance on quantitative indicators poses a limitation with severe repercussions for sustainability assessment. Their quantitative nature means that issues measured qualitatively are less likely to be integrated into sustainable development assessments, regardless of their relevance for sustainability. As mentioned earlier, it is possible not only to communicate information in qualitative terms but also to process qualitative information by using indicators. Especially in the social sciences, indicators based on qualitatively obtained data (e.g., surveys of happiness, compliance, or agreement) are increasingly important. These data are not easily interpreted and are even more difficult to update in a robust fashion. Their integration with quantitative data remains a critical methodological challenge. The feasibility and reliability criteria for indicators, whether quantitative or qualitative, relate strictly to the scientific quality of the acquisition, reliability, and treatment of the data from which they are derived.

For a number of issues that are hard to address directly with an adequate indicator (e.g., missing data, insufficient knowledge of interactions), proxy or substitute indicators are widely used. Two broad kinds of proxy indicators can be identified: proxies as representations of complex systems (e.g., number of bird species instead of local ecosystem biodiversity) and proxies as metaphors (e.g., treaty signature instead of degree of implementation). If the first type can be very useful for communicating complex issues, the second type of proxy indicator is prone to oversimplification and value-laden assessments. For instance, bird presence has been used as proxy for certain insect populations or even for biodiversity as a whole (in the United Kingdom), suicide rates serve as proxy for a series of social issues, and GDP is used as a proxy for welfare. None of the indicators comprehensively represents the issue it is a proxy for, but each one should at least move in the same direction as that issue and thereby usefully detect and signal general changes. However, proxy indicators are much less suitable for identifying the precise dynamics of change and possible policy intervention levers. Like any indicator, proxies can be difficult to interpret and lead to wrong conclusions about the actual state of the system, as can be illustrated with the indicator "protected areas as percentage of total land area." On one hand, the higher the percentage of protected land, the stronger the policy implementation probably is. However, the higher the percentage, the more areas could have been proven to need protection, which means that the former conservation policy failed or that human activities have unacceptably high levels of impacts.

With reference to the policymaking cycle, indicators are occasionally characterized as input or output indicators. Input indicators are measurements of the procedural or substantive means engaged by policy actors to influence a condition (e.g., ratio of budgets assigned to control compliance to environmental legislation, taxes levied according to the polluter-pays principle). They are meant to provide the different types of policy actors (e.g., enterprises, consumers, politicians, lobbyists, civil servants) with an insight into the existence, potential, and performance of policy levers or societal responses. At the other end of the causal chain, output indicators measure the evolution of the identified problem itself (e.g., the state of an ecosystem). It is generally acknowledged that policy actors need such output indicators to increase their awareness of the problems but that input indicators are more appealing to them when they define policies and responses because input indicators hint more directly at the necessary implementation schemes, levers for change, or behavior adaptations.

Finally, with reference to causalities and change, two general types of indicators exist: state or stock indicators and rate or flow indicators. Especially in the environmental domain, state or stock indicators are of major importance in assessing the evolutions of systems with regard to their limits (e.g., amount of biodiversity for a given ecosystem). However, for most issues it is impossible to determine the sustainable level of the relevant stock (e.g., the necessary amount of biodiversity for the given ecosystem). In order to avoid the inherent difficulty of defining limits of acceptability (or carrying capacities), many indicator initiatives focus on the development of rate or flow indicators. Furthermore, rate or flow indicators are often more policy relevant in the short term and more attractive in a political business cycle (typically of 4–5 years). In this period, the direction and intensity of flows (e.g., CO^2 emissions) can be influenced by policy and behavior in the short or medium term, whereas stocks or states are often characterized by inertia. On a sustainability time scale of decades or centuries, ignoring stock-related indicators in science and policy will hide fundamental system properties from users. Given current knowledge, improving the articulation between stock and rate indicators is a widely underestimated necessity for sustainability assessments.

Criteria for Methodological Strength of Indicators

The development of indicators is a matter of concessions and compromise. The quality of indicators reflects the developers' dexterity in responding to and anticipating a number of constraints. The way an indicator is used and performs depends not only on how it responds to individual criteria applicable at the level of the individual indicator but also on whether it responds in a balanced way to the sum of criteria as an interconnection of constraints.

Criteria can be developed on a number of levels, from technical quality criteria intervening at the level of the statistical nature of the data to quality criteria related to the usability of the indicators. The errors in developing indicators are inextricably linked to the danger of measuring the wrong issues perfectly or the correct issues inadequately.

Constructing information on integrated and complex issues is bound by our imperfect understanding of reality (i.e., constraint by the measurable), and for issues where reasonable understanding exists, developers of indicators are bound by what is actually measured (i.e., data availability) and how it is measured (i.e., data quality). Epistemologically, our understanding of reality is limited by knowledge gaps and our inability to measure, simplify, and compare many of the complex factors involved in sustainable development. Technically, any representation of reality is limited by the quality, accessibility, and reproducibility of the background data. Societally, integrated assessments that are concerned with multidimensional issues are influenced by the way the issues involved are interpreted.

To help indicator developers overcome the constraints encountered during their work, relevant metacriteria can be set out to make the constraints on the level of the indicator selection and construction more transparent (Table 3.1):

Purpose: This refers to why an indicator exists, the appropriateness of scale, and the accuracy with which it links its purpose to the general concept of sustainable development. The quality of an indicator lies in the way it addresses its purpose and provides clear information on the state or trend of some aspect of sustainable development. The appropriateness of the scale at which an indicator is usable is a secondary consideration. Although some indicators may be used intentionally for cross-scale comparisons, generally the scale at which an indicator is to be used must be clearly defined. (See Box 3.1.)

Measurability: This refers to how the values in an indicator are measured and the extent to which it measures reality. Even though indicators necessarily limit themselves to the sphere of the measurable, their link to reality is imperfect to varying degrees because they use sample measures taken on specific days or at specific times and locations. They are also limited by the way the raw data reflecting reality are translated into quantitative data and measures. (See Box 3.2.)

Representativeness: This refers to the completeness and adequacy with which an indicator measures or expresses the phenomena with which it is concerned. Although many aspects of human–environment interactions and their associated complex feedback loops are not completely scientifically understood, enough is known about many issues for the scientific accuracy with which indicators represent reality to be assessed.

Reliability and feasibility: This reflects the truth and reproducibility of indicators and their robustness in statistical terms and the ability to develop the indicator in practical terms of data and cost-effectiveness. The quality of an indicator depends on the data from which it is derived and from the indicator construction method, which may increase uncertainty. Gaps in monitoring of human–environment interactions and deficiencies in the spatial coverage of many global and local data sets constrain the quality of many indicators. Ideally indicators should be built on existing data sets, but many desirable indicators lack suitable background data. Collection of new data and repeated data collection exercises may be costly and lengthy enterprises.

Communicability: This is the extent to which indicators are understood and the effectiveness with which they convey their purpose and meaning to their target audiences. Fundamentally, indicators are communication tools. An indicator that fails to do this is redundant. Because sustainable development is a multistakeholder project, indicators must be meaningful to many different actors (e.g., citizens, policymakers, decision makers). Thus the capacity of an indicator to reach its target audience ultimately determines its communicability and contribution to sustainable development.

Table 3.1. Case study: Applying assessment criteria to three economic headline indicators.

GDP per Capita

Purpose	The virtues that help GDP per capita retain its utility in sustainability assessments include its widespread acceptability and implementation in national policy as a measure of and influence on the level or vigor of exchange transactions and its influence on political survival.
Measurability	Like any other highly aggregated measure used to represent the long-term quality of human activities, GDP per capita is far from perfect, and it must be considered in tandem with related developments represented by key social indicators such as life expectancy at birth, choice or freedom measures, security, access to education, and subjective well-being.

Table 3.1. Case study: Applying assessment criteria to three economic headline indicators (*continued*).

Representativeness	The limitations of GDP per capita as a comprehensive measure of economic welfare of a nation's citizens stem primarily from problems associated with capturing and distinguishing all relevant sources of welfare (even within a given period). In addition, GDP levels do not embody information on the level of wealth, long-term sustainability, and quality of life because GDP represents an income flow for a short time and does not reveal whether this income was derived from qualitative or quantitative gains in productive capital stock (including natural resources) or from depletion of existing assets that will jeopardize future economic sustainability. There have been laudable attempts to adjust GDP per capita for environmental and other capital losses, unpaid labor, defensive expenditure, and so forth, but interpretive difficulties and their methodological inadequacies linked to data needs, questionable construct validity, and economic valuation limit their utility as viable headline indicators for the economic domain.
Feasibility	GDP is supported by a long history of well-developed underlying data and methods.
Communicability	This "social product" has many positive influences in terms of the choices, diversity, and access to resources that make up other major facets of long-term sustainability (e.g., adequate nutrition, health services, dematerialization, and related eco-efficiency–enhancing technologies and infrastructural and strategic change). However, trends in GDP per capita should be complemented by simultaneous consideration of the various components of overall expenditure that have significant effects on sustainability (e.g., social products committed to social indicators such as education, preventive eco-efficiency, natural capital protection and augmentation, and, arguably, the nature of production and consumption, examined in environmental pressure terms).

Income as a Measure for the Distribution of GDP

Purpose	In addition to an account of the overall economic value added that is represented by GDP, another indicator is necessary that addresses the distribution within the socioeconomic system and thus shows the amount of income per person or household. Concerning sustainability, household income provides necessary information on the distribution of wealth among the population.

(continued)

Table 3.1. Case study: Applying assessment criteria to three economic headline indicators (*continued*).

Measurability	GDP per capita often is used as proxy for individual income. However, GDP is not evenly distributed among the population. Thus, other indicators are necessary to cover distributional issues (e.g., income per household).
Representativeness	Income data are difficult to acquire in economies where there is a large informal sector, and the data obtained will not adequately represent reality.
Feasibility	See above.
Communicability	The limitations of income data to represent distributional issues are of diminishing importance because they are readily understood by a majority of users (unlike those of more complex indicators such as the Gini coefficient).

Material and Energy Intensities

Purpose	Given that human activity relies on a natural system with finite resources, sustainable economic development has to account for environmental limits (both sink and source). Material and energy intensities relate economic data to indices of natural resource use (and waste output) per unit of economic output. These hybrid approaches are at the basis of the material and energy intensity measures, which can provide good indications of natural resource or eco-efficiency needed for dematerialization processes.
Measurability	Economic indicators are not defined solely as those that use monetary values. In recent years, there has been a substantial shift in interest (in the environment–economics nexus) toward depicting the economy in terms of the production, exchange, and demand of physical resources and goods providing welfare services. For environmental demands, the key factor becomes the material and energy flows (often effective consumption) generated by the overall economy and specific types or classes of economic activity and output (and associated technologies, inputs, and waste outputs).
Representativeness	Dematerialization indicators and material or energy intensities are being used more often. However, any dematerialization observed can be derived in two different ways: through real reduction of resource input per unit of GDP or simply outsourcing of material-

(continued)

Table 3.1. Case study: Applying assessment criteria to three economic headline indicators (*continued*).

	or energy-intensive processes to other countries. The latter does not result in dematerialization on the global scale but only for the specific socioeconomic system observed. Any interpretation of dematerialization therefore has to take into account the global scale and thus consider effects on the international level.
Feasibility	See above.
Communicability	Users of indicators do not necessarily think about energy, water, and material budgets in the same way as they think about monetary budgets. However, the mechanisms underlying material and energy intensity indicators are similar to thinking in terms of monetary values and therefore can be readily understood.

Box 3.1. Limiting purpose and scale: The EEA and DEFRA experiences.

The European Environment Agency (EEA, www.eea.eu.int) is implementing assessment and reporting systems with the objective of providing timely and responsive information on the state and trends of the environment and related pressures and impacts. The principal purpose of their indicator systems is to inform about the environment, an objective they meet successfully. The link to a formalized sustainable development (SD) system is not given, however, even though the agency is attempting to be consistent with SD systems at the international level. Therefore, the EEA can contribute only indicators pointing in the direction of sustainability from the environmental perspective. By focusing on the European level and the environmental domain, EEA enhances feasibility, reliability, and representativeness.

The UK-Department for Environmental Food and Rural Affairs (UK-DEFRA) (www.sustainable-development.gov.uk) developed headline indicators of sustainable development for the United Kingdom with the objective of translating issues of public concern into a small number of indicators. These issues have been identified by public consultation with the intention to link them to the concept of SD and quality of life in general. The headline indicators "are intended to raise public awareness of SD, to focus public attention on what SD means, and to give a broad overview of progress." By focusing on a very small number of indicators and linking directly to public opinion, DEFRA enhances communicability.

Box 3.2. Satisfying measurability without neglecting communicability: The Wellbeing of Nations

The indexes of the *Wellbeing of Nations: A Country-by-Country Index of Quality of Life and the Environment* are developed along a hierarchical system of indicators, subelements, elements, dimensions, and subsystems. Two subsystems are divided into five elements of measurement each:

- People: health and population, wealth, knowledge and culture, community, and equity
- Ecosystem: land, water, air, species and genes, and resource use

For the sake of aggregation, a normalization method (i.e., the relationship to best and worst observed values) is used to obtain performance scores on a 0–100 scale. Different methods of weighting are used on the level of the subelements and elements, whereas the two subsystems are assigned identical weights. Four indices were calculated: the Human Wellbeing Index (HWI, i.e., the people subsystem), the Ecosystem Wellbeing Index (EWI, i.e., the ecosystem subsystem), the Wellbeing Index (WI, i.e., representing the average of EWI and HWI), and the Wellbeing/Stress Index (i.e., ratio of human well-being to ecosystem stress).

The intention of the Wellbeing of Nations is to compare countries and emphasize the trade-offs between the two subsystems (people and ecosystems) on which countries build their development. It ranked Sweden as the best performing among 180 countries (in 2001) although it considers Sweden an ecosystem-deficit country because its people dimension's excellent performance is obtained at a high environmental cost.

For more information, see

IUCN Web site: www.iucn.org/info_and_news/press/wbon.html.
Robert Prescott-Allen (2001). *The Wellbeing of Nations: A Country-by-Country Index of Quality of Life and the Environment.* Washington, DC: Island Press.

Pocedural Issues: User Involvement in Formulating and eveloping Indicators

Involvement of stakeholders and users is often assumed to be a prerequisite of SD indicators. However, because we do not fully understand how to promote stakeholder involvement in a useful and efficient manner, we need critical analyses of the potentials and limits of participatory and deliberation processes. The key reasons for user involvement include the following:

- Integrating local knowledge with technical and scientific expertise to improve understanding of societal and ecosystems aspects of sustainable development
- Improving decision-making processes by allowing users or the public to influence the topics, components, and nature of indicators and the relative weightings given to different components
- Using stakeholders as a counterbalance to the influence of implicit values in experts' selection of indicators and widening the ownership of assessment instruments and hence the political commitment to action on the SD issues emerging from the participation process

The two major ways of achieving stakeholder and civil society participation are as follows:

- Involving the public (or part of the public) on a near-representative or random basis in order to increase the democratic value of the indicators. This involvement can be realized at two levels: involvement of the public to define, select, and prioritize the issues to be addressed with indicators and involvement of the public to select indicators to be integrated into a set.
- Involving users to increase the efficiency of decision-making processes (audience targeted) and thus reforming the process of making decisions about sustainable development. One of the goals of indicators is to achieve greater transparency of information and open up decision-making processes (e.g., by allowing shared knowledge to be spread).

Whereas involvement by users (e.g., policymakers, civil servants, business actors) is widely considered and implemented as a basic condition of many "science for sustainable development" initiatives, the wider public has been considered less critical. Whatever the involvement mechanism, the feasibility and value of public involvement are also determined by the scale of the indicator initiative. Comprehensive democratic public involvement synthesizing local knowledge would not work well for global-scale indicator initiatives. However, such involvement is highly appropriate at the local level (see Chapter 19, this volume), where expert opinions are likely to differ from the preoccupations and understanding of individual communities (diminishing the subsequent usability of the indicators). Local involvement for local indicators, as envisaged in Local Agenda 21, increases public ownership of indicators, improves their communication, makes measured issues more relevant to the local population, increases the reliability of indicator systems, and facilitates access to nonformalized data such as local knowledge that can complement weak formal data on many phenomena at the local level. Nevertheless, the difficulty of conceiving public and user involvement at global level reflects the difficulty of developing indicators that address cross-scale issues; these issues reveal a high level of difficulty at the level of both technical and participatory development. Because of the difficulties and resources needed to organize successful participatory processes, stakeholder involvement processes should not be initiated without clear objectives.

Communicating to Different Audiences

The ideal indicator would be one that communicates for a specific purpose to a range of audiences. This may not be achievable, given the diversity of stakeholders. Conceiving tools for policymaking entails different approaches to indicator construction, issue selection, and depth of information provided than are needed in developing instruments to provide general information for citizens.

Despite widespread acknowledgment of the problems of indicator acceptance and use, the ways in which indicator methods and design should be improved to overcome these difficulties remain unclear. Indicator developers tend to ignore these constraints, preferring to work toward enhancing the capacities of audiences to comprehend complex information. Although capacity building is the long-term goal, immediate efforts are needed to develop robust means of communicating the messages carried by indicator initiatives in comprehensive ways. Despite the usefulness of simplified communication interfaces (e.g., dashboards, barometers, headline indicators, and indices), the majority of indicator development activity still concentrates on improving the quality of individual indicators, with sometimes marginal usability for decision making.

Conclusions

Developing indicators for sustainable development entails a series of methodological trade-offs at many different levels during the process of defining, constructing, and communicating indicators. Defining a consistent yet useful framework of hierarchically interlinked levels of data, indicators, and indices is the first step toward developing unambiguous indicators for decision making. In the context of trade-offs between technical feasibility, societal usability, and systemic consistency, aggregation techniques should be considered one way to reconcile form and function of indicators.

The robustness of the methodological choices being made should be scrutinized. In this respect, indicators can be assessed with five quality criteria: purpose, measurability, representativeness, feasibility and reliability, and communicability. Again, no indicator can be perfect on all five criteria. However, in order to develop into useful and robust decision-making tools, indicators should not lack any of these basic qualities.

Methodological challenges in indicator development extend beyond the mere technical data-determined issues to the need for policy relevance. If we want to prevent indicators for sustainable development from being confined to sterile expert and stakeholder discussions (and eventually to fall into oblivion, as did the social indicator movement of the 1970s), we must continue to improve our understanding of the impact of indicators on decision making at the level of a society. The effective monitoring of the performance of specific indicators will partially determine further adaptations of the multidimensional decision framework necessary for sustainable development. The flexibility of indicators responses to the challenges of usability will determine the strengths of the individual indicators being developed today.

Literature Cited

Adriaanse, A. 1993. *Environmental policy performance indicators: A study on the development of indicators for environmental policy in the Netherlands.* The Hague, the Netherlands: SDU Publishers.

Atkinson, A. B., B. Cantillon, E. Marlier, and B. Nolan. 2002. *Social indicators: The EU and social inclusion.* Oxford: Oxford University Press.

4

Ensuring Policy Relevance

Louise Rickard, Jochen Jesinghaus, Christof Amann, Gisbert Glaser, Stephen Hall, Marion Cheatle, Alain Ayong Le Kama, Erich Lippert, Jacqueline McGlade, Kenneth Ruffing, and Edwin Zaccai

> How much more tidy the world would be if there were some experts who, in possession of the appropriate applied science, could tell us what to do in the cause of sustainability.—Funtowicz et al. (1999)

In a study of sustainable development indicators, Parris and Kates (2003) argue that indicator developers display a surprising degree of political naïveté, which is illustrated by the gap between the stated aim of informing decision making and the weak efforts made to ensure that the indicators are designed to achieve this. Technical experts often make only vague reference to concepts such as policy relevance, policy process, and policy impact and tend not to give careful consideration to the components of these issues.

An example of this would be the United Nations Environment Programme (UNEP) publication of the Global Biodiversity Assessment (GBA), produced in 1995. The GBA contained two chapters on ecosystem functioning, one coauthored by thirty-three and the other by sixty-six international scientists. In addition, there were many other contributors to these chapters, both of which were extensively and internationally peer reviewed (Loreau et al. 2002). However, despite initial high hopes, the GBA "sank like a lead balloon" (Kaiser 2000:1677, cited in Cash and Clark 2001). Some authors believe that although the GBA represented the scientists' views of what they thought the important problems were, based on literature review, they did not take into account the issues policymakers faced (Loreau et al. 2002) and therefore did not address the needs of potential users (Cash and Clark 2001).

Simply stating that indicators should guide the decisions made by decision makers leaves many unanswered questions about the type of decision, who these decision makers actually are, and on what basis they are empowered to make decisions. Political naïveté on the part of experts may also reflect a more fundamental disparity between the prevalent societal belief system, in which clear lines of demarcation exist between science and politics, and the more complex and fuzzy science–policy interface that exists in reality. It is in this fuzzy zone that many sustainable development initiatives are created and launched.

The policy process itself is changing as policymakers seek to implement overarching principles, such as sustainable development, in the face of scientific uncertainty and complexity (Funtowicz et al. 1999). The way in which contemporary society discusses problems and their possible solutions may also be very different from that of the past, when environmental problems could often be addressed by stand-alone thematic policies. Defining specific issues becomes more difficult as greater awareness of cause and effect links environmental problems to human behaviors that are determined by a range of societal and economic driving forces. Bludhorn (2002) illustrates this when he writes that specific problems may have lost some of their identity in the traditional sense by merging into the larger pool of conflicting social interests, values, and preferences. An example of this would be the European biofuel policy, which has the global aim of reducing CO_2 emissions and improving energy security; local debate in member states has tended to focus on the economic aspects such as industry support and the replacement of agricultural subsidies. This will have an effect on the practical application of the biofuel policy (e.g., which plants are grown, which market strategies are supported) and consequently its environmental impact.

As policy becomes increasingly integrated and complex, it is accompanied by rising demands for transparency and openness. How an indicator is developed can affect its legitimacy and credibility as much as its timeliness and relevance to the wider political process. It is easy to overstate the role of sustainable development indicators (SDIs) and indicators as a whole. They are tools for informing decision making, but even the best indicators may not be able to influence decision-making processes if the area addressed is outside the political priority issues. Public concern is a key driver in advancing policy issues, and the media are instrumental in raising public awareness.

This chapter investigates the processes that lie behind indicator creation and use and seeks to answer the following questions: What makes an indicator or indicator set successful? Can specific factors be found that contribute to its positive impact on the policy arena?

Policy Processes

SDIs are intended mostly for use in the wider political arena, whether at the local, national, or global level. Targeting the external political process wisely therefore is essential, although the indicator development process itself also has a role to play.

In very generic terms policy has a life span: It starts with the realization that there is a need for policy to tackle a certain issue, followed by the design and implementation of that policy. This is often followed by an evaluation, leading to revisions and possibly, in the long term, to the phasing out of the policy as the problem is resolved (Chapter 8, this volume). Information that arrives at the wrong time in the evolution of an issue can fail to influence action (Cash et al. 2003).

In the widest sense the issue identification and framing for sustainable development have already taken place, but for measurements of progress to be possible, the framing must become more focused. An analogy could be measuring progress in reducing the environmental impacts of transport (Chapter 8, this volume). Although there was general consensus that action was needed to tackle this problem, the issue had to be broken down further before progress toward a set of indicators could be made. This was done by a reframing of the wider issue ("What are the environmental impacts of transport?") into subquestions that then allowed progress to be measured (Box 4.1).

The process of framing the issue is critical because this is where scientific and technological expertise meets the stakeholders and the political process, and synergistic activity to create a common work program is needed. If the framing stage is well grounded in a participatory process, the indicators themselves are likely to be more acceptable and credible.

The policy life cycle highlights the need for flexibility in the design of indicator sets. However, there is a risk that this desired flexibility may come at the expense of stability and familiarity, both of which are important in the development of successful indicators. A single indicator will not remain policy relevant unless it specifically meets a need in the policy life cycle that may exist for a long time, such as routine monitoring.

Box 4.1. Seven key questions on transport and the environment in the European Union.

- Is the environmental performance of the transport sector improving?
- Are we getting better at managing transport demand and at improving the modal split?
- Are spatial and transport planning becoming better coordinated so as to match transport demand to the needs of access?
- Are we optimizing the use of existing transport infrastructure capacity and moving toward a better-balanced intermodal transport system?
- Are we moving toward a fairer and more efficient pricing system, which ensures that external costs are internalized?
- How rapidly are improved technologies being implemented, and how efficiently are vehicles being used?
- How effectively are environmental management and monitoring tools being used to support policy and decision making?

Source: EEA (2000).

Political priorities tend to change in response to objective or subjective judgments about what voters want and to events and actions at local, national, or global scales. The flux of political priorities is likely to be most pronounced where the subject is not supported by the specific needs of legislation or by an established wider process or framework. Large international institutional bodies and processes may have an element of stability and continuity that similar national or local activities do not have.

Kates et al. (2000) suggest that research focus in developed countries tends to be more global in orientation and more theory driven than in developing countries, where emphasis may be on local issues and processes. In the area of sustainable development, assessments often fail to include the concerns and perspectives of developing country citizens, sometimes intentionally but more often as an artifact of unrepresentative participation by developing countries (Cash et al. 2002). If actors believe that their views and concerns were not considered, the resulting assessments may not have the desired policy impact, even if they are relevant.

Indicator Development

Indicator developers use frameworks to provide a common language and perspective on the issue and its solution. This facilitates indicator development, particularly when many different actors are involved. The way in which issues are framed becomes important in the interpretation and deeper analyses of the results because the frameworks are the assumptions and rationales on which the indicator is based and should be made available to those wanting to interpret the indicators. Understanding these assumptions and the frameworks is essential in order to compare and discuss indicators from different institutions because they may be based on different frameworks. For the majority of users, however, showing the frameworks themselves, or categories from such frameworks, would only add an unnecessary degree of complexity that might distract them from the results.

Institutions that work on sustainable development need to have one foot in the politics of problem definition, responsive to issues of appropriate participation and representation, and the other foot in the world of science and technology, responsive to issues of expertise and quality control (Clark 2003). Clark writes that perhaps the strongest message to come out of the Johannesburg summit was that the research community needs to complement its historical role in identifying problems of sustainability with a greater willingness to join other organizations in finding practical solutions to those problems and that institutions that spend most of their time doing pure science or pure politics are not likely to be as successful as boundary-spanning institutions (e.g., those providing scientific assessments or regional decision support). Boundary-spanning institutions that consciously manage and balance the multiple boundaries within a system (e.g., between disciplines, between organizational levels, and between different forms of knowledge) tend to be more effective than other institutions in creating information that can influence policymaking (Cash et al. 2002).

The three criteria of credibility, legitimacy, and salience are key attributes for characterizing the effectiveness of sustainable development indicators (Cash et al. 2002; Parris and Kates 2003) where *credibility* refers to the scientific and technical adequacy of the measurement system, *legitimacy* refers to the process of fair dealing with the divergent values and beliefs of stakeholders, and *salience* refers to the relevance of the indicator to decision makers. The indicator development process itself is responsible for ensuring at least the first two of these criteria.

Resources (e.g., equipment, monitoring, data, research, and knowledge) vary substantially between developed and developing countries. The socioeconomic, environmental, and knowledge dichotomies between the two hemispheres may be exacerbated by this resource distribution (Clark 2003). Finding ways to bridge this resource gap is essential for equitable representation, both geographically and in terms of recognizing and framing important issues. Equitable representation increases the legitimacy and credibility of both the process and the final product.

A capacity for mobilizing and using science and technology is also an essential component of strategies promoting sustainable development (Cash et al. 2003). Generating adequate scientific capacity and institutional support in developing countries is particularly urgent in order to enhance resilience in regions that are vulnerable to the multiple stresses that arise from rapid, simultaneous changes in social and environmental systems (Kates et al. 2000).

However, scientific capacity alone is not sufficient for the purposes of producing credible SDIs. Instead, capacity building is needed, with emphasis placed on supporting the wider processes that ensure legitimacy and credibility of the indicator development process.

Effective capacity building places emphasis on the key components of communication, translation, and facilitation (mediation) (Cash et al. 2003). Providing for adequate communication between stakeholders is essential, as is ensuring that mutual understanding is possible. Communication is often hindered by jargon, language differences, experiences, and presumptions about what constitutes a persuasive argument. Facilitation or mediation further enhances transparency of the process by bringing all perspectives to the table, defining the rules of conduct and procedure, and establishing decision-making criteria.

A serious commitment by institutions to managing the boundaries between expertise and decision making will help link knowledge to action. Establishing accountability to key actors across the boundary and using joint outputs to foster cohesion and commitment to the process are also helpful in developing capacity for sustainable development.

Indicator legitimacy and acceptability hinge on recognition of the plurality of legitimate perspectives (Funtowitz et al. 1999). Where there are complex issues, the qualities of the decision-making process itself are critical, and processes designed to open the dialogue between stakeholders rather than diluting the authority of science are key to creating a broad base of consensus. It has been suggested that the role of indicators is to serve as aids to this dialogue and decision making (Funtowitz et al. 1999).

The value of a specific indicator set varies between users and situations. The users should be able to influence the choice of the indicators that they will have to use. Sometimes this local choice will result in a loss of comparability as different groups and processes elect to use different indicators. This can be acceptable when the main purpose of indicators is to promote effective decision making. On the other hand, when the main purpose is comparability, then more importance should be given to standardization. It is not always possible to have both.

Participation in the Process

Stakeholder groups often include young people, nongovernment organizations (NGOs), local authorities, scientists and technical experts, trade unions, farmers, business and industry, indigenous people, and women. These are the nine categories of key stakeholders and civil society groups listed for Agenda 21.

In general, the importance of stakeholders in securing legitimacy of the process increases as the complexity of the studied issue increases, to the point where conclusions cannot be reached by scientific facts alone, and some value-based judgment is needed. Funtowitz et al. (1999) write that the guiding principle must ensure quality in the process (more than in the end product). The traditional process of scientific researchers engaging in dialogue with peer researchers must be replaced by a broader recognition of multiple values and multiple truths, where dialogue with all stakeholders becomes the norm. Managing such a participative process can be difficult, and for indicators the participatory process may vary widely from one case to another. Finding relevant indicators for water sanitation, for instance (once it is agreed, through a broader participatory process, that water sanitation must be included), is a topic in which inputs from experts is likely to be needed.

Communication

At least two separate strands of communication can be identified: how best to communicate the content or substance and how best to package and market the product. Communicating the content is linked to the indicator development process, taking the users' needs into account and consulting appropriately with relevant stakeholders.

The format of the publication describing the indicators also must be suitable for the intended audience. Given the digital divide that exists both within societies and between countries, a Web-based publication might not be sufficient. In particular, there may be a divide between the resources available to stakeholders and potential audiences in developed and developing countries, which may affect the format of the output. The timing of publication can also be very important, at least for media coverage. For example, the Environmental Sustainability Index (ESI) reports are published to coincide with the annual meeting of the World Economic Forum.

The loose definition of the phrase *policy impact* in SDI development circles makes its attainment difficult to measure. The key role of indicators is to measure progress, so if legitimacy is established through a transparent indicator development process, then the measure of progress shown by the indicator will be more widely accepted. The legitimacy of the institution proposing indicators probably is as important as the transparency of the process used to determine them.

However, even if the indicator development process is transparent, the resulting progress report may not be of any consequence (it may be ignored or overshadowed by other events), or the reverse may occur. In the case of the fifteen headline indicators in the UK set, the policy impact was far greater than anticipated. This was assisted by the unexpected link made by a media cartoon between bird population size and human happiness or welfare (published in the *Financial Times*, November 24, 1998). This cartoon caught the imagination of the general public. Although the indicator developers themselves did not promote this link, it made a lasting impact, and the association has persisted, giving the indicator public and policy relevance far greater than the measured variable itself.

Indicator Users

At least ten categories of users, with very different needs, can be identified. These users can be clustered into three main groups. The first needs very simple, structured information and includes voters, the broader (nonspecialist) media, and decision makers. The second category needs an intermediate level of detail and simplicity and includes local government, policy implementers and checkers, NGOs, research funding bodies, and industry. The third category needs technical information: policymakers, academics, and some NGOs.

It is useful to distinguish between decision makers and policymakers. Decision makers typically are ministers and parliamentarians, policy implementers and checkers such as regulatory bodies, environmental protection agencies, and NGOs that can deal with technical details and disseminate information to the public. Policymakers are a distinct, expert group of scientists, economists, and social scientists who design policy portfolios, evaluate policy alternatives, and construct and evaluate indicators of sustainability. This category also includes the non–politically appointed civil servants who undertake some of the underpinning research and brief ministers.

Insights into the Use of a Global Report: The Case of GEO-2000[1]

UNEP has published three reports to date in the comprehensive Global Environment Outlook (GEO) series—GEO-1 in 1997, GEO-2000 in 1999, and GEO-3 in 2002—and GEO-4 will be published in 2007. The objectives of the main GEO report series are to produce a comprehensive, policy-relevant overview of the state of the global environment that incorporates global and regional perspectives and to provide an outlook for the future. Indicators are used to encapsulate and convey information on the state and trends of high-priority environmental issues and related pressures, impacts, and policy responses, to illustrate global and regional dimensions, and to support outlook analyses. UNEP uses a participatory, consultive process for GEO integrated environmental assessment and reporting.

Senior decision makers and their advisors, especially those in ministries of environment, are the primary audiences targeted by GEO reports. Secondary target groups include international environmental organizations and NGOs, the academic community, other UN agencies, the media, and concerned members of the general public (Table 4.1).

Monitoring and evaluation of the use of GEO-2000 have provided insight into the composition of the GEO user community and how it has used the report (Figure 4.1).

Table 4.1. User categories and needs.

Voters	• Indicators relevant to voters would help them identify actions that they can take and actions that government should take. • The indicators must be applicable and relevant at an individual or local level and conceptually clear. • The indicators should also be few in number, simple, and unambiguous, with no technical and methodological information.
Media	• Journalists need clear bundling of information. • They also need information ("sound bites") that they can use to pad out other stories. • The data should be unambiguous and simple, with clear messages and assessments (notes to editors, interpretation guidelines, limitations) that enable journalists to make statements about whether a trend is stabilizing, worsening, or improving.
Decision makers	• Decision makers need simple information that provides an overview but with some assessment and possibly some analysis that highlights areas where action should be taken. • Targets are important.
Local government	• Local governments need to be able to disaggregate the information in order to target policy appropriately. • They need the indicators and methods to be applicable and relevant in different settings for towns, cities, and municipalities.

(continued)

Table 4.1. User categories and needs (*continued*).

Policy implementers and checkers	• Policy implementers and checkers need a wide range of indicators that are clearly defined and stable in terms of methods and data requirements and can be used to monitor progress over time. • They need good guidelines and clearly formulated targets, objectives, and policy effectiveness indicators.
NGOs	• NGOs need information for use in campaigns to raise public awareness and lobby politicians. • They might need a wide range of indicators, with assessment and some analysis; this should include access to technical documentation, guidelines, and possibly data (these might be made available on the Web).
Industry	• Industry probably needs indicators that provide engagement incentives and use appropriate language (e.g., *eco-efficiency, cost effectiveness, sector-specific and pressure indicators*). • It needs indicators that can anticipate future trends (for investments needs) and costs.
Policymakers, developers, and designers	• Policymakers need a comprehensive set of many indicators to inform specific areas of policy. • Indicators are likely to already be in use outside the SD area, so the specific need may be for an SD set or a focus on the interlinkages between the separate pillars. • There might be a need for links to outlooks and scenarios and to costs when designing policies. • The indicators should link to existing indicators and data if possible.
Academics	• Academics need very specific data for research, as inputs to studies and models and for use in evaluating and developing methods. • They are also likely to need the detailed assessments, analyses, and reasoning behind the analyses.
Research funding bodies	• These users may need a set of indicators as a basis on which to evaluate whether to select further project proposals for funding. • They may also need information on data availability, the conceptual basis of indicators, methodology, feasibility, and reliability.

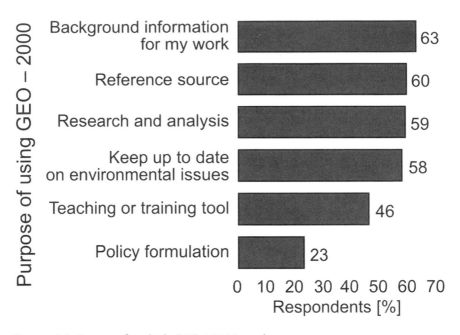

Figure 4.1. Purposes for which GEO-2000 is used.

The largest categories of GEO-2000 users were members of the research community and information compilers, the policy development and decision-making community, and other environmental information depositories and distributors.

The results of the reader survey indicate that GEO-2000 was read not only by the target audiences but also by a variety of people in other types of organizations and positions. The largest number of respondents was from the education sector, accounting for almost 38 percent of the total respondents. Sixty percent of respondents in this group held professional or faculty positions. Fourteen percent of respondents belonged to ministries of environment and related government bodies. The largest number of respondents (33 percent) worked as professional staff or faculty members, closely followed by senior management and other decision makers (28 percent). This example shows that it is possible, with the same report, to reach different types of audiences using different layers of communication.

An indicator set must be flexible to provide maximum policy relevance, but this flexibility must be weighed against the risk of losing familiarity and continuity. Composite indicators and indices are made up of a combination of separate variables, often with different weighting. They can be used to summarize complex or multidimensional issues and provide the big picture for policymakers (Saisana and Tarantola 2002; Nardo et al. 2005). The separate underlying variables in composite indicators constitute a pool of background information and provide flexibility, and the summary index or composite ensures continuity and familiarity.

Headline indicators fulfill a similar function because the small and familiar headline set can be extracted from a larger pool of underlying variables. However, the use of headline sets might increase management needs because the headline indicators must be reviewed and changed in response to changing policy needs. The larger pool of underlying variables would also need to be maintained and updated rather than allowed to decline in quality. The main successes in the use of headline sets occur when the larger pool is made up of indicators that have other uses and therefore are automatically maintained (e.g., as part of a core set of national statistics) and when the headline set itself is reviewed and updated regularly. A key example of this process occurs in the United Kingdom, but this is also the pattern used by some organizations, such as the European Environment Agency (EEA), where the larger set is used for reports and assessments that are not specifically targeted at policymakers (e.g., general public, students, and NGOs). This dual purpose ensures the maintenance of a larger base than the small, flexible, and policy-relevant core set (EEA 2005).

Distortion is an additional concern for indicator sets: A set of indicators cannot adequately represent reality if the composition is skewed or biased in some way. A skewed headline set might contain more indicators that show improving trends than those that show a lack of progress or deteriorating trends. Similarly, an index or composite indicator might be skewed if the choice of variables or the weighting given to each is biased in some way. A wide participatory process would help in preventing and responding to such criticisms.

Data that are clearly needed and used to populate a stable indicator system are a helpful driver for creating stable data flows and structured, balanced monitoring systems. There is a debate about whether to publish indicators that contain data known to be of poor quality (e.g., with substantial data gaps), but although there must be a lower quality limit beyond which the indicator becomes distorted and misleading, publishing a poor-quality indicator often acts as a driver for an improvement in data quality. The 2001–2002 ESI is an example that was published using the best available but often low-quality data, resulting in some rapid improvements of data flows and data quality.

The evolution of methods also provides difficult choices: Policy relevance and wider acceptability mean that an indicator should be believable. So if the methods are improving, the indicator may need to change to reflect this improvement. Although changing the methods of an existing indicator may increase its relevance and acceptability, it may create difficulties in comparability over time, especially where definitional changes are made in underlying data (e.g., municipal waste).

Meadows (1998) states that an environmental indicator becomes a sustainability indicator with the addition of time, limit, or target. Indicators become especially powerful tools for policy when they relate to political targets, thus adding the element of performance; for example, the Kyoto Protocol, with its precise reduction targets, could not exist without the indicator "CO_2 emissions."

Three distinct types of targets can be identified:

• Political or hard targets
• Soft targets such as those for sustainability reference values, minimum viable populations, and thresholds
• Benchmarks

Hard targets are set through political processes and usually are beyond the scope of the expert indicator producers. However, the nature of these targets can play a key role in their use by the indicator producers. A conceptual target such as "halting the loss of biodiversity" may be an essential policy driver, providing a focus for research and indicator development, but it must be broken down to more accessible and specific subsidiary targets for inclusion in indicator exercises. Even apparently concrete targets such as "decoupling transport demand from gross domestic product" will raise many questions about the definition and the measurement of progress toward this target. In general, vague or qualitative hard targets in need of clarification or definition can be identified and highlighted by the indicator community, although it is the responsibility of the political process to set and refine such targets. Drawing attention to the lack of targets may be a key role of indicator producers, and identifying vague targets can provide an opportunity to define them in a more precise fashion.

Soft targets, such as a sustainable reference value or a minimum viable population size, could be used more fully by the indicator community. Although the identification and use of soft targets are often associated with scientific debate and differing opinions, the use of such targets can both highlight the inadequacy of the political targets and raise awareness of the complexities and uncertainties inherent in environmental systems. In addition, soft targets may offer an opportunity for analysis and interpretation of the fuzzy areas that hard quantitative targets do not provide.

Benchmarking is a widely used way to add context to indicators. The mean value of neighboring countries is often used as a benchmark, as are the Organisation for Economic Co-operation and Development and the EU-15 and EU-25 averages. Benchmarking has a key role to play at local and national levels, helping to create a race for the top, whereas at the international level the use of benchmarking and best practice examples as illustrative targets could be further developed. Benchmarking is distinct from ranking in its focus on comparisons with a few selected countries rather than a list of ranks. A benchmark is more powerful if the comparison made is acceptable and relevant. Benchmarking against the average for a group of countries may not have the desired effect if the mean does not provide an example of best (or better) practice.

Ranking on the basis of an indicator set or composite indicator is appealing: For the national media, the international ranking of countries often makes headlines. However, such assessments often are irrelevant to policymakers: Ranking based on relative performance means that a country may be successful and be ranked highly one year but then may be ranked low the next year not as a result of a decline but because the rate of improvement slows and even stops once a country has reached the top. This occurred in the United Kingdom, which was ranked sixteenth in the ESI in 2001 and ranked only ninety-first in 2002. Because the public was not aware of any environmental catastrophe between those two assessments that in their minds would explain such a drop, the index lost credibility in the United Kingdom and no longer has impact on policy or the media.

Conclusions

In a global setting, the external policy process is very diverse, and SDIs form only one aspect of the information available to decision makers. Without public and media visibility, indicators are unlikely to have major impact on political processes, but organizing an indicator development process that is targeted to the public together and emphasizes communication of the results may increase its impact.

If stakeholder involvement plays a key role at each stage in the process and, most crucially, in the framing stage, the indicators will have legitimacy and credibility that make them valuable for many stakeholders. Such indicators can act as tools for measuring the progress of existing policies and steering further action.

Indicator developers therefore need to embrace inclusive participatory processes, ensuring adequate and appropriate representation of the diverse stakeholder groups and taking into account the resource divides that may exist between them. However, participation should not come at the expense of scientific validity. There may be a case for expert methodological discussions following wide agreement on a particular topic to be addressed.

Many SDI initiatives are salient and scientifically sound but do not have the legitimacy associated with a wide base of a good process. This reflects the trade-offs between processes and products and between salience, credibility, and legitimacy. The most perfect technical SDI might not be useful, and the most useful SDI might not be perfect. Understanding that even the best technical efforts of indicator developers may not be sufficient to ensure that an indicator has policy impact and reflecting this understanding in the emphasis given to the indicator development process may be the best way to increase the chances of having a positive impact on the policy arena.

Note

1. This section is based on a GEO user profile and impact study commissioned and carried out in 2000 by an independent firm of consultants on GEO-1 and GEO-2000, reader survey questionnaires that were distributed inside the hard copies of the GEO-2000 report and made available electronically in all languages on the GEO-2000 Web page, and traffic on the GEO Web site.

Literature Cited

Bludhorn, I. 2002. *Post-ecologism and the politics of simulation.* Paper presented at the 30th European Consortium of Political Research Joint Sessions of Workshops in Turin (Italy), Workshop 10: "The End of Environmentalism?," March 22–27.

Cash, D., and W. Clark. 2001. *From science to policy: Assessing the assessment process.* Faculty Research Working Paper RWP01-045. Cambridge, MA: John F. Kennedy School of Government, Harvard University.

Cash, D., W. Clark, F. Alcock, N. Dickson, N. Eckley, and J. Jäger. 2002. *Salience, credibility, legitimacy and boundaries: Linking research, assessment and decision making.* Faculty Research Working Paper RWP02-046. Cambridge, MA: John F. Kennedy School of Government, Harvard University.

Cash, D. W., W. C. Clark, F. Alcock, N. M. Dickson, N. Eckley, D. H. Guston, J. Jaeger, and R. B. Mitchell. 2003. Knowledge systems for sustainable development. *Proceedings of the National Academy of Sciences of the United States of America* 100(14):8086–8091.

Clark, W. 2003. *Institutional needs for sustainability science.* Posted to the Initiative on Science and Technology for Sustainability (ISTS) Web site, http://sustsci.harvard.edu/ists/docs/clark_governance4ss_030905.pdf September 24, 2003.

EEA. 2000. *Are we moving in the right direction? Indicators on transport and environmental integration in the EU: TERM 2000.* European Environment Agency Environmental Issues Report 12. Copenhagen: European Environment Agency.

EEA. 2005. *EEA core set of indicators: Guide.* European Environment Agency Technical Report no. 1/2005. Copenhagen: European Environment Agency.

Funtowicz, S. O., J. Martinez-Allier, G. Munda, and J. Ravetz. 1999. *Information tools for environmental policy under conditions of complexity.* EEA Environmental Issues Series no. 9. Copenhagen: European Environment Agency.

Kates, R. W., W. C. Clark, R. Corell, J. M. Hall, C. C. Jaeger, I. Lowe, J. J. McCarthy, H. J. Schellnhuber, B. Bolin, N. M. Dickson, S. Faucheux, G. C. Gallopin, A. Gruebler, B. Huntley, J. Jäger, N. S. Jodha, R. E. Kasperson, A. Mabogunje, P. Matson, H. Mooney, B. Moore III, T. O'Riordan, and U. Svedin. 2000. *Sustainability science. Research and Assessment Systems for Sustainability Program.* Discussion Paper 2000-33. Cambridge, MA: Environment and Natural Resources Program, Belfer Center for Science and International Affairs, Kennedy School of Government, Harvard University. [Modified version published in *Science* 292:641–642, April 2001.]

Loreau, M., S. Naeem, and P. Inchausti. 2002. *Biodiversity and ecosystem functioning: Synthesis and perspectives.* Oxford: Oxford University Press.

Meadows, D. 1998. *Indicators and information systems for sustainable development.* Report to the Balaton Group. Hartland, VT: Sustainability Institute.

Nardo, M., M. Saisana, A. Saltelli, S. Tarantola, A. Hoffman, and E. Giovanni. 2005. *Handbook on constructing composite indicators: Methodology and user guide.* OECD statistics working paper and Joint Research Centre report JT00188147. Ispra, Italy: Joint Research Center of the European Commission.

Parris, T. M., and R. W. Kates. 2003. Characterising and measuring sustainable development. *Annual Review of Environmental Resources* 28(13.1):13–28.

Saisana, M., and S. Tarantola. 2002. *State-of-the-art report on current methodologies and practices for composite indicator development.* Joint Research Centre report, EUR 20408EN. Ispra, Italy: Joint Research Center of the European Commission.

PART II
General Approaches

Arthur Lyon Dahl

The multiple dimensions of sustainability do not lend themselves to a single approach or type of analysis. The SCOPE project therefore assembled a diverse group of experts from many fields and perspectives and asked them to provide inputs to the assessment process. These background papers have been reworked to enrich the debate on sustainability assessment. The general approaches described in this part raise very broad issues.

In Chapter 5 Jesinghaus is intentionally controversial, showing how profound an impact indicators can have on our thinking and decision making, often today in an undesired and inappropriate way. He highlights the power of gross domestic product (GDP) as an indicator and its distorting effects. He then describes the Dashboard of Sustainability, a useful tool he has devised for the simple graphic presentation of complex indicator sets, demonstrating the messages that can be communicated. Following on the failure of the sustainable development indicator set of the Commission on Sustainable Development to achieve wide implementation, he holds out some hope that the indicators for the Millennium Development Goals may do better. However, there is a major gap in indicators for governance. He challenges us to keep trying for the acceptance of an effective measure of sustainable development.

Chapter 6 gives an economist's view on assessment of sustainability. Zylicz argues that sustainable development can be addressed in modern economic theory, at least by indicating whether the economy is developing in a sustainable way. Accepting the difficulty that comes from the GDP assumption that welfare is confined to the consumption of marketed goods and services, he shows that the broader economic concept of utility can capture all dimensions of sustainability, including any of several welfare functions. The great strength of GDP, and the challenge for any alternative measure, is its independence from arbitrary valuations. Among the economic approaches that have been tried are a green GDP and natural capital accounts. He proposes a green net domestic product as a better measure that avoids arbitrariness, although there is still the problem of determining alternative valuations of nonmarket goods and services. The economic approach also does not address the different social philosophies of income distribution.

For an alternative political science approach, Spangenberg focuses in Chapter 7 on the importance of the institutional dimension of sustainability. This is often marginalized in the North, where institutional frameworks are already strong, but it is recognized in development discourse. However, most efforts address the sustainable development of institutions, not institutions for sustainable development able to integrate all dimensions of sustainability. The chapter provides clear definitions of institutional sustainability at the macro level to assist in defining indicators, especially the boundary between social and institutional dimensions. It then proposes a methodology for developing a coherent set of institutional indicators.

These three chapters provide complementary reflections on the failings of present economic indicators and the challenges of the institutional or governance dimension to achieving sustainability.

5

Indicators: Boring Statistics or the Key to Sustainable Development?

Jochen Jesinghaus*

At first sight, indicators seem to be addressed to experts, economists, and statisticians, and yet they are part of our daily life. For example, whenever I watch the evening news on television, I am informed about at least two indicators: First, I am told how the Dow Jones Index evolved on that day, and then the weather report informs me about the next day's temperature.

The weather report helps me decide how to dress the next day, but the Dow Jones Index is absolutely useless for me because I do not own American shares. Nonetheless, I am forced to digest Wall Street's news. Of course, I could switch the television off, but a few seconds after the Dow Jones Index is given, the movie starts. There is hardly enough time to pour myself a drink, so I will probably just stay there and listen. Statistics show that ordinary people spend several hours watching TV every day. Even if we watch the news only once a day, this still implies that after 1 year, we have seen how the Dow Jones scored about three hundred times. During our professional life (i.e., about 40 years), people like you and me will have been told more than ten thousand times that a falling Dow Jones is bad, and a rising Dow Jones is good. After a while, we start believing the message. Maybe it is sheer coincidence, but on January 7, 2003, U.S. President Bush announced a tax cut on dividends aimed at helping the dwindling stock markets. This generous present to a small minority of shareholders will cost more than US$600 billion in the next 10 years. Who will pay this bill in the end?

The daily Dow Jones brainwash is a recent phenomenon, and I hope it will never get full control over people and politics. But there is an even more powerful indicator, the gross domestic product (GDP) growth rate, that does indeed strongly influence the democratic debates in our societies.

*This chapter reflects the strictly personal opinions of the author. Comments should be sent to Jochen.Jesinghaus@jrc.it.

Of course, it is almost impossible to prove this statement because no head of government has ever declared in public, "I made this decision, aimed at accelerating economic growth, because I wanted to be reelected, and I knew my voters would judge me on the basis of the GDP growth rate." Yet it is equally impossible to find a newspaper that does not contain at least one article lamenting the critical economic situation and urging the government to take measures to accelerate economic growth, preferably by lowering the tax burden of small and medium enterprises.

Certainly, the wealth of a nation depends to a great extent on its economic output (and that is essentially what GDP measures: output valued at market prices), but there is also a broad scientific consensus that GDP should not be misused—and unfortunately this is still common practice in the media—as a way to measure the well-being of our societies.

Under pressure from the media, governments are pushed to follow the "more growth is better" message of GDP. But do we really want to get richer,[1] even if the price is destruction of the environment, poverty for the South, violation of human rights, gender inequality, and child labor?

The answer should be a clear "No," at least for somebody who has no television and does not read newspapers. However, most citizens do have television, are brainwashed, and consequently do vote for political parties and governments that behave as if "more growth" were the only important goal of Western democracy. No prime minister in the world can ignore the GDP growth rate.

There is an obvious solution to end this distortion of the political agenda: the *abolition* of GDP. If the statistical offices stopped publishing the GDP growth rate, the media pressure would end, and our politicians would start explaining to us in detail how their decisions make our lives better. Unfortunately, although there is abundant literature on why GDP is a flawed measure of economic success, it continues to be used as the most important policy performance barometer: A government may do plenty of good and intelligent things to increase the true welfare of its citizens, but if the GDP growth rate is −2 percent, it will definitely lose the next election. Given its role in politics, including some legitimate uses in economic policy, it is very unlikely that statistical offices will ever stop publishing GDP.

However, we could balance the power of GDP, and of its almost equally powerful companions inflation and unemployment rate, by redefining government performance on the basis of a comprehensive set of indicators covering all the important goals of society.

The Dashboard of Sustainability

Fortunately, this idea is not new, so a lot of the legwork has already been done. There have been several attempts to produce indices aiming at a replacement of GDP, such as the Human Development Index (HDI), published by the United Nations Development Programme (UNDP), together with the annual Human Development Report.[2] However, very few have the ambition to give comprehensive coverage of today's most important political debates. Perhaps the most advanced example is the Dashboard of Sustainability, developed by a small group of indicator program leaders called Consultative Group on Sustainable Development Indices (CGSDI, www.iisd.org/cgsdi/).

A car driver, an Airbus pilot, and the captain of a cruise ship all have a dashboard in front of them, with an impressive array of instruments that help them make their decisions. Likewise, the captains of nations need tools to steer our modern societies into the twenty-first century, and in a participatory democracy, citizens insist on looking over the shoulder of the captain in order to understand, comment on, and criticize the decisions of their governments.

Currently, only a handful of indicators, namely the rates of GDP growth, unemployment, and inflation, are communicated to citizens. However, judging government performance with only three indicators is like traveling with a captain who tells the passengers, "As long as there is fuel on board and the compass points into the right direction, everything is OK."

The complexity of decision making in the twenty-first century calls for much better decision support tools. The Dashboard of Sustainability presents sets of indicators in a simple pie chart format based on the following three principles:

• The size of a segment reflects the relative importance of the issue described by the indicator; for example, in Figure 5.1 the theme "Economy" has a weight of 45 percent.[3]

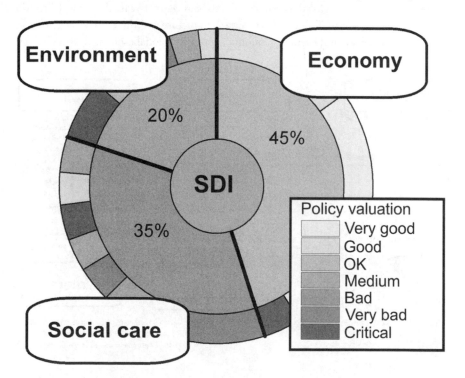

Figure 5.1. Communication language of the Dashboard of Sustainability.

- A color code signals performance relative to others: green means "good," yellow is "average," and red means "bad." On the Internet, colors work fine, but because this book is in black and white, the scale from red to green is replaced by a gray scale in which "red" is the darkest and "green" the lightest shade.
- The central circle (Sustainable Development Index [SDI]) summarizes the information of the component indicators. The inner circles are aggregations of the individual indicators.

The CGSDI sees this tool as an attempt to help launch process of putting indicators at the service of democracy. Given the professional background of the group's members, this is an SDI with special focus on environmental aspects.

Assuming that such an SDI were in place and that the media had fully adopted this new way of judging government performance, the political cycle might work as illustrated in Figure 5.2.

Of course, the diagram cannot capture the full complexity of politics in a media society, and many other factors contribute to voters' decisions. But this political cycle exists already, unfortunately reduced to a "performance index" with only three components: economic growth (measured as GDP), unemployment, and inflation.

In an attempt to demonstrate that a full SDI would give a much richer picture of politics, the CGSDI had prepared, just in time for the Johannesburg World Summit on Sustainable Development, a global assessment for the ten elapsed years since the 1992 Rio Summit. The indicator set follows the Agenda 21 structure and therefore is organized in four clusters: social, environmental, economic, and institutional aspects. In this chapter,

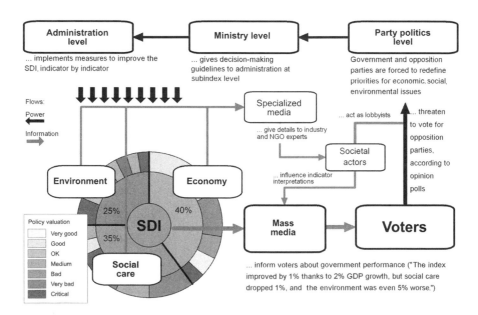

Figure 5.2. Policy cycle in a media society using an SDI.

some key results for the least developed countries (LDCs), compared with the rest of the world, are presented.

The social cluster consists of nineteen indicators (Figure 5.3). The overall social situation for the LDCs is very dim: Most indicators are deep in the red (i.e., dark) zone, with a few exceptions:

- The income distribution, measured as Gini coefficient, is good (i.e., even the rich in these countries are relatively poor); South America has by far the worst gap between rich and poor.
- Unemployment rates are at an acceptable level (but doubts about the validity of the underlying figures might be raised).

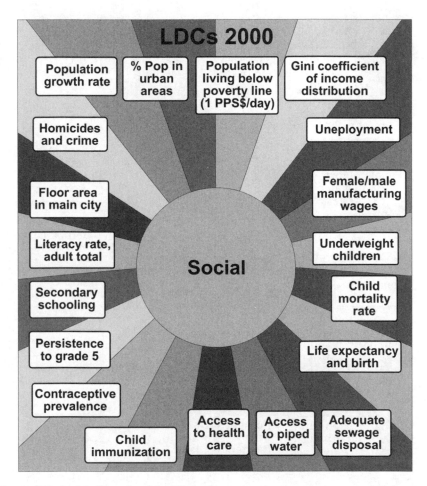

Figure 5.3. Social pillar of sustainable development.

• The ratio between female and male salaries is average; the situation is worse in Asia and, again, South America is at the bottom of the list.
• Crime rates and urbanization[4] are still "green" (i.e., light) compared with those of the other country groups.

The environmental cluster consists of twenty indicators and shows a very mixed picture (Figure 5.4):

• The LDCs score very well for problems associated with a high living standard, such as CO_2 and chlorofluorocarbon (CFC) emissions, fertilizer use, and pesticide use.
• Their ecosystems and fauna are judged more positively by these indicators, but not many of these areas are protected.

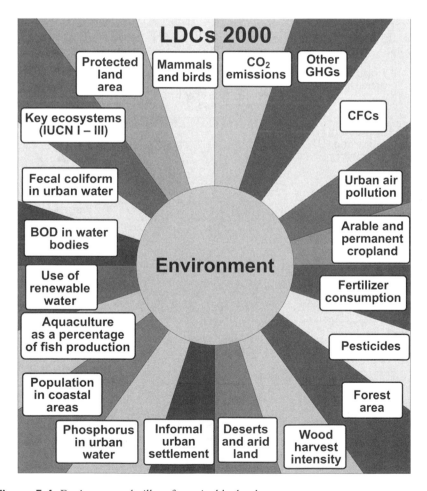

Figure 5.4. Environmental pillar of sustainable development.

- Urban air quality is not so good (but is much worse in Asia).
- Forests are in bad shape, maybe also because firewood is still an important source of energy.
- Finally, the LDCs face serious water problems (phosphorus, biological oxygen demand [BOD]), despite a "green" (light, i.e., "good") rating for use of renewable water.

The economic cluster features fourteen indicators and, not surprisingly, is deep in the "red" (i.e., dark) zone for most indicators (Figure 5.5). The LDCs are ranked last for income per capita, investment, current account balance, external debt, development aid received, renewable energy resources, and energy efficiency of GDP, adequate solid waste disposal, and (third last) waste recycling.

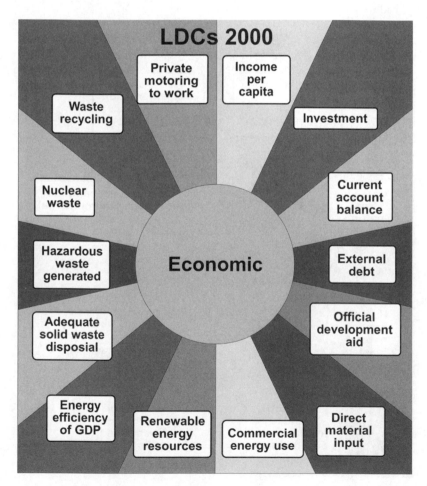

Figure 5.5. Economic pillar of sustainable development.

The picture is more positive for private motoring to work, hazardous waste generated, direct material input (DMI), and commercial energy use,[5] issues that typically go along with high Western living standards.

For economists, this indicator set looks strange: Recycling and renewable energy are rarely found in the business pages of our newspapers.

Given the importance of governance in today's politics, the institutional cluster is very interesting but also disappointing (Figure 5.6). Do the number of telephone lines, access to the Internet, and the availability of SD strategies and indicators really tell us something about the efficiency and quality of institutions?

The Commission on Sustainable Development (CSD) indicator set enjoys high legitimation because it comes from the United Nations. However, the institutional cluster in particular reflects also the delicate political climate in which the UN expert groups had to design their set.

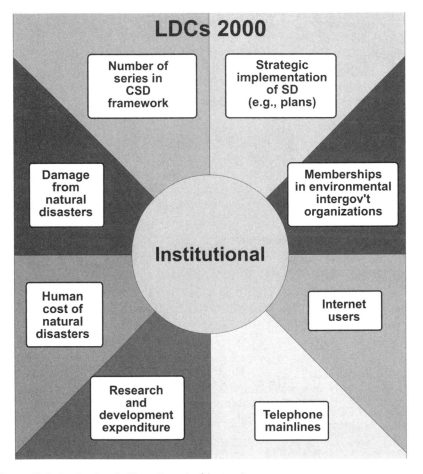

Figure 5.6. Institutional pillar of sustainable development.

The institutional results are similar to the others, with LDCs almost always at the bottom of the ranking, except for the indicator "national sustainable development strategies."

The overall assessment of sustainability for the year 2000 shows the European Union on top (old EU-15), with 750 of 1,000 possible points, followed by the three North American Free Trade Agreement states (United States, Canada, and Mexico), the Organization for Economic Co-operation and Development (OECD), and Europe (Figure 5.7).[6] At a distance, Asia and South America follow, and the continent of the poor, Africa, is close to the bottom. Among the forty-nine LDCs, characterized by low income, weak human resources, and a low level of economic diversification, are two thirds of the African countries, plus Afghanistan, Bangladesh, Bhutan, Cambodia, Haiti, Laos, Myanmar, Nepal, and Yemen. In the context of the globalization debate, LDCs have a particularly important role because they can be both victims of globalization, because of their extreme vulnerability, and globalization winners if they manage to develop their natural and human resources in a sustainable way.

A powerful communication function of the Dashboard is the map view (Figure 5.8). A trained eye needs only milliseconds to see where the problems are. In sub-Saharan Africa, Kenya, Ghana, and Cameroon are positive surprises, whereas Somalia, Niger, Angola, Liberia, Eritrea, and Sierra Leone are all deep in the "red" (i.e., dark) zone.

Unfortunately, a trend analysis shows that the gap is becoming wider: The European Union, already on top of the ranking for the situation in 2000, is also the country group that shows the best trend between 1990 and 2000. On the bottom end, the developing countries and the smaller subgroup of LDCs appear to have no chance of catching up. However, a more detailed analysis would reveal that the LDCs made considerable progress in social problems; they rank second among nine country groups, mainly because of improvements in infant mortality, illiteracy, and access to safe water. However, interpreting that as a result of good development cooperation (the LDCs enjoy certain preferences) would take a more detailed and more robust data set.

The Dashboard of Sustainability contains the sixty indicators proposed by the UN Commission on Sustainable Development and covers almost two hundred countries (available at esl.jrc.it/envind/mdg.htm). The software is free for indicator developers and has been applied to many other indicator sets, of which more than a dozen are downloadable. Among these are an almost complete set of Millennium Development Goal (MDG) indicators, a set of World Bank governance indicators, and the Environmental Sustainability Index (ESI) produced by Columbia and Yale Universities for the World Economic Forum.

Are Sustainable Development Indicators Still "Mainstream"?

Most observers quote the 1972 Stockholm UN Conference on Human Environment as the starting point of the sustainable development process, which had its outstanding events with the 1992 UN Conference on Environment and Development and the

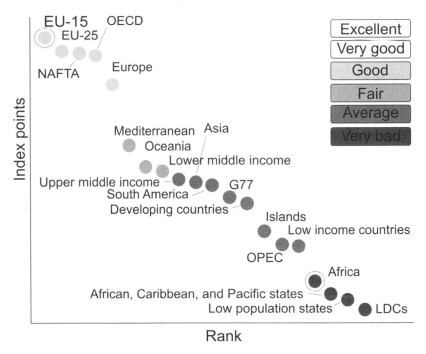

Figure 5.7. Global picture: Sustainable development by country groups.

2002 World Summit on Sustainable Development in Johannesburg. The indicators listed in this chapter were developed in the context of this process by experts convened by the UN CSD, and nominated by UN lead agencies and national governments. According to UN DESA,[7] the UN CSD set is "the result of an intensive effort of collaboration between governments, international organizations, academic institutions, non-governmental organizations and individual experts."

However, the CSD set has never really taken off. In 1997, Eurostat (the Statistical Office of the European Communities) published a booklet with fifty-four CSD indicators, followed by a second edition in 2001.[8] To my knowledge, these are the only international official publications based on the CSD set. Although the set represents a reasonable coverage of sustainable development themes, it faced stiff resistance from some member states of the G77 (the group of developing countries in the UN system), and at CSD-9 (the session dedicated to information for decision making) it was decided that the indicators should be used on a strictly voluntary, country-owned basis, a decision that was not conducive to a focused effort on increasing the quality and avail-

Sustainable Development Index

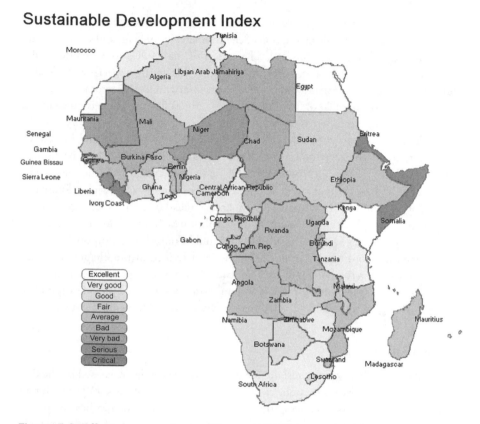

Figure 5.8. Effective communication: The map of Africa.

ability of the data. However, it would be unfair to put the blame for the failure of this set on the developing countries; indeed, very few OECD states have tested the set, and no government adopted it in the end. Consequently, as of today, the only common indicators for OECD countries are still GDP growth, unemployment rates, and inflation.

The good news is that there is hope for developing countries: After the Millennium Summit, a set of forty-eight indicators related to the eight MDGs has been agreed upon by the main actors in development politics.[9] The quality and availability of data for these indicators are not overwhelming, but they have quickly become popular: A Google search for the phrase "MDG indicators" yielded 544 effective hits in March 2006, compared with only 257 for "CSD indicators."[10] Even more impressive is the availability of Excel tables for all MDG indicators on the Web sites of the United Nations Statistics Division, the UNDP, and the World Bank, a sign that the MDG set is being taken seriously by the UN system.

The bad news is that the MDG indicators, though enjoying a broad political legitimation, have two important gaps. First, the environmental pillar receives far too little attention (only eight indicators, compared to twenty in the CSD set). That is not enough even for a very crude description of environmental problems, and sooner or later this lack of detail must be addressed. The review of the MDGs in September 2005 would have been an excellent opportunity to launch this debate, especially for Goal 7, but this summit was overshadowed by debates on security and UN reform.[11]

Second, the fourth pillar of SD, institutions, is almost absent. There is no technical reason for this vacuum; there is abundant literature on governance indicators, and the World Bank site in particular offers data for a wide range of countries and indicators.

A more comprehensive indicator set could be constructed from three prominent sources: the UN CSD set, the forty-eight MDG indicators, and a governance index. The latter could be based on an index developed at the World Bank (widely known as KKZ, for its authors, Kauffman, Kraay, and Zoido-Lobaton), composed of six subindices called "Voice and Accountability," "Political Stability," "Government Effectiveness," "Regulatory Quality," "Rule of Law," and "Control of Corruption."

Rearranged according to the four-pillar model described in this chapter and cautiously reduced to a total of about sixty indicators, such a merged CSD–MDG–KKZ set would build on the most advanced global indicator processes and provide a complete picture for judging sustainable development at the level of UN member states.

Conclusions

Many of the problems on the sustainable development agenda are caused by the distorting effect of indicators on politics, particularly the abuse of the GDP growth rate as a measure of success. Balancing this power by embedding such indicators into a comprehensive performance index covering the four pillars of sustainable development (economic, environmental, social, and institutional) would enable citizens to better judge governments' actions. Consequently, and even more important, such an index would permit politicians to do reasonable things without fear of being punished by the media's oversimplifying message, "This threatens economic growth and employment."

Replacing GDP in its role as political lead indicator will not be easy, and great intellectual and financial efforts will be needed. However, progress in recent years has been encouraging: The data situation has improved, and with tools such as the Dashboard of Sustainability, in its applications to the UN CSD and MDG indicator sets, we are able to communicate complex messages to the public. Sustainable development must become measurable if we want operational political targets; would there be a Kyoto Protocol if we didn't know how to measure CO_2 emissions?

Notes

1. Strangely enough, high GDP growth is rarely interpreted as, "We got richer, wow!" In most cases, the interpretation is, "That's good for employment." Such abuse

of GDP growth as an unemployment forecasting indicator is very popular in the media but is misleading because it says nothing about long-term unemployment.

2. Available at hdr.undp.org/.

3. The figure is illustrative. The Dashboard software allows up to ten main segments and more than two hundred indicators.

4. High urbanization gets a negative score. In richer countries, urbanization is not necessarily bad; for example, people in the countryside use their cars more often. In poor countries, though, the modern infrastructure needed to run a mega-city such as Lagos or Manila puts a heavy economic and resource burden on the hinterland. A conclusive assessment of the pros and cons of urbanization in poor countries is lacking.

5. High energy use gets negative scores in light of dwindling resources and rising prices. It might be argued that in poor countries commercial energy use entails an improvement of living conditions, but where should we draw the line between "good" and "bad" energy? At the level of India, China, or Mexico? Why not Canada? Also, living conditions (e.g., diseases, infant mortality) have their own indicators, so using energy use as a proxy would mean double-counting. However, I would add "percentage of population with access to electricity" as a positive energy indicator if data were available.

6. The y axis of Figure 5.7 shows calculated performance points, and the x axis displays the country's position in the ranking.

7. See *Indicators of sustainable development: Guidelines and methodologies*, United Nations Department of Economic and Social Affairs, www.un.org/esa/sustdev/natlinfo/indicators/indisd/indisd-mg2001.pdf.

8. Available from Office for Official Publications of the European Communities, Luxembourg, 2001, ISBN 92-894-1101-5; also available at www.eu-datashop.de/veroeffe/EN/thema8/entwickl.htm.

9. *Road map towards the implementation of the United Nations Millennium Declaration*, A/56/326, available at http://unpan1.un.org/intradoc/groups/public/documents/un/unpan004152.pdf.

10. Google "omitted some entries very similar to the 257 already displayed." For "MDG indicators," Google claims 38,700 hits, but only 544 pages are found.

11. The UN Division for Sustainable Development recently started reflections on how to better integrate the CSD and MDG indicator sets; see "Indicators of Sustainable Development: Proposals for a Way Forward," by László Pintér, Peter Hardi, and Peter Bartelmus, available at www.un.org/esa/sustdev/natlinfo/indicators/egm Indicators/crp2.pdf.

6

Sustainability Indicators: An Economist's View

Tomasz Zylicz

Many environmentalists and social critics see economics as a discipline that is inconsistent with what most friends of the earth would like to see. Yet I argue in this chapter that sustainability can be fruitfully addressed on the grounds of modern economic theories. The concept of a greened domestic product can be developed and contrasted with alternative measures based on noneconomic indicators. Whereas the latter may address specific issues, the former reflects the idea of a social welfare function whose objective is to capture the overall predicament of a society. One can easily ridicule the idea of a single indicator by pointing at situations such as airplane flight safety or a patient's health; they cannot be meaningfully characterized by a single measure. Indeed, the more complex a research object is, the more indicators are necessary to capture its predicament. Nevertheless, often one would like to raise simple yes-or-no questions such as "Is this economy developing in a sustainable way?"

The classic definition asserts that sustainable development meets the needs of the present without compromising the ability of future generations to meet their own needs (WCED 1987). Although it would be difficult to challenge the logic of this succinct description, it is equally difficult to make it operational (Pezzey 1989). Economists have developed the concept of utility to capture the essence of satisfying human needs. This goes beyond what is routinely applied in statistical analyses based on gross domestic product (GDP) accounting. The latter assume that welfare is confined to the consumption of material goods and services available on the market. And yet not everything that determines human well-being can be bought in the market. First, many services are provided directly by households to their members. Second, some goods or services are provided directly by the natural environment. Third, human well-being is also determined by psychological factors such as a subjective feeling of justice, social cohesion, and a sense of purpose. Unlike the market value of goods and services consumed, utility can in principle reflect all these considerations. It also captures the phenomena that escape market valuation.

The main point, however, is not to swap the GDP for alternative indices too quickly. A major advantage of the GDP is its independence from arbitrary valuations. Quantities are recorded statistically, and prices are set by markets. Therefore, there is no room for arbitrary manipulations by researchers. In contrast, many other welfare indices are subject to arbitrariness, even though some researchers may not be aware of it. For instance, a typical non–GDP-based welfare index is a composition of a GDP-like number (e.g., the consumption of some material goods) and a series of other numbers reflecting social welfare (e.g., the availability of medical treatment, longevity, green areas per capita) (Moldan and Billharz 1997). Although many or all of such factors are indeed important elements of welfare, the totals finally reached depend on the number of factors included and the choice of measurement units. Any choice of units implies assigning weights that a researcher attaches (perhaps unconsciously) to a given factor.

Of course, there are procedures aimed at freeing such composite indices from arbitrariness, but they have no scientific foundation. Any expert judgment and method of statistical grouping or standardization may provide a researcher with the comfort of feeling not guilty of conscious manipulation, but in fact they are not objective (Kobus 2002).

The only alternative welfare indices that can be defended as consistent with economic theory are those based on GDP. Because many exercises of this type were carried out specifically in order to take environmental considerations into account, the concept of a greened GDP was formed. The latter is a GDP in which certain corrections were added to account for the environmental factors that are not adequately reflected in the standard GDP.

Yet another approach is based on the critical natural capital concept (Ekins et al. 2003). According to this, there are limits to substitution between various elements of welfare: The lack of a resource cannot be compensated by the abundance of something else. The purpose of studying critical natural capital is to identify certain minimum or safe levels of natural resources essential for sustainability. This is a promising area of study, but it will take a long time before the approach produces widely accepted sets of indicators (Ekins 2003).

An Economic Idea of Welfare

In economic theory the individual well-being of a consumer i is represented by a utility function $u_i(x_i, G)$, where x_i stands for the individual consumption of so-called private goods (i.e., those that are individually acquired, perhaps in different quantities by every consumer), and G is the consumption of so-called public goods (by definition identical for all consumers). Since the mid-twentieth century, economic theory has applied the Bergson–Samuelson function of economic welfare $W(x_1, \ldots, x_k, G)$, where the consumers are numbered from 1 to k. No conditions are imposed on W other than if all the consumers prefer one situation over another, W must indicate this universally preferred situation as superior. This concept can be formalized in the following way. If for every $i = 1,...,k u_i(x'_?,G') \geq u_i(x_?,G)$, then also $W(x'_1,...,x'_k,G') \geq W(x_1,...,x_k,G)$. The general definition of the Bergson–Samuelson function does not indicate the effect on overall welfare when some consumers are better off while others are worse off. One has to make additional assumptions in order to draw further conclusions.

Many specific social welfare functions are considered in economic theory. Three of them are particularly well studied (see any advanced text of microeconomics, e.g., Mas-Collel et al. 1995:825–828):

Bentham function (utilitarian)

$$W(x_1,...,x_k,G) = \alpha_1 u_1(x_1,G) + ... + \alpha_k u_k(x_k,G), \text{ where } \alpha_1,...,\alpha_k \geq 0$$

Rawls function

$$W(x_1,...,x_k,G) = min_i [u_1(x_1,G),...,u_k(x_k,G)]$$

Nietzsche function

$$W(x_1,...,x_k,G) = max_i [u_1(x_1,G),...,u_k(x_k,G)]$$

These functions differ in what they assume about how social welfare depends on the welfare of individual consumers.

Bentham's function asserts that social welfare increases even when a consumer j has experienced a decrease of utility as long as a consumer m has experienced a higher increase of utility (taking into account the relation α_j/α_m). If one is guided by the Bentham function with identical coefficients α ($\alpha_1 = ... = \alpha_k$), that is, no consumer is favored over another, then the social welfare is maximized when resources are allocated to the consumers who value them most. In contrast, the Rawls function asserts that social welfare is identical to the welfare of the worst-off consumers. According to Rawls, welfare changes among those who are well off do not affect the social welfare. The latter increases only when the welfare of the very worst-off consumers improves. Finally, the Nietzsche function (referring to the nonegalitarian convictions of the philosopher) asserts that social welfare is identical with welfare of the best-off consumers. If the Nietzsche function were preferred, resources would have to be allocated in order to improve the situation of the most privileged ones, even at the expense of the least privileged ones.

These examples of welfare functions demonstrate how flexible economic theory can be in analyzing social and economic changes. It can provide analytical tools irrespective of political convictions of the analyst. Bergson–Samuelson social welfare functions can model any system of resource allocation, both purely egalitarian and the opposite.

Economic theory can accommodate any relationship between the natural environment and social welfare. Thus, one can study how the environment affects the satisfaction derived from both individually consumed goods (x_i) and public ones (G). In particular, economic theory does not take an a priori position with respect to privatization or socialization of certain services provided by natural resources. For instance, the demand for clean water can be satisfied either by improving the quality of natural aquifers (i.e., providing public goods) or by developing the market for bottled water (i.e., providing private goods). The first way reaches all consumers because, by definition, everybody has access to a public good. The second way reaches only those who put the highest value on clean water first because the demand is satisfied through individual purchases; therefore, not all consumers need to purchase the same amount.

The most challenging questions refer to welfare impacts of the environment in the future. Some decisions are based on present preferences despite the fact that they will affect future generations of consumers whose preferences are not known. For instance, land use decisions, particularly those that determine proportions between built-up areas and natural ecosystems, are made without knowledge of whether they are consistent with the future generations' preferences.

Economists ponder whether the environment is a luxury good, that is, a good for which demand grows faster than consumers' income. There is some evidence indicating that indeed this might be the case. Nevertheless, a fully satisfactory and universal answer will remain unknown because we cannot predict the preferences of future generations.

Greening the Conventional GDP

We now turn back to the concept of GDP. Its simplest definition denotes the value of all newly produced goods and services. It was contemplated by economists for centuries, but the idea was not formalized until the 1930s. Its purpose was to quantify the global demand in an economy so that business cycles could be better controlled. It performed in this role so well that both economists and lay citizens started to accept it as an overall indicator of economic activity and welfare, despite the fact that this was not its original purpose.

Abusing GDP and pretending it indicates what it could not triggers criticism from an environmental point of view also. Critics say, "GDP counts what does not count, and it does not count what counts," suggesting that it is not an adequate measure of welfare. For instance, if there is an oil spill, GDP may increase as a result of rescue and recovery actions. What is most disturbing is that GDP does not depend on the state of the environment. This state can either improve or deteriorate without any impact on the GDP. Even worse, the production of some goods (e.g., soundproof windows) may grow, increasing the GDP, while welfare (because of the higher noise outdoors) goes down. A similar effect can be observed when GDP grows in response to the increasing production of pesticides applied to counteract the declining resilience of ecosystems.

The difference between gross and net product is also important. Theoretically, the distinction is easy. Gross product contains all newly produced goods, including those that will substitute for depreciated and scrapped capital. However, the eliminated capital reduces the wealth of society. Therefore, from a welfare point of view, depreciated capital should be subtracted from the newly produced investment goods. The result would then be called net domestic product (NDP). Hence, NDP contains only the investment goods that do not simply restore the used-up capital but increase it. Although NDP seems to be a much better indicator of material welfare than the GDP, the latter is more widely used because economists do not trust depreciation statistics. Anybody familiar with bookkeeping knows how arbitrary write-off rules can be; they do not have to reflect actual depreciation of capital. For that reason many economists prefer using gross indicators because they are not affected by the write-off regulations that are not always reliable.

Public opinion demands a quantitative indicator of economic activity, and GDP has been widely accepted in this role. Consequently, many environmentalists have continued efforts to green it. Ideas emerged to exclude from GDP so-called protective goods, which do not really improve welfare but rather protect against its loss from environmental disruption. For instance, actions to prevent or remediate oil spills would be excluded from GDP. However, this is not a satisfactory solution because classifying products into distinct categories of those that improve social welfare and those that prevent welfare loss must be arbitrary. There would be a category of ambiguous products (e.g., computers) that could both improve the welfare and protect against its loss. That is why any correction rules based on common wisdom should be looked at with caution (Dasgupta and Mäler 2001).

Only in the last two decades of the twentieth century did economists develop a consistent idea of how to amend the GDP (and NDP) concept in order to take the environment into account systematically. The idea is to estimate a hypothetical Bentham welfare function W with a linear approximation. The original W function could be nonlinear in the consumption of both private and public goods, rendering it difficult to operationalize even if its formula were known. This formula is actually not known, which makes the situation even worse. Nevertheless, subject to some mathematical assumptions (that will not be discussed here), this unknown function can be approximated by a linear one constructed as the sum of products of quantities and equilibrium prices (Aronsson 2000):

Domestic product $= p_1 q_1 + \ldots + p_n q_n$,

where $p_1, \ldots, p_n$ stand for the prices of the n goods and services produced in the economy, and $q_1, \ldots, q_n$ denotes their quantities.

Thus an appropriately defined NDP can be interpreted as a linear approximation of an unknown welfare function. The problem then boils down to what products to include and what prices to apply. One can further demonstrate that in the case of many goods pure market prices (net of subsidies and taxes) are sufficient. Only with respect to nonmarket goods, including environmental protection, do alternative valuations need to be sought.

Thus an adequately greened NDP has a similar form to its traditional prototype except that certain goods are systematically excluded from summation (if they do not contribute to welfare), certain goods are systematically added (if they do contribute to welfare but they are neither sold nor bought on the market), and prices are not always market ones. The starting point is a greened GDP that is the sum of consumption, savings, and the environmental services that are consumed directly (not bought on the market). The greened NDP is then calculated by switching from gross to net values (i.e., instead of all savings, one has to take into account only those that increased the value of capital), and from the value of environmental services one has to subtract environmental damages (i.e., the cost incurred in order to compensate for environmental amenities lost).

Particular attention should be given to the value of capital and net savings. In modern economic theory, capital consists of three major components: human-made, natural, and human. All three can depreciate, and all three can be enhanced through investment. The depreciation of natural capital results from activities such as raw material extraction, biodiversity loss, and environmental disruption. The greened NDP thus ignores revenues from selling capital (including natural resources). According to modern economic theory, not every revenue considered an "income" in common language is a true income (Aronsson 2000:585). At least since John Hicks's (1939) fundamental work, economic income has been understood as a flow of revenues that can be sustained in the future. In other words, cash from selling a house is not income because this is a one-time transaction, and the owner simply changes the form of his or her wealth. In contrast, renting out a house may generate income.

In the environmental protection context, income is what comes from a sustainable use of resources. If the use of resources is an extractive one, it diminishes the natural capital whose depreciation should be accounted for. The greened NDP concept captures the essence of Hicksian income and hence it also brings the notion of sustainability.

To sum up, the NDP greened in the way outlined here differs from the traditional one in three aspects. First, it accounts for direct consumption of environmental services. Second, it adds investment in natural resources or subtracts their depreciation. Third, it subtracts environmental damages. The revised indicator reflects changes in social welfare better than GDP, but still this is just an approximation of the true value.

There have been numerous practical attempts to calculate a greened GDP. One of the best-known exercises is the Index of Sustainable Economic Welfare (ISEW), defined and computed for the United States by Daly and Cobb (1989). The bottom line of that study was that—contrary to conventional GDP statistics—American sustainable economic welfare has stabilized at the level of the 1970s, and it does not grow. If a greened NDP decreases, or if it increases at a rate lower than the corresponding conventional one does, it means that the economy develops in an unsustainable way. In other words, the present generation tries to meet its needs by compromising the ability of future generations to meet their own needs. Referring to our previous example, if one sells the house (which presumably could have served two or three more generations), one increases his or her economic welfare at the expense of the welfare of children. Daly and Cobb calculations suggest that this is what has happened in the United States.

Similar computations were carried out for Poland in the 1990s (Gil and Sleszynski 2003). As expected, the pace of growth measured with ISEW was lower than that measured with GDP. However, the Polish study demonstrates both theoretical and practical weaknesses of the ISEW method. The latter are caused by the lack of data necessary to calculate costs of environmental degradation and benefits from its recovery. In the absence of other data, the method allows approximation of these values with costs of environmental protection. This is highly questionable because changes in these costs do not necessarily reflect changes in environmental damages. In fact, such an assumption can be defended only where environmental policies are set at a socially optimum level. A more fundamental problem with the ISEW method is that raw GDP data (adjusted as in the greening process described earlier) are then multiplied by the income concentration index. The less egalitarian income distribution, the lower the ISEW. This reflects the assumption made in the general theory of Bergson–Samuelson social welfare function W; the theory envisages that not only the sum but also the distribution of individual wealth may affect W. Nevertheless, applying this specific measure of income concentration to ISEW is arbitrary, and it reflects a certain social philosophy adhered to by its authors. In particular, it often reflects an assumption that individual utility functions are concave. Although this general approach can be defended, specific forms of these functions are arbitrary.

Conclusions

Greening the GDP provides an opportunity to improve the measure of social welfare without adopting arbitrary assumptions about relative weights of various physical aspects of human well-being. This is not the only alternative economists have explored, having realized that traditional GDP misrepresents welfare. There is also a strong tendency to analyze physical accounts as more adequate for representing the quality and intensity of economic activities (Ayres 1998). The approach is best manifested in the critical natural capital concept. Although other indices are interesting and sometimes justified, the greened NDP has the advantage of being closest to what politicians and policy analysts are familiar with.

A major theoretical advantage of the greened NDP is its affinity with the concept of sustainable development. By definition, this indicator is sensitive to instances of natural capital depletion. Likewise, it reacts positively to investments in environmental quality. Therefore, its changes provide good indications of what makes the society improve or worsen its long-term perspectives for economic growth. Additionally, by comparison with the conventional GDP, it indicates whether the present generation lives sustainably or whether it eats up the capital to be handed over to the next one. If the greened NDP is lower than the conventional one, it means that the present generation compromises the ability of future generations to meet their own needs. If it is the other way around, then the present generation accumulates wealth that can be used by its successors. Consequently, the greened NDP is not a measure of present welfare. A low or declining index signals that the present generation lives at the expense of future generations, but its own welfare can be high and perhaps even growing.

There are many issues to be sorted out before a greened NDP can be calculated. One of the most challenging ones is pricing nonmarket goods such as clean air, soil, and biodiversity. Economists refer to hypothetical or surrogate markets in order to estimate the prices that cannot be observed in actual transactions (Shechter 2000). For instance, in order to approximate the value people attach to silence, one can analyze differences in real estate prices caused by differences in the noise in their neighborhoods. Likewise, in order to approximate the value people attach to the preservation of species, analysts develop scenarios of hypothetical protective actions that may be taken and ask people how much they would be willing to pay for these actions to be carried out. Values revealed by such exercises are questionable, but a lot of progress has been achieved in the last several decades in soliciting them and making them more credible.

An important question remains: whether sustainability can be assessed with just one indicator. It is a complex, multifaceted, and multidisciplinary issue and therefore can never be reflected fully by a single number. Therefore, a set of sustainability indicators rather than a unique figure should be sought. Nevertheless, economists tend to develop aggregate measures to approximate people's general perception of certain phenomena. Because sustainability is likely to remain high on the policy agenda, fundamental yes-or-no questions have to be expected. If we were to answer whether a given economy develops sustainably, then—in addition to reviewing an entire set of sustainability indicators, some of which may suggest movement in different directions—we should refer to a composite index. Many such indices can be conceived, but most of them are arbitrary in the choice of elements and the weighting system. The greened NDP concept offers a solution to the arbitrariness problem and comes closest to what many people identify as sustainable welfare.

Literature Cited

Aronsson, T. 2000. Social accounting and national welfare measures. Pp. 564–601 in *Principles of environmental and resource economics. A guide for students and decision-makers*, 2nd ed., edited by H. Folmer and H. Landis Gabel. Aldershot, UK: Edward Elgar.

Ayres, R. U. 1998. Rationale for a physical account of economic activities. Pp. 1–20 in *Managing a material world. Perspectives in industrial ecology*, edited by P. Vellinga, F. Berkhout, and J. Gupta. Dordrecht, the Netherlands: Kluwer.

Daly, H. E., and J. B. Cobb Jr. 1989. *For the common good. Redirecting the economy toward community, the environment and a sustainable future*. Boston: Beacon.

Dasgupta, P., and K.-G. Mäler. 2001. Wealth as a criterion for sustainable development. *World Economics* 2(3):19–44.

Ekins, P. 2003. Identifying critical natural capital. Conclusions about critical natural capital. *Ecological Economics* 44:277–292.

Ekins, P., C. Folke, and R. De Groot. 2003. Identifying critical natural capital. *Ecological Economics* 44:159–163.

Gil, S., and J. Sleszynski. 2003. An index of sustainable economic welfare for Poland. *Sustainable Development* 11:47–55.

Hicks, J. R. 1939. *Value and capital,* 2nd ed. Oxford, UK: Clarendon.

Kobus, D. 2002. *The evaluation of progress towards achieving sustainable development in countries in transition.* Ph.D. thesis, University of Manchester.

Mas-Colell, A., M. D. Whinston, and J. R. Green. 1995. *Microeconomic theory.* Oxford: Oxford University Press.

Moldan, B., and S. Billharz. 1997. *Sustainability indicators. Report of the project on indicators of sustainable development.* SCOPE 58. Chichester, UK: Wiley.

Pezzey, J. 1989. *Economic analysis of sustainable growth and sustainable development.* Environment Department Working Paper no. 15. Washington, DC: World Bank.

Shechter, M. 2000. Valuing the environment. Pp. 72–103 in *Principles of environmental and resource economics. A guide for students and decision-makers,* 2nd ed., edited by H. Folmer and H. Landis Gabel. Aldershot, UK: Edward Elgar.

WCED (World Commission on Environment and Development). 1987. *Our common future.* Oxford: Oxford University Press.

7

The Institutional Dimension of Sustainable Development

Joachim H. Spangenberg

Sustainable development is a complex concept, as several chapters in this volume illustrate. It is essentially a normative concept calling for a decent quality of life for all the earth's citizens now and in future, to be provided within the limits of the environment's carrying capacity. Its strategic core approach is the delimitation of responsibilities in space and time and the integration of policy domains for coherent strategies. This includes environmental objectives (respecting ecological limits), social standards (dignified life), economic conditions (competitiveness, often also growth), and institutional desiderata (e.g., participation, empowerment of communities and women, peace and justice). All four domains are addressed in the key documents such as the Brundtland Report, Agenda 21, the Rio Declaration, or the Johannesburg plan of implementation, and monitoring all four dimensions has been a sine qua non ever since the United Nations published its first set of sustainable development indicators (UND-PCSD 1996). At least in principle, the same applies to the results of other major UN conferences (e.g., the Beijing summit on gender issues, the Cairo one on population, Istanbul for community development, and Copenhagen for social development).

Because the description as a combination of separate dimensions misses the integrative character of sustainability as much as the dynamics of the development process, a better basis for understanding may be the description as a metasystem, based on the coevolution of four independent but permanently interacting subsystems. For each of them, the internal conditions for permanent reproduction must be secured for development to be sustainable, and their mutual influence must not undermine this reproductive capability. Besides its specific mechanisms and structures, the institutional system in a sustainable development concept has its own set of normative objectives, frequently discussed in all reference documents.

Unfortunately, some key elements of the concept have been underemphasized or even lost in the current framing of the political debate by antietatism, free markets, deregulation, and economic relativism, according to which by definition every scarcity is relative. Elements lost include the existence of environmental (and social) limits (ignoring the full text of the Brundtland Commission's definition including them and leading to a definition of sustainability as an "organizing principle of discourses" with no restrictions on the possible outcomes), the need for integration (resulting in the metaphor of three pillars separate from each other), and the character of institutions as a fourth dimension in its own right. It is often doubted or ignored; institutions are considered as an element supporting sustainable development but not part of it (particularly in governance discourse), or they are subsumed in the social dimension.

The complexity of the concept of four coevolving systems and their delimitation and integration as categorical imperatives exceeds the limits of the steering capacity of current (increasingly deregulated) institutional settings. Good governance programs find the task challenging and are tempted to "pragmatically simplify" the concept (e.g., by ignoring imperatives and by merging or externalizing certain dimensions). As in system analysis, the analyst is free to define the system boundaries ("this *can* be done, but *should* it?").

This chapter explains why the institutional system should be singled out as a separate dimension of sustainable development, describes the incoherent use of the terminology so far, presents a refined definition of this dimension, and finally suggests a procedure for deriving institutional sustainability indicators.

Institutions as a Dimension of Sustainability

Institutions are defined differently by different disciplines; an in-depth analysis shows that a definition from political science is the most appropriate one in the sustainable development context. For instance, sociology and economics analyze two different directions of interaction (humans on organizations and vice versa), and historical analysis refers to organizations and cultural rules. In contrast, political science focuses on what is essential for a normative concept, the conditions or rules of decision making, including the most familiar kind of institutions (i.e., organizations) but complementing them with mechanisms and orientations (Spangenberg 2002).

Institutions have been described as essential to sustainable development because of their indispensable role in implementing social, economic, and environmental objectives, but they have been denied a role as a dimension in its own right. With this view, however, although the serving character is rightly emphasized (in a coevolutionary setting, each dimension serves all others permanently), independent institutional objectives are lacking, and the specific characteristics of the institutional system (e.g., its development dynamics and lead sciences) are neglected.

More recently, institutional issues have been subsumed under the "governance for sustainability" discourse, as indeed institutions and governance are partly overlapping themes. Furthermore, despite initially different research questions and methods (institutionalism starting from the structure, with governance more prone to the mechanisms), in both research communities the trend has been to recognize that structure and dynamics must be analyzed jointly (R. Kemp, personal communication, 2005; Spangenberg and Giljum 2005). However, the analytical categories are still quite different, although organizations and mechanisms cover both disciplines. Orientations (e.g., culture, value systems) are more often neglected in governance analysis, despite their role as a key constraint for modifying mechanisms and restructuring organizations for the sake of sustainable development.

Finally, governance for sustainable development is usually described as a process external to sustainable development itself; otherwise, it would be a prominent part of the institutional dimension. However, this way the development aspect is underemphasized, and the institutional objectives such as equality, justice (including gender), and human rights are considered not as constitutive to sustainable development but as part of the governance processes supporting it. In the best cases, some of these objectives are internalized by integration into another dimension (usually the social one), but then the internal dynamic of sustainable development is still deprived of important institutional aspects.

Furthermore, merging the institutional and the social dimension is no remedy to the complexity challenge. As the interaction of agency and structure, of humans and society goes on, it must be analyzed within one dimension, which does not make it easier to describe. Rather, this implies a multilevel description of the merged socioinstitutional dimension.

The extended social dimension thus becomes a kind of a residual category, too complex to be integrated with the other dimensions on equal footing because it needs internal disaggregation to be operational. Thus, instead of mixing the social and institutional dimension into one category, in order to strengthen the role of social concerns a more structured view is preferable. Giddens calls this the duality of structure, referring to the interdependence of agency and structure, which are conceptually distinct but inseparable: Social structures enable as well as constrain human action (Jackson 2003). Once adopted, such distinct structure and agency concepts can guard against reductive social theory and help to unveil the full complexity necessary for substantial sustainability strategies. Consequently, because of the different actors, response systems, and lead sciences (humanities vs. social sciences), it makes sense to keep the difference between the two dimensions or capital stocks in mind.

For this purpose, and in analogy to the fruitful distinction between human and social capital in economics, we have defined the social dimension or human capital as the stock of personal (i.e., intraindividual) assets and capabilities, such as health, knowledge, dedication, experiences, and skills, their socialization and habits, attitudes, and orientations (the preferences of individuals as customers and citizens). It includes the agents and their human capital. The human capital stock is an aggregate or macro-level description of the sum of these individual assets and attitudes (whereas the rules governing them are part of the institutional dimension). Social objectives are focused on self-determination, the individual quality of life, and the ability to sustain oneself and all dependents on one's salary.

In contrast to that, the institutional dimension comprises the systems of interindividual rules structuring the activity of agents (i.e., the social capital of the society). Like the other dimensions, the institutional one has core objectives of its own (e.g., participation, access, gender justice; Table 7.1) but also interlinkage objectives (i.e., it enables but also limits the implementation of social, individual, economic, and environmental objectives).

Like any division, the one suggested here is artificial, but it is plausible because the separate systems thus defined are characterized by a different functional logic, different time structures, different normative objectives (self-fulfillment and quality of life vs. justice and cohesion), and different lead disciplines analyzing them (humanities vs. social sciences). Furthermore, using these definitions permits us to make the body of economic research on capital stocks usable for sustainable development analysis, such as the insight that both human and social capital are essential factors in the production of wealth, often more so than the human-made capital (Serageldin and Steer 1996). Processwise, keeping the human and social and the societal and institutional dimensions and capital stocks apart helps in distinguishing policies and outcomes (necessitating different indicators and monitoring systems for policy implementation and policy success), as in the case of education: School enrollment (institutional) is crucial, but it provides limited information about schooling success (human capital formation, social), as the Organization for Economic Co-operation and Development Program for International Student Assessment (PISA) studies have shown.

More generally, trade-offs between individual and social benefits and differences between societal efforts and individual achievements become more accessible if described as a linkage of two dimensions rather than an internal contradiction within one. One example of significant political relevance is the different concepts of justice as an indispensable element of sustainable development emerging from the social and the institutional perspective. Following the institutionalist concept of justice developed by Rawls (1971), a society can be considered just if the current situation has been the result of just rules, regardless of the actual pattern of distribution. Against this, Sen (1999) and others argue that just institutions are not enough, that justice is an individual (i.e., social) rather than an institutional phenomenon, depending on the capabilities of individuals to realize the potentials provided by the institutions.

This example illustrates the interaction of institutions (i.e., rules enabling and restricting human behavior) and behavior itself, which, in addition to external rules, is determined by a wide range of intrinsic factors and motivations. Individual preferences and orientations are part of the social or individual dimension, even if shared by many others, but once they become an element of distinction between social groups, they are an institutional phenomenon.[1] Like other institutions such as visions of societal devel-

Table 7.1. Selected social objectives and criteria.

Institutional Sustainability	Social Sustainability
Social security (systems, orientation)	Having basic material security, meeting needs
Public health (systems, orientation)	Enjoying physical and psychological safety
Social integration (mechanisms, orientation)	Participating in social activities
Participation opportunities	Being politically active and empowered
Gender equity	Benefiting from gender mainstreaming
Justice and welfare orientation	Enjoying and practicing solidarity
Freedom concerning the way of life	Being able to choose an individual way of life

Sources: Empacher and Wehling (1999), Littig (2001), Kopfmüller et al. (2001).

opment, *leitbilder*,[2] norms, relations, but also organizations and property rights, they are formed as social responses to challenges from the societal and the natural environment and the cooperation problems that emerge while we cope with them (Weisbuch 2000).

Following these lines of thought, the criteria for social sustainability from some major research projects can be divided into social and institutional ones. To illustrate the distinction, the corresponding institutional and social criteria from a number of studies are listed in Table 7.1.

The demarcation line between the institutional and the economic dimension or subsystem must be clarified as well. In the terminology applied here, the economic subsystem is made up by economic stocks and processes, inputs, and outputs, whereas the rules governing the economy—not the individual decisions—are institutional. Thus property rights, markets, and the public support for a competition-based system are institutional conditions of a market economy.

State of the Debate on Institutional Sustainability in Theory

Despite the fact that hardly any sustainable development program occurs without governance reforms, changes in legal mechanisms, and other institutional adjustments, the institutional setting has not been systematically integrated into sustainability planning. In the North, the dominant perception of sustainable development considers reconciling the environmental and the economic dimension as the key sustainability challenge (Figure 7.1). Humans and their actions, and the rules regulating them (i.e., the social and the institutional dimensions), are understood as secondary elements of sustainability, moderating the interaction of economic and environmental processes. Sometimes they are even reduced to means of managing the side effects of environmental policies (OECD 2001), which in turn are defined economically, as the internalization of external costs. However, without a change in orientations, central elements of

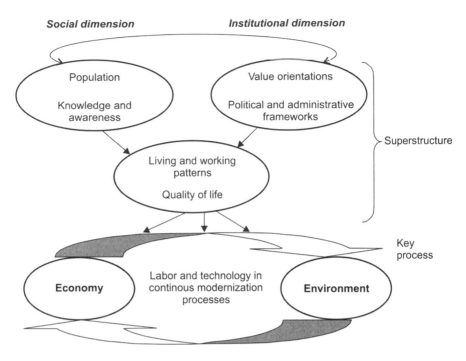

Figure 7.1. Northern perception of sustainable development (Huber 1995, modified).

sustainable development such as the shift toward sustainable patterns of production and consumption will not materialize. Similarly, without appropriate mechanisms, neither the greening of the economy nor gender mainstreaming, community empowerment, and civil society participation will be implemented, although they are key demands of Agenda 21.

This biased perception reflects the situation of countries where social and institutional problems are considered less pressing than environmental ones, whereas in countries with prevailing poverty different priorities apply. Here the main emphasis in development planning, national and by donor organizations, often is on the interaction of the social and economic dimensions (e.g., poverty eradication, literacy promotion, public health improvements), with the environment often playing a secondary role and institutions again as a moderator. However, as the Arab Human Development Report illustrates (without reference to the sustainable development concept), it is societal orientations and political decision-making mechanisms that hinder successful development, not the lack of investment capital or organizational matters (DGVN 2005).

State of the Debate on Institutional Sustainability in Practice

In development circles, the crucial role of institutional sustainability has long been recognized, in particular on the micro level, although with a narrow definition. This situation is at least partly the result of the systematic neglect of social structure in the dominant neoclassical economic paradigm, which has sidelined institutional and evolutionary economics in the North but is not as hegemonic in development economics. In development practice, the usual understanding of *institutional* is synonymous with *organizational*, and *sustainability*, following a long tradition in development politics, is usually understood to mean financial or fiscal sustainability. As a consequence, the focus is on the sustainability of specific institutions (i.e., their long-term viability and effectiveness), understood as sufficient availability of effective administration and funds, called sustainable institutional capacities and financial sustainability. Such micro-level discourses refer to the institutional capability and financial sustainability of

- All kinds of development projects and organizations
- The results of local and regional self-organization (e.g., institutional capacities of nongovernment organizations [NGOs], financial sustainability of the management of microcredits)
- Public institutions (e.g., sustainable health systems, financial services, local institutions)

In each case, the focus is on the *sustainable development of institutions* (i.e., the durable functioning of organizations); however, few deal with *institutions for sustainable development* (i.e., the institutional setting in a broader sense, including mechanisms and orientations), which would be needed for the development process as a whole to be sustainable. This indicates a missed opportunity because the restructuring of organizations and legal and economic mechanisms often is mandatory in development cooperation and financial support programs (e.g., the structural adjustments). However, such conditions are oriented toward other, mostly economic objectives, instead of orienting development toward sustainability: Social and environmental demands are taken into account in specific cases but not permanently and on equal footing.

One positive example in this context is the institutional indicators for housing and sustainable settlements, again on the local level, from New Zealand. They distinguish

- The roles and responsibilities of actors
- Institutional integration
- Process organizations and management

Another example is the use of indicators in public policies in several Brazilian cities, where local policies with social and economic characteristics have been developed and implemented using political institutional indicators (Ferreira 2000).

However, usually no link to the macro level is discussed in reference to institutional sustainability. This is a pity because sustainable development is essentially a macro phenomenon dealing with the global environment, global poverty, and the generations to come. In both cases, because the independent role of institutions is ignored and reduced to organizational and financial soundness on the micro level, their contribution to sustainable development is neglected. They play an increasing role in capacity building, but efforts for empowerment, capacity building and technology transfer cover only a subset of institutional development needs. Therefore, it seems important to more clearly define institutional sustainability on the macro level. Such clear definitions are no end in themselves; they will be essential for deriving indicators, that is, for making the concept operational. At least partly because of the lack of clear definitions, indicators referring to the institutional dimension are rarely applied on the macro level; some of those suggested by the UN Department of Policy Co-ordination and Sustainable Development were already dismissed in the test phase of the first set of indicators without being replaced by more suitable ones.

Institutions and Sustainability: Refining the Definitions

Institutions as in this chapter comprise not only formal and informal organizations but all systems of interindividual rules structuring the activity of agents (Czada 1995). This definition of institutions as interpersonal systems of rules governing decision making and comprising organizations, institutional mechanisms, and orientations has been proven fruitful for analyzing the mechanisms fostering or hindering sustainable development (Hans-Böckler-Stiftung 2001).

As derived from political science, the definition rightly focuses on the sustainability-relevant aspects of the role of organizations, decision-making processes, and orientations, the impacts they have, and the consequences they cause or contribute to. A sustainable institutional setting is a specific state of the institutional system favoring sustainable development. This definition requires an externally set norm to distinguish between sustainable and unsustainable constellations, derived from Agenda 21 or the World Summit on Sustainable Development Plan of Action, for example.

In general terms, institutions for sustainable development are the rules that

> structure the choice of action of individual or corporate and other collective actors within a society to the benefit of sustainable development. This includes organisations, which influence all actors or groups of actors in a society, if they directly or through these actors have a significant impact on society as a whole, and mechanisms and orientations (both implicit or explicit systems of rules), which apply to all actors, or groups of actors in a society, and systems of rules which apply to specific collective actors, if these rules through these actors have a significant impact on the sustainable development of the society as a whole.[3] (from Spangenberg 2002, modified)

Corporate actors here refers to constituted political actors (governments, NGOs, unions, associations) with at least a minimal organization. In contrast, *collective actors* refers to groups of individuals who potentially might act similarly and simultaneously because of comparable interests and preferences in specific circumstances but without an organizational structure. Whereas corporate actors are themselves institutions, collective actors are not, although their influence (e.g., in market processes) is significant.

Understanding Coevolution

Such a system of institutions for sustainability is sustainable in itself, that is, it delivers a sustained contribution to the institutional objectives (e.g., as set out in Agenda 21; core institutional criterion), and it creates the opportunity space and favorable conditions for the economic, social, and environmental development processes to be sustainable (interlinkage criterion). Thus, as a result of the coevolution of the subsystems, in order to be sustainable the institutional system must not only deliver on the institutional objectives but also meet economic, social, and environmental demands. In other words, it must be socially, economically, and environmentally sustainable as well. Although this may look confusing at first glance, the idea behind it is simple: For each of the four subsystems to be sustainable, it is not enough to serve itself; it must also meet the sustainability demands of the other dimensions.

For instance, one key element of sustainable development as defined by Agenda 21 is the right to a dignified standard of living for all citizens. Whereas such a right is part of the system of rules governing human interaction and thus an institutional criterion, the dignified life and the well-being of the population as such are essentially social phenomena. Obviously, realizing this objective is dependent on and influenced by the prevailing institutions, and if they support the realization of the social objective, they can be considered socially sustainable institutions. For example, a thriving economy may be economically sustainable, but as long as it is based on overexploitation of the environment and salaries below the poverty line, it must be considered socially and environmentally unsustainable.

So in more general terms, we have to distinguish the dimension and the objectives, and to evaluate its sustainability we must subject each dimension to a multicriterion assessment. In this context, it is necessary not only to focus on the institutions as such but also to ask what the institutional criteria are for the sustainability of the economic, social, and environmental system.[4] For instance, because institutions such as public organizations depend on a functioning economic system that provides resources such as tax income, an economy not providing such resources must be judged as institutionally unsustainable. Similarly, because public trust is an essential institution for any democratic state, an economic system permitting mechanisms such as tax evasion and corruption beyond a minimum accepted by society are institutionally unsustainable. Human cultures (including the political cultures) are institutions providing orientations, structured by mechanisms, and facilitated by organizations, but they are not possible without the active and often voluntary work of skillful and dedicated individuals, a social dimension phenomenon. Similarly, social cohesion and peaceful conflict solution depend on such people. To be considered institutionally sustainable, the social subsystem must deliver on these institutional demands.[5] These examples illustrate that the relationship between the dimensions is not one of delivery based on demand and supply but a system in coevolution.

Toward Institutional Sustainability Indicators

Because the four dimensions and the corresponding four sets of criteria are omnipresent in human life, sustainable development can be understood as a group of specific constellations (expressed by the criteria) in all four dimensions, characterized by the fact that their synergistic interaction permits and even creates a variety of feasible pathways for continued existence and reproduction of the overall system (Bossel 1998).

Thus, although industrial societies can be characterized as productive societies, sustainability calls for reproductive societies, including the need to permanently reproduce the institutional and societal as well as the social and human dimension of each society. Substantial sustainability will have to take both into account. However, the complex, nonlinear interaction of institutions with each other and with the other dimensions, the impossibility of listing and counting all of them, and the fact that the same effect can be produced by widely varying institutional settings render fruitless any attempt to test the sustainability of the institutional system and to derive indicators based on simplified, causality-based analytical systems or to analyze the institutions one by one regarding their appropriateness for specific purposes (such an analysis would not even help us understand the functioning of the institutional system as an emergent property on a higher system level). Instead, indicators must be derived for the institutional system as a whole, based on the explicit or implicit objectives and targets of the sustainable development paradigm, and designed to measure the performance of the institutional system in implementing them. The following six steps lead to a comprehensive set of institutional indicators by applying the differentiation outlined here.

First, all institutions present in the reference documents (organizations, mechanisms, and orientations) and the purposes they are referred to are identified. The result is a systematic list of all institutional aspects in the reference documents as the basis for further analysis.

Second, the institutional aspects are classified according to their objectives, resulting in indicators for the institutional sustainability of the social, economic, and environmental dimension and those for sustainable institutions (core institutional indicators).

In a third step, the purposes should be cross-checked with the existing sets of sustainability indicators to find out whether they have already been covered. This way, a number of prevailing indicators can be identified that measure the effectiveness of institutions by assessing the implementation of their purposes.

Based on this analysis, in a fourth step, complementary indicators can be suggested, based on measuring institutional sustainability as the effectiveness of implementation of the purposes of institutions.

Having exhausted the explicitly mentioned purposes, from step five on the indicator development must be based on the implicit ones. This refers to institutions and institutional purposes that have not been explicitly defined in the reference documents or for which the scope of purposes mentioned clearly is only a fraction of the functions the institution has in reality. Once the purposes are plausibly derived from the objectives, actors, and institutions mentioned, the corresponding indicators can be developed as described earlier.

In step six the comprehensiveness of the list of purposes is tested against the sustainability objectives mentioned in the same context. If the objectives are not covered by the purposes mentioned or developed so far, further amendments to the purpose list are to be derived based on the objectives, together with the corresponding indicators.

With this step, the total of implicit and explicit purposes of institutions in the reference documents is covered and indicators developed. However, because there probably will be objectives and institutions not mentioned in the references but important for sustainable development, a final test is recommendable, checking additional relevant documents such as UN decisions, international conventions, and conference results in a similar fashion to identify gaps regarding important institutional aspects in the basic documents used for the analysis. This kind of sensitivity analysis is disputable because "important for sustainable development" is a criterion that—beyond the official documents mentioned—will always depend on subjective assessments. Nonetheless, the feasibility of this approach has been demonstrated by stepwise analysis of the institutional imperatives in Agenda 21 and development of indicators suitable to monitor their implementation (Spangenberg 2002; Spangenberg et al. 2002); the result of this exercise is documented in Appendix 7.1.

This appendix is not intended to be the final set of institutional indicators because the discourse on methods in the scientific community is still in its infancy, but it may serve as food for thought. For instance, it illustrates that many aspects can be covered by data mining and introducing unusual cross-references. However, other indicator suggestions face severe data poverty, but it is instructive to see which domains (e.g., gender issues) have been neglected in data collection. In this sense, the state of data availability itself can be used as an indicator pointing to institutional issues that have been neglected in the sustainable development discourse.

Finally, the indicator formulation in itself is only the first step of a longer journey. The plethora of indicators suggested here is systematically derived, and no smaller set can be chosen by simple cherry picking without losing the systematic character. Instead, additional work is needed to derive indicator hierarchies (headline indicators for strategy development, policy-level indicators, and implementation-level indicators) in a manner reflecting the experiences with the UN sustainable development indicators and the new set suggested by the European Commission for the revised EU sustainability strategy.

Appendix 7.1. Core institutional indicators, suggestions, and sources.

DECENTRALIZATION AND ACCOUNTABILITY

Share of local authorities in total public expenditure	New
Number of elected members in parliaments and councils per 100,000 citizens	New
Percentage of population involved in locally managed credit systems	Established
Locally managed credit systems as share of national volume of commercial loans	New
Share of municipalities that implement local Agenda 21	Established
Share of population that takes part in local Agenda 21 processes*	New

PUBLIC POLICIES AND CIVIL SOCIETY EMPOWERMENT

Percentage of GDP spent on environment and development policies	New
Share of development plans including environmental impact assessments and social and economic acceptability assessments	Established
Percentage of environment and development expertise in government consultancy, plus gender shares thereof*	New
Ratio of full-time paid and voluntary sustainability and development experts in government, business, academia, and NGOs to total staff by gender*	New
Financial support for NGOs as percentage of total subsidies	New
Number of people involved in work for NGOs*	Germany (1999)
Number of court cases on claims of violating sustainability legislation per billion dollars GDP	New
Share of NGO-initiated cases	New
Share of national and regional development plans under legal scrutiny because of NGO initiatives	New
Share of NGOs entitled to file suit	New

EDUCATION AND RESEARCH

Percentage of research expenditures for sustainability, including share of gender-sensitive research and development	New
Percentage of interdisciplinary policy-relevant research in total research and development budget	New
Percentage of public–private partnership expenditure in sustainability-related research and development	New
Share of private funding in research for sustainability	New

*Characterized by severe data poverty.

Appendix 7.1. Core institutional indicators, suggestions, and sources (*continued*).

Percentage of sustainability-related education in schools and adult education or time budget spent in grades 5–8 on environmental "syndromes"*	Germany (1999)
Percentage of teachers taking part in training for sustainability education per annum.	New
Share of adult population taking part in adult education programs (full and part time)	Established
Share of university professors researching traditional methods of knowledge as related to share of indigenous people in the total population	New
Average number of languages spoken per person	New

GENDER RELATED

Similar constitutional and legal rights for women and men in the areas of electoral rights, inheritance, contractual relations, divorce, and choice of profession as percentage of limitations on these rights	Established (ordinal indicators)
Share of measures to secure baby food quality in drinking water investments*	New
Share of water infrastructure plans based on women's day-to-day water use analysis*	New
Relationship of average incomes in production and reproduction work	New
Share of women earning more than their partners and the share of men doing so*	New
Gender-sensitive control mechanisms in legislation and implementation	New
Share of gender-specific data collection and interpretation as a share of total data collection with reference to population groups*	New
Share of gender-sensitive research in the research budget per discipline	New
Percentage of female experts in expert databases	New
Share of data collection work based on problem definitions developed from the everyday life experience of women, particularly in agricultural, water management, and health care research and planning*	New

*Characterized by severe data poverty.

Appendix 7.1. Core institutional indicators, suggestions, and sources (*continued*).

Share of women in the 2 top levels of the 10 biggest companies, in public administration, in national NGOs and interest groups, in parliament and government, and among professors	Established
Share of these institutions with 50% or more women in the 2 top levels	New
Participants and budget share of top-level training courses specifically for women	New
Average frequency and expenditure for effectiveness assessment of plans to reduce gender inequality in main organizations*	New
Share of staff in charge of analyzing conditions of and progress in reducing gender inequality	New
Share of men in top positions with demonstrated qualifications in reproductive and care work (e.g., having taken educational time off)*	New
Share of official information publications specifically dedicated to gender issues	Established
Share of research expenditure for these links in economics, policy sciences, environmental sciences, and sociology and in the national research budget*	New
Frequency of budget lines including these links as a purpose or criterion for eligibility in total institutions that support funding of the ministries for research, economics, environment, and development	New

HUMAN RIGHTS

Violations of Human Rights Charter (including social rights)	Established
Government ratification of 8 international conventions related to fundamental human rights	UNDPCSD (1996)
Number of people and percentage of population living in absolute and relative poverty	Established
Sufficient shelter and nutrition (percentage of population)	Established

Interlinkage Indicators: The Social Dimension as an Example

The Social Interlinkage

HEALTH ISSUES

Percentage of people with basic health training	Established

*Characterized by severe data poverty.

Appendix 7.1. Core institutional indicators, suggestions, and sources (*continued*).

Percentage of Service Delivery Point (SDP) at the primary health care level offering three or more integrated reproductive health services	UNFPA (1998)
Contraceptive prevalence rate	UNFPA (1998)
Percentage of births assisted by health personnel trained in midwifery	UNFPA (1998)
Percentage of population with access to primary health care services	UNFPA (1998)
Maternal mortality ratio	UNFPA (1998)
Number of nurses and doctors/1,000 inhabitants	UNDP (1994)
Body mass index	Germany (1999)
Share of smokers in population	Germany (1999)
Share of GDP spent on preventive health care	Established
Water expenditure as percentage of disposable income of households	UNDESA (1998)

EMPLOYMENT AND INCOME ISSUES

Percentage of population employed	Established
Ratio of average female wage to male wage	Established
Ratio of top 1%, 3%, and 20% of private income to bottom 20% of private income	Established
Average real tax paid by top 20% of private income earners in comparison to national average tax paid*	New
Spending on recreation as share of disposable income by gender	UNDESA (1998)
Time spent on leisure, paid and unpaid work, and travel by gender	UNDESA (1998)
Employees represented by elected councils or comparable institutions in the workplace	ILO (1993)
Share of elected representative bodies with competencies for environment and development	ILO (1993)
Share of elected representative bodies with codecision rights for employment policies	ILO (1993)
Share of elected representative bodies with codecision rights for industrial strategies	ILO (1993)
Share of workers covered by collective framework contracts (employers and trade unions)	ILO (1993)

*Characterized by severe data poverty.

Appendix 7.1. Core institutional indicators, suggestions, and sources (*continued*).

Sustainability Beyond Agenda 21

VULNERABILITY INDICATORS

Peripherality and accessibility: distance to main trading partners	Crowards (1999)
Export concentration: share of main products	Crowards (1999)
Convergence of export destination: share of recipients	Crowards (1999)
Dependence on import energy: share of total consumption	Crowards (1999)
External finance and capital: share of total investment	Crowards (1999)
Share of imported food in national food consumption	Established

INDICATOR ON DISASTER PREPAREDNESS

Share of population trained in first aid*	Lass and Reusswig (1997)
Trained helpers in disaster protection (percentage of the population)	Lass and Reusswig (1997)
Expenditures for disaster prevention (share of GDP)	Lass and Reusswig (1997)
Frequency of risk assessments and contingency plans in business*	New

Peace

Share of defense spending in national budget	Established
Share of armaments in total industrial exports	Established
Share of armaments in total industrial production	Established
Peace research expenditure	Established
Time share of conflict management and de-escalation training in the total education of police and armed forces*	New

Notes

1. With the motivation of consumption in affluent countries shifting from the satisfaction of needs to the symbolic functions of goods (positional and identity functions), both dimensions are deeply intermingled in everyday life.

2. This German word is also used in English to describe a normative vision of a desirable state, defined as the joint long-term common ground of what is desirable and what can be realistically expected.

*Characterized by severe data poverty.

3. Social entities and general systems of rules are included in this definition because both shape human behavior. Furthermore, including both simultaneously reflects the fact that agents and structures have a dialectic relationship, and both can play a decisive role. Which of the elements—agents or structures—actually determines the outcome in a specific situation is determined case by case.

4. In dealing with this question, economics fails completely. Although it can express the four dimensions as capital stocks (the economic, social, environmental, and institutional dimension correspond to human-made, human, natural, and social capital, respectively), it cannot distinguish different, cumulative qualities within such capital stocks.

5. Similarly, environmental and economic criteria can be defined. A nature protection law is an institution, and if well designed and effectively implemented, it constitutes an element of the environmental sustainability of the system of institutions. This applies also to orientations toward dematerialization, policies for material flow reduction, and organizations such as efficient environmental NGOs and benevolent environmental authorities. The protection of property (private and public) and the enforceability of contracts are some of the best-known conditions for a system of institutions to be economically sustainable.

Literature Cited

Bossel, H. 1998. *Earth at a crossroads: Paths to a sustainable future.* Cambridge: Cambridge University Press.

Crowards, T. 1999. *An Economic Vulnerability Index for developing countries, with special reference to the Caribbean.* Bridgetown, Barbados: Caribbean Development Bank.

Czada, R. 1995. Institutionelle Theorien der Politik. Pp. 205–213 in *Lexikon der Politik*, Vol. 1, edited by D. Nohlen and H.-O. Schultze. Munich: Hanser.

DGVN (Deutsche Gesellschaft für die Vereinten Nationen). 2005. *Arabischer Bericht über die menschliche Entwicklung 2004. Auf dem Weg zur Freiheit in der arabischen Welt* [abridged edition in German, English, French, and Arabic]. Berlin: DGVN.

Empacher, C., and P. Wehling. 1999. Indikatoren sozialer Nachhaltigkeit. *ISOE Diskussionspapiere.*

Ferreira, L. d. C. 2000. *Political-institutional indicators: Creating and conciliating public demands.* Paper presented at the IASA 2000 conference, Florida, USA.

Germany, Federal Government. 1999. *Erprobung der CSD Nachhaltigkeitsindikatoren in Deutschland. Zwischenbericht der Bundesregierung.* Bonn: Federal Ministry of the Environment.

Hans-Böckler-Stiftung. 2001. *Pathways towards a sustainable future.* Düsseldorf: Hans-Böckler-Stiftung.

Huber, J. 1995. *Nachhaltige Entwicklung.* Berlin: Edition Sigma.

ILO. 1993. *Trade unions and environmentally sustainable development*. Geneva: International Labour Organisation.

Jackson, W. A. 2003. Social structure in economic theory. *Journal of Economic Issues* 3:727–746.

Kopfmüller, J., V. Brandl, J. Jörissen, M. Paetau, G. Banse, R. Coenen, and A. Grunwald. 2001. *Nachhaltige Entwicklung integrativ betrachtet*. Berlin: Edition Sigma.

Lass, W., and F. Reusswig. 1997. Konzeptionelle Weiterentwicklung der Nachhaltigkeitsindikatoren zur Thematik Konsummuster: Kapitel 4 der Agenda 21. UBA Texte 36/99, Vol. 3, pp. 1–68 in *Konzeptionelle Weiterentwicklung der Nachhaltigkeitsindikatoren der UN Commission on Sustainable Development*. Berlin: Umweltbundesamt UBA.

Littig, B. 2001. *Zur sozialen Dimension nachhaltiger Entwicklung*. Vienna: Strategy Group Sustainability.

OECD. 2001. *Analytic report on sustainable development* SG/SD(2001)1–14. Paris: OECD.

Rawls, J. 1971. *A theory of justice*. Cambridge, MA: Harvard University Press.

Sen, A. 1999. *Development as freedom*. New York: A. A. Knopf.

Serageldin, I., and A. Steer. 1996. Sustainability and the wealth of nations: First steps in an ongoing journey, *Environmentally Sustainable Development Studies and Monographs Series*, Vol. 5. Washington, DC: World Bank.

Spangenberg, J. H. 2002. Institutional sustainability indicators: An analysis of the institutions in Agenda 21 and a draft set of indicators for monitoring their effectivity. *Sustainable Development* 10(2):103–115.

Spangenberg, J. H., and S. Giljum (guest eds.). 2005. Special issue on governance for sustainable development. *International Journal of Sustainable Development* 8(1/2):1–150.

Spangenberg, J. H., S. Pfahl, and K. Deller. 2002. The institutional content of Agenda 21. *Ecological Indicators* 2(1):61–77.

UNDESA (United Nations Department of Economic and Social Affairs). 1998. *Measuring changes in consumption and production patterns. A set of indicators*. New York: United Nations.

UNDP (United Nations Development Programme). 1994. *Human development report. An agenda for the social summit*. Oxford: Oxford University Press.

UNDPCSD. 1996. *Indicators of sustainable development, framework and methodologies*. New York: United Nations.

UNFPA (United Nations Population Fund). 1998. *Indicators for population and reproductive health programmes*. New York: United Nations.

Weisbuch, G. 2000. Environment and institutions: A complex dynamic systems approach. *Ecological Economics* 35(3):381–391.

PART III
Methodological Aspects
Arthur Lyon Dahl

Many of the challenges in assessing sustainability are methodological. How do we turn the concept of sustainability into a framework or model and then identify indicators that describe its essential properties in a way that is easily understandable? Although this part raises a number of critical issues, it emphasizes an underlying theme of the need for more integrating indicators of system sustainability.

The first two chapters distill the experience of the European Environment Agency (EEA), which has become a leader in indicator development. Chapter 8 describes its evolving use of frameworks for environmental assessments and indicators, from simple arrangements of indicators to more cross-cutting frameworks based on scenarios and models. Starting with the traditional driving force, pressure, state, impact, and response framework, with its static list of indicators, and its adaptation by the World Health Organization to incorporate multicausal effects of exposures, the EEA has moved toward a more sophisticated typology of indicators in the policy life cycle, using indicators to define problems, measure performance and efficiency, monitor policy effectiveness, and assess total welfare. The authors propose six steps toward a common indicator development process, aiming for more consistency and reliability in indicators.

In Chapter 9, Stanners et al. look critically at frameworks to achieve environmental policy integration in other sectors of government. Models of sustainable development in common use, such as the three-pillar economic, social, and environmental framework, are misleading because it is not possible just to add together different sets of policy objectives. There is an unjustified assumption of independence and commutability and a tendency to overlook interlinkages between the pillars. Such frameworks are too simple to guide indicator selection. The authors call for new impact assessment methods and indicators for the synergistic interlinkages and overlaps between economic, social, and environmental policies. A more holistic reporting on sustainable development, acknowledging the interdependence of the socioeconomic system and the environment, would show sustainability as an emergent property of the whole system. As a step in this direction, the EEA has identified eight key features of sustainable development that can be used as a checklist or guideline to test the relevance of indicators in the context of environmental assessment. They are also developing methods for evaluating complex scientific evidence to bridge science and policy (e.g., in application of the precautionary principle).

In Chapter 10, Dahl reviews the use of indicators in integrated assessments at the international level. After suggesting a scientific approach to the definition of sustainability, he compares the many assessments that simply use indicators as illustrations and a more statistical approach where data sets are compiled and analyzed to generate part or all of the assessment. The latter approach raises particular challenges of data adequacy, the selection of indicators, and their weighting, which leave them open to criticism. Methodologically, none of the present assessment processes has succeeded in addressing the challenges of integration or the definition of indicators of whole system sustainability. Nor have they adequately considered their susceptibility to underlying assumptions, values, or worldviews, although there is some progress in this direction. Other challenges concern policy relevance and legitimacy, with the difficulty of bridging the short-term perspective of policymakers and the long-term requirements for sustainability. In addition, people have a strong preference for indicators that reflect their own values and perspectives rather than those that are most objective. Dahl identifies a number of research needs, including linkage indicators, new data sets from global observing systems, indicators of less tangible dimensions such as governance, science, culture, values, and spirituality, and measures of intergenerational sustainability.

Finally, in response to the issues raised in the preceding chapters, in Chapter 11 Grosskurth and Rotmans propose a concept for an indicator of the sustainability of systems. Given the difficulty in understanding the complex dynamics of a whole human–environment system, this indicator would focus on the whole system structure rather than its parts. It would start with a conceptual model of the real-world system to be assessed, arranging stocks, flows, and actors in an influence diagram and defining the first-order influences as positive or negative. Because the model must include normative choices as to what is desirable, these should be made through a consultative stakeholder process. The model makes it possible to identify inconsistencies or conflicts, where progress in one area would undermine the system elsewhere. A qualitative system sustainability index can then be calculated based on the proportion of total flows containing inconsistencies. To improve sustainability, some inconsistencies could be corrected by changing the structure of the system, but ultimately it would be necessary to choose between inconsistencies, giving up some goals to achieve others. The indicator thus would be able to define realistic policy options.

8

Frameworks for Environmental Assessment and Indicators at the EEA

David Stanners, Peder Bosch, Ann Dom, Peter Gabrielsen, David Gee, Jock Martin, Louise Rickard, and Jean-Louis Weber

Over the past 10 years the European Environment Agency (EEA) has published assessments and indicators on most European environmental issues. These assessments and indicators are changing to reflect the increasingly cross-cutting nature of new environmental issues such as water management, biodiversity and ecosystem services, climate change and biofuels, health, and chemicals. Assessments are also needed to capture changes across the enlarged European Union (EU)—which covers more socially, economically, and biogeographically diverse countries—to cover longer time spans, and to include more scenario analyses and models. These new and increasingly demanding challenges put a spotlight on the manner and underlying assumptions of knowledge creation.

In this context, this chapter presents some key EEA frameworks that underpin the approaches taken to build environmental data, information, and indicators. These frameworks have already proved useful to the EEA and others and appear to be robust. However, to help improve and extend their application to complex and persistent environmental problems, we welcome extended peer review as a step toward their improvement.

Why do we need frameworks? Applying frameworks to analyze and structure information helps us move from data to information and on to the structured knowledge needed to elucidate environmental and sustainability issues and to design effective responses. However, experience shows that available knowledge is not systematically put to use in policy: "Policy-makers only take that knowledge in consideration that does not cause too great tension with their values. . . . These values are embedded in 'policy frames' or 'policy theories.' Knowledge that does not fit into these policy theories is not agreeable and will be discarded" (Veld 't 2004:83).

Therefore, the purpose of these frameworks is to help improve the organization, structuring, and analysis of environmental information, to increase the use of information and the consistency of its handling, to minimize mishandling, and to help avoid gaps in analysis and assessments. "If the principal actors do not agree about the problem definition, the values that are at stake and the knowledge that is thought to be relevant, we consider the problem unstructured" (Veld't 2004:83). Thus, if we gain agreement on frameworks, information generated based on them has a greater chance of acceptance, improving the effectiveness of associated indicators and assessments. Work in this area contributes to the framing of complex environmental problems and helps policymakers frame sound and effective policy measures.

The DPSIR Analytical Framework

To structure thinking about the interplay between the environment and socioeconomic activities, the EEA uses the driving force, pressure, state, impact, and response (DPSIR) framework, a slightly extended version of the well-known Organisation for Economic Co-operation and Development (OECD) model (Figure 8.1). This is used to help design assessments, identify indicators, and communicate results and can support improved environmental monitoring and information collection.

According to the DPSIR system analysis view, social and economic developments drive changes that exert pressure on the environment; consequently, changes occur in the state of the environment. This leads to impacts on, for example, human health, ecosystem functioning, materials (such as historic buildings), and the economy, where *impacts* refers to information on the relevance of the changes in the state of the envi-

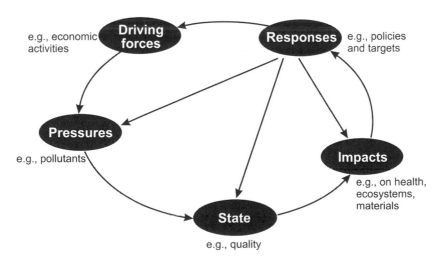

Figure 8.1. DPSIR framework for reporting on environmental issues (courtesy of the EEA).

ronment. Finally societal responses are made that can affect earlier parts of the system directly or indirectly. Many assessments and sets of environmental indicators used by national and international bodies refer to or use directly this DPSIR framework or a subset or extension of it (see the EEA's core set of indicators [CSI]).[1]

The first indicator framework commonly known is the stress–response framework, developed by two scientists working at Statistics Canada, Anthony Friend and David Rapport (personal communication, 1979). Their STress Response Environmental Statistical System (STRESS) framework was based on ecosystem behavior distinguishing between environmental stress (pressures on the ecosystem), the state of the ecosystem, and the ecosystem response (e.g., algal blooms in reaction to higher availability of nutrients). However, the original ideas encompassed all kinds of responses.

When the STRESS framework was presented to the OECD, the ecosystem response was taken out in order to make the concept acceptable to the OECD. The rephrasing of *response* to stand only for societal response led to the OECD pressure, state, response (PSR) model. *Pressures* encompassed all releases or abstractions by human activities of substances, radiation and other physical disturbances, and species in or from the environment. *State* was initially limited to the concentrations of substances and distribution of species.

Because environmental statisticians dealt not only with PSR categories, an early DPSIR model came into use at various statistical offices in the early 1990s as an organizing principle for environment statistics. This framework for statistics described human activities, pressures, state of the environment, impacts on ecosystems, human health and materials, and responses. The Dobris Assessment (EEA 1995a) was also built on this idea.

With the development of the large environmental models Regional Air Pollution INformation and Simulation Model (RAINS) and Integrated Model to Assess the Global Environment (IMAGE) by the International Institute for Applied System Analysis (IIASA) and the Dutch National Institute for Public Health and the Environment (RIVM), the DPSIR model became further formalized, with a precise differentiation between driving forces, pressures, the resulting state of systems, the impacts (including economic), and policy responses. However, it was the EEA that made the simplified DPSIR framework more widely known in Europe. The RIVM report "A general strategy for integrated environmental assessment at the EEA" (EEA 1995b) provided the analytical basis for the DPSIR framework. It was accepted by the EEA Management Board at that time as the basis for integrated environmental assessment.

Over the past 20 years, the analytical framework has developed from a tool to describe natural ecosystems under stress to an overall framework for analyzing many different environmental problems. Furthermore, the DPSIR model has not only been useful as a framework for analyzing environmental problems and identifying indicators. It has also been important for establishing the wide scope of work necessary for effective environmental assessments: When in its early years of operation pressure was being put on the EEA to confine itself to working on the "state of the environment," the DPSIR framework provided an effective tool to legitimize work on driving forces and responses.

From a policy point of view, there is a clear need for information and indicators on all parts of the DPSIR chain:

Indicators for *driving forces* describe the social, demographic, and economic developments in societies and the corresponding changes in lifestyles and overall levels of consumption and production patterns. Primary driving forces are population growth and developments in the needs and activities of individuals. These primary driving forces provoke changes in the overall levels of production and consumption. Through these changes in production and consumption, the driving forces exert pressures on the environment.

Pressure indicators describe developments in release of substances (emissions), physical and biological agents, the use of resources, and the use of land. The pressures exerted by society are transported and transformed in a variety of natural processes to manifest themselves in changes in environmental conditions. Examples of pressure indicators are CO_2 emissions by sector, the use of materials for construction, and the amount of land used for roads.

State indicators give a description of the quantity and quality of physical phenomena (e.g., temperature), biological phenomena (e.g., fish stocks), and chemical phenomena (e.g., atmospheric CO_2 concentrations) in a certain area. For example, state indicators may describe the forest and wildlife resources present, the concentration of phosphorus and sulfur in lakes, or the level of noise in the neighborhood of airports.

Impact indicators are used to describe the relevance of changes in the state of the environment. They are often compared against a threshold or may be measurements of exposure. Examples include frequency of fish kills in a river or the percentage of population receiving drinking water below quality standards.

Response indicators refer to responses by groups and individuals in society and government attempts to prevent, compensate, ameliorate, or adapt to changes in the state of the environment. Some societal responses may be regarded as negative driving forces because they aim to redirect prevailing trends in consumption and production patterns. Other responses aim at raising the efficiency of products and processes by stimulating the development and penetration of clean technologies. Examples of response indicators are the relative amount of cars with catalytic converters and recycling rates of domestic waste. An often-used broad response indicator is that describing environmental expenditures.

To use this framework to look at the dynamics of the system means that we have to understand what happens in the links between D, P, S, I, and R (Figure 8.2). For example, eco-efficiency indicators such as emission coefficients and energy productivity show what happens between driving forces and pressures. This kind of information

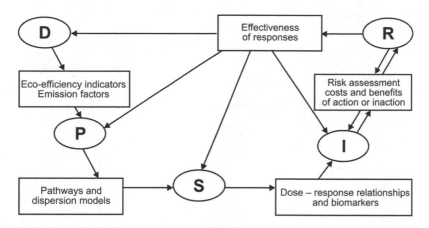

Figure 8.2. DPSIR links and associated information flows (courtesy of the EEA).

helps us answer such questions as "Are we succeeding in making shifts in the economy, such as decoupling?" and "Are we making technological progress?" The combination in one diagram of the pressure (release of nutrients from agriculture) and the state (development of nitrate concentration in surface waters) tells a story of time delay in natural processes and the possible "time bombs" created in the environment. A focus on links generates the need for new information flows (EEA 1999a).

To help better address the effects of human exposure to environmental factors, the World Health Organization (WHO 2002) has extended DPSIR to the DPSEEA model (Figure 8.3). How people react to environmental exposures depend in part on their individual makeup (e.g., their genetics, health, fitness, and age), where they live, frequency of exposure, and what they have been exposed to before. The effects of exposure therefore are the result of a multicausal chain of risks and probabilities. By adding an extra step in the chain between state and response, the DPSEEA framework attempts to capture the multicausal effects of exposure (see also Chapter 9). Although the effects of human exposures are not readily reduced to a simple linear cause-and-effect framework, the DPSEEA model is helping to guide the development of environmental health indicators to support the development of effective policies to protect human health and the environment and to measure their effectiveness (WHO 2004).

The DPSIR Framework and the Policy Life Cycle

When designing indicator lists, conscious use should be made of the DPSIR framework and the policy life cycle (Figure 8.4). For problems that are at the beginning of their policy life cycle (i.e., the stage of issue identification), indicators on the state of the environment and on impacts play a major role (Figure 8.5). In theory, sentinel indicators could play an important role giving advance warning of alarming developments in the

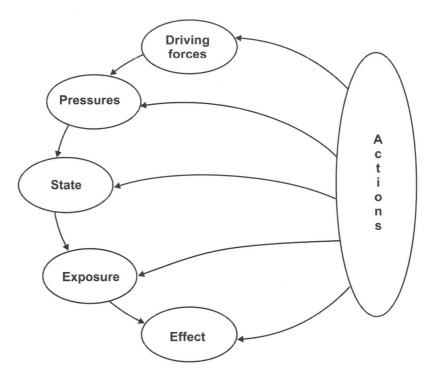

Figure 8.3. DPSEEA model of environmental health (WHO 2002).

state of the environment to allow precautionary measures to be taken. However, few such indicators have been identified that are reliable and would command the attention of decision makers. The best-known cases of state indicators that give rise to policy reactions are those showing the sudden decline of selected species (e.g., fish in acidified Scandinavian lakes, seals in the Dutch Waddensea), surface water quality (e.g., salt in the river Rhine, which was used for irrigation in horticulture), and air quality in cities (e.g., summer smog in Paris and Athens).

This function of state indicators is limited in time: As soon as a problem is politically accepted and measures are being designed, the attention shifts to pressure and driving force indicators. Nevertheless, there is a long period in which state and impact indicators support the process of getting political acceptance of policy responses. Greenhouse gas policies provide clear examples in which indicators of climate change impacts such as extreme weather events (heat waves, floods, and storms), the number of hot summers, average temperatures, the movement of treelines, and species distribution are being used to gather political support for the Kyoto Protocol. Such indicators rise in importance when political opposition increases.

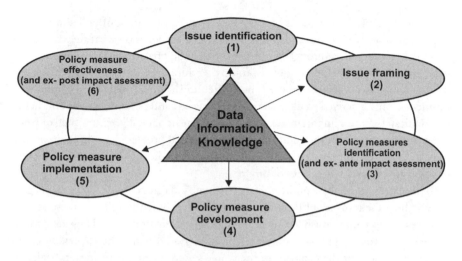

Figure 8.4. Main stages in the policy life cycle, supported by data, information, and knowledge (courtesy of the EEA).

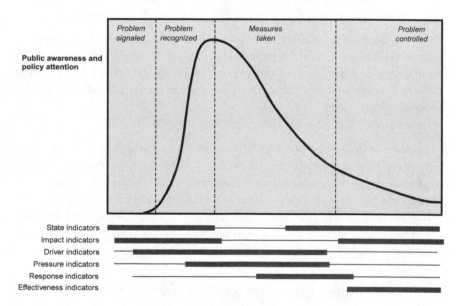

Figure 8.5. Indicator use in the policy life cycle (courtesy of the EEA).

In the next and longer stages of the policy cycle (formulation of policy responses, implementation of measures and control), policymakers focus on what they can influence: the driving forces through volume measures, the pressures with technical measures, and responses with educational projects. Performance indicators on changes in driving forces and pressures are used most often in this phase. The state of the environment is only a derived result of activities in society, and policy reactions and hence state indicators are less important, except in management of biodiversity as such or when organisms play a role in the solution of environmental problems. In these situations, indicators such as biomass production, forests as carbon dioxide sinks, and forest composition are important measures of progress.

In the last, control phase of the policy cycle, state indicators become important again for watching the recovery of the environment, and a limited number of these indicators are used to continuously monitor the state of the environment. They are accompanied by an equally limited number of indicators on driving forces, pressures, and responses to monitor the behavior of the whole system. As implementation begins to demand effort and resources, impact indicators are again needed to remind people why efforts are needed and to reveal improvements. Effectiveness indicators then come into play to assess outcomes of the policy.

A Typology of Indicator Designs

The DPSIR framework has analytical significance for indicators in a policy context. In such a context, environmental indicators are used for three major purposes:

- To supply information on environmental problems, in order to enable policymakers to evaluate their seriousness (this is especially important for new and emerging issues)
- To support policy development and priority setting by highlighting key factors or places in the cause-and-effect chain that cause pressure on the environment and that policy can target
- To monitor the effectiveness of policy responses

Regardless of its position in the DPSIR system, an indicator should always convey a clear message, based on relevant variables (Box 8.1). The indicator typology outlined here aims to provide a classification to aid indicator design. As a means of structuring and analyzing indicators and their related environment–society interconnections, the typology can be used to analyze existing indicators to check their coverage and suitability and can also help to identify possible gaps, pinpoint indicator requirements, and support indicator construction.

Descriptive Indicators (Type A): "What's Happening?"

Descriptive indicators can be used for all elements of DPSIR, although they are seen most commonly as state, pressure, or impact indicators. They can be represented as

---

Box 8.1. What is an indicator?

Indicators always simplify a complex reality, focusing on certain aspects that are regarded as relevant and for which data are available. Indicators are meaningful only as part of a framework or story. Indicators are a necessary part of the stream of information we use to understand the world, make decisions, and plan our actions.

Indicators are communication tools that

- Simplify complex issues, making them accessible to a wider, nonexpert audience.
- Can encourage decision making by pointing to clear steps in the causal chain where it can be broken.
- Inform and empower policymakers and laypeople by creating a means for the measurement of progress in tackling environmental progress.

Indicators cannot replace scientific studies of cause and effect. They are presentations of associations and links between variables. When we choose to present variables together as part of an indicator, we make an explicit assumption of the connection between them. Indicators therefore can never replace statistical analyses of data or the development and testing of sound hypotheses.

Source: EEA.

---

numbers, in pie or bar charts, on maps or other forms, and in line graphs, which are commonly used to present trends in a variable over time, such as the cadmium content of blue mussels, the number of indigenous species in biogeographic regions, or the share of organic farming in an agricultural area (Figure 8.6).

If descriptive indicators are presented in absolute terms, such as "mg/kg dry matter," the relevance of the numbers given is often difficult for a nonexpert to assess. Comparison with another relevant variable (as in Figure 8.6) or as a performance indicator often improves their communication value.

Performance Indicators (Type B): "Does It Matter?" ("Are We Reaching Targets?")

Performance indicators may use the same variables as descriptive indicators but are connected with target values. They measure the distance between the current environmental situation and the desired situation (target): "distance to target" assessment. Performance indicators are relevant if specific groups or institutions can be held accountable for changes in environmental pressures or states. They are typically state, pressure, or impact indicators that clearly link to policy responses.

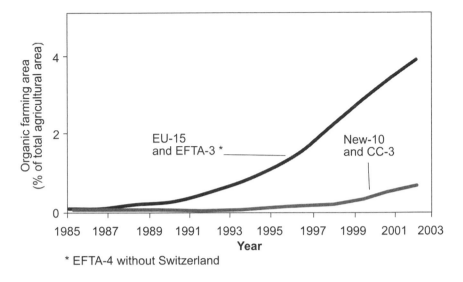

Figure 8.6. Example of a descriptive indicator: Share of organic farming in total agricultural area (courtesy of the Institute of Rural Sciences, University of Wales, Aberystwyth).

Most countries and international bodies develop performance indicators on the basis of nationally or internationally accepted policy targets or tentative approximations of sustainability levels. A typical presentation of a performance indicator is shown in Figure 8.7.

Efficiency Indicators (Type C): "Are We Improving?"

Efficiency indicators relate drivers to pressures. They provide insight into the efficiency of products and processes in terms of resources, emissions, and waste per unit output. The environmental efficiency of a nation may be described in terms of the level of emissions and waste generated per unit of gross domestic product (GDP). The energy efficiency of cars may be described as the volume of fuel used per person per mile traveled.

An absolute decoupling of environmental pressure from economic development is necessary for sustainable development. Most relevant for policymaking, therefore, are indicators that show the most direct relationships between environmental pressures and human activities. For reasons of clarity, these indicators are best presented with separate lines rather than as a ratio. Figure 8.8 gives a good example for the energy supply sector. The diverging lines for energy consumption and GDP indicate increasing eco-effi-

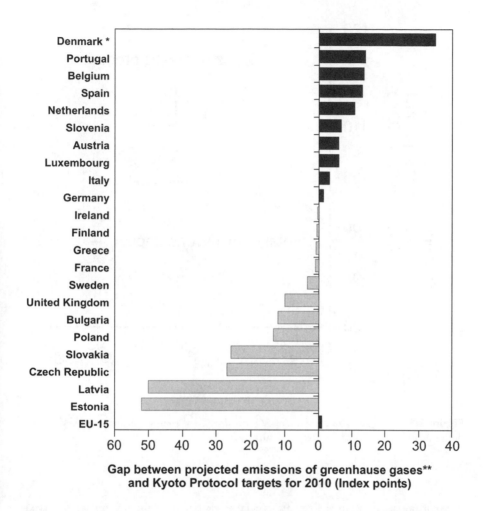

Figure 8.7 Example of a performance indicator: Projected progress toward Kyoto Protocol targets (courtesy of the United Nations Framework Convention on Climate Change UNFCCC, DG Environment, European Commission).

ciency. Presented in this way, eco-efficiency indicators combine pressure and driving force indicators in one graph.

Policy Effectiveness Indicators (Type D): "Are the Measures Working?"

Policy effectiveness indicators relate the actual change of environmental variables to policy efforts. Thus, they are a link between response indicators and driving force, pressure,

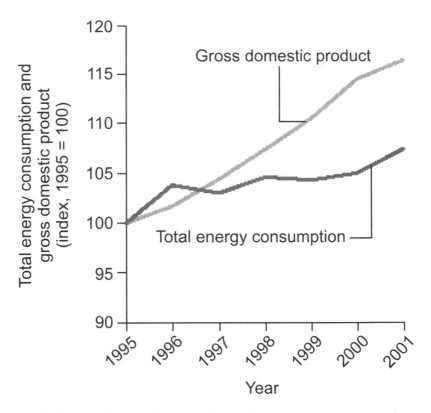

Figure 8.8. Example of an eco-efficiency indicator: Total energy consumption and gross domestic product, EU-25 (courtesy of Eurostat).

state, or impact indicators. They are crucial in determining the reasons for observed developments. The Dutch yearly environmental indicator report (RIVM 2000) contains several examples of this type of indicator. The first examples for the EU have been published in EEA's *Environmental Signals* reports (EEA 2001a, 2002).

Whereas for the previously mentioned indicators an assessment text is necessary to communicate the background information on the reasons behind the development of an indicator, for policy effectiveness indicators much of this information is included in the graph. The production of this type of indicator takes a large amount of quantitative data and expert knowledge. With the expected increase in national and European capacities to carry out policy analysis, it is likely that this type of indicator will develop from the current model, which links with technical measures (e.g., decrease in sulfur emissions in Figure 8.9), to a model that indicates the link with the policy decisions that started off the technological changes.

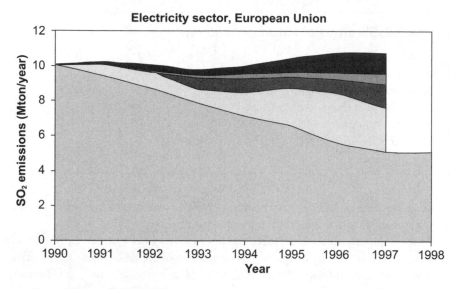

Figure 8.9. Example of a policy effectiveness indicator: Reduction of sulfur dioxide emissions in the electricity sector, EU (courtesy of the EEA).

Total Welfare Indicators (Type E): "Are We on the Whole Better Off?"

In any discussion of sustainability and human welfare, the balance between economic, social, and environmental development is crucial. For an integral assessment, some measure of total sustainability is needed in the form of a green GDP. The Index of Sustainable Economic Welfare (ISEW) is one such example that also includes measures of inequalities and of nonpaid work.

Toward a Common Indicator Development Process

Although the frameworks and typologies described in this chapter are useful tools for building indicators, the process chosen for building indicators can also have an important influence on the relevance, effectiveness, and scientific underpinning of the indicators. Based initially on EEA's experience with developing the Transport and Environment Reporting Mechanism (TERM)[2] (EEA 1999b and 2001b) and its CSI, six important steps have been identified for an effective indicator-building process (Box 8.2).

Beginning the indicator development process with agreement on a story establishes a clear and explicit understanding of the purpose of the indicators. The indicator story must be closely linked to relevant policies, strategies, and related objectives and should address causes, measures, and links with other policies and societal developments. In

---

Box 8.2. Six steps of indicator building.

1. Agree on a story.
2. List policy questions.
3. Select indicators (ideal and actual).
4. Define and compile data.
5. Interpret indicators.
6. Modify, adapt, update, and iterate conclusions.

Source: EEA.

---

addition, the story should describe relevant scientific knowledge, including factors such as multicausality, critical thresholds, and uncertainties.

To develop ownership and increase relevance, the story must be developed with all relevant stakeholders. The design of the story involves the description of the stakeholders' views about the issue, the limits of the problem being addressed, and how they think it should be solved. Such an approach brings out the hopes, beliefs and ethical standpoints of the stakeholders, including those of the policymakers who design the policies that the indicators are intended to track, improving the relevance of the resulting indicators. An example storyline for the environment–transport problem is summarized in Box 8.3.

Once a clear story is established, it is important to make explicit the relevant policymakers' questions. Ideally there should be a balance in questions related to causes, effects, and solutions to the problem. Box 8.4 lists the main questions of the environment–transport storyline.

Box 8.3. Description of the transport problem in the EU.

- Growing greenhouse gas emissions from the transport sector jeopardize the achievement of the EU's emission reduction target under the Kyoto Protocol.

- Impacts on air quality, noise nuisance, and the increasing fragmentation of the EU's territory are equally worrying.

- Transport growth, which remains closely linked to economic growth, and the shift toward roads and aviation are the main drivers behind this development.

- Technology and fuel improvements are only partly effective in reducing impacts.

- They must be complemented with measures to restrain the growth in transport and to redress the modal balance.

Source: EEA.

With the first two steps complete, defining and selecting indicators becomes a clearer and more focused exercise. When indicators for complex cross-cutting issues (e.g., measuring the positive and negative impacts of biofuels on the environment) are being developed, specific integrated frameworks must be built for assessing the broad, cross-sectoral environmental impacts to ensure that all important factors are taken into account. Indeed, even for less complex issues an explicit framework or model of relevant processes is useful to steer indicator development. The DPSIR framework can be a useful basis for such models.

To be effective, indicators must be selected that come close to answering the policy questions, taking into account the relevant environmental, societal, and economic interactions described in the framework or model for that issue and the relevant policy levers (i.e., the policy measures that could have an effect on the issue). We can improve the indicators by making connections between the type of policy questions and the type of indicators used to provide answers, as defined in the indicator typology. To ensure relevance, it is important not only to consider indicators for which data are currently available but also to identify ideal indicators that may have new requirements.

Because indicators are often constructed using a combination of data sets (e.g., map-based indicators derived from geospatially referenced data made up of multiple data layers combined in complex algorithms), it is necessary to define the algorithm of

---

Box 8.4. Seven key questions on transport and the environment in the EU.

- Is the environmental performance of the transport sector improving?

- Are we getting better at managing transport demand and improving the modal split?

- Are spatial and transport planning becoming better coordinated so as to match transport demand to the needs of access?

- Are we optimizing the use of existing transport infrastructure capacity and moving toward a better-balanced intermodal transport system?

- Are we moving toward a fairer and more efficient pricing system, which ensures that external costs are internalized?

- How rapidly are improved technologies being implemented, and how efficiently are vehicles being used?

- How effectively are environmental management and monitoring tools being used to support policy and decision making?

Source: EEA.

indicator construction in the third step and unravel the data requirements before data collection in the fourth step.

Once produced, we must interpret the indicators, explaining why they are developing as they are and linking them back to the story and policy questions. This must be done in connection with other information using relevant literature, more detailed studies, and comparisons with other available data and indicators. The various factors steering the development of an indicator should be distinguished as much as possible (e.g., natural processes, changes in the size and structure of the economy or society, and changes deliberately brought about by environmental policies). Specific regional phenomena influencing the indicator should be highlighted, such as strong economic growth or differences in welfare.

The last step consists of making conclusions about the whole set of indicators, communicating them to the network of people making or influencing decisions, and preparing an improved indicator set for the next round of reporting.

Using common processes and frameworks for developing indicators will not necessarily result in a common set of indicators. Common processes, frameworks, and typologies are guides for the identification and development of indicators. They support a scientific, systematized approach, help enforce consistency with existing knowledge, and help provide balance in outcomes, including highlighting gaps. Each indicator-building process may require different indicators, but within a certain scope (and at different scales) the frameworks and typologies can be more universal. New frameworks may be needed or existing ones extended as the extent and purpose of the indicators vary, such as between environment and health issues (e.g., DPSIR and DPSEEA).

Consistency of indicators is important within a certain field for practical reasons, including data availability, coordination, and efficiency of data collection and processing. Consistent indicators can also be more effective and reliable communication tools because over time they become familiar and long-term trends can be built up. For all of these reasons, consistency and reliability favor a small core set of indicators, because the fewer the indicators, the more recognizable and manageable they are. However, a small core set does not have the flexibility of a larger indicator set for covering a full cause-and-effect framework. Also, there is a risk that as issues evolve and their scientific understanding improves, a small indicator set will stagnate unless regularly reviewed, updated, or expanded. To understand and manage this tension between stability and flexibility of indicator sets and to develop the necessary trade-offs, suitable processes must be established and run with the appropriate stakeholders. It is here that the common processes, frameworks, and typologies presented in this chapter are useful for enforcing consistent approaches and ensuring that the indicator development and selection process falls within scientific understanding and acceptable norms.

Conclusion

Indicators can be powerful tools in the communication of environmental issues to policymakers. They serve a useful function in simplifying complex issues, steering policymaking, and measuring environmental and policy progress. However, although the simplicity of indicators makes them powerful communication tools, it also represents their limitation. Determining what constitutes sustainability—environmentally, socially, and economically—and comparing current developments against these goals requires indicators to capture multidimensional trade-offs and comparisons in a single two-dimensional graphic.

Although indicators can provide the common language and the accepted yardstick for benchmarking between different countries, regions, or municipalities, they can also be misleading in their simplicity. The theoretical basis for indicator selection therefore must be modified continuously to capture current developments and maintain policy relevance.

Notes

1. The CSI, launched by EEA in March 2004 (eea.europa.eu/coreset), is intended to provide a stable and manageable basis for indicator reporting by EEA, to provide a means of prioritizing improvements in data quality from country level to aggregated European level, to enable streamlined contributions to other indicator initiatives (e.g., structural indicators), and to strengthen the environmental dimension in the sustainability debate.

2. The aim of TERM was to develop indicators to plot progress with the integration of environment into EU transport policies as part of the EU Cardiff process (CEC 2004).

Literature Cited

CEC. 2004. *Integrating environmental considerations into other policy areas: A stocktaking of the Cardiff process.* Commission Working Document COM(2004) 394 final. Brussels: Commission of the European Communities.

EEA. 1995a. *Europe's environment: The Dobris assessment.* Copenhagen: European Environment Agency.

EEA. 1995b. *A general strategy for integrated environmental assessment at the EEA.* An unpublished report for the EEA by RIVM.

EEA. 1999a. *Making sustainability accountable: Eco-efficiency, resource productivity and innovation.* Topic Report no. 11/1999. Copenhagen: European Environment Agency.

EEA. 1999b. *Towards a Transport and Environment Reporting Mechanism (TERM) for the EU: Part I and II.* Technical Report no. 18. Copenhagen: European Environment Agency.

EEA. 2001a. *Environmental signals 2001.* Environmental Assessment Report no. 8. Copenhagen: European Environment Agency.

EEA. 2001b. *TERM 2001: Indicators tracking transport and environment integration in the European Union.* Environmental Issue Report no. 23. Copenhagen: European Environment Agency.

EEA. 2002. *Environmental signals 2002.* Environmental Assessment Report no. 9. Copenhagen: European Environment Agency.

RIVM. 2000. *Milieubalans 2000. Het Nederlandse milieu verklaard* [*Environmental balance 2000, The Dutch environment explained*]. Alphen aan den Rijn, the Netherlands: Rijksinstituut voor Volksgezondheid en Milieu/Samson bv.

Veld 't in, R. 2004. The importance of being a boundary worker. In *Book of Abstracts* for "Bridging the Gap: Information for Action," Dublin, Ireland, April 28–30.

WHO. 2002. *Environmental health indicators for the WHO European region: Update of methodology.* Geneva: WHO report EUR/02/5039762.

WHO. 2004. *Environmental health indicators for Europe: A pilot indicator-based report.* Geneva: World Health Organization Regional Office for Europe.

9

Frameworks for Policy Integration Indicators, for Sustainable Development, and for Evaluating Complex Scientific Evidence

David Stanners, Ann Dom, David Gee, Jock Martin, Teresa Ribeiro, Louise Rickard, and Jean-Louis Weber

To assess sustainable development (SD), new approaches are needed to deal with the issues of system complexity, uncertainty, and ignorance. The necessary information must be condensed and made accessible to a wide and diverse audience ranging from policymakers, decision makers, and citizens who are striving to apply both precaution and prevention. These new and increasingly demanding challenges put a spotlight on the manner and underlying assumptions of knowledge creation. This chapter reviews some key approaches to building sustainability indicators, underlying models, and frameworks for evaluating complex evidence, all needed for a thorough appraisal of progress toward SD. The chapter begins by analyzing policy integration indicators, a key approach to addressing unsustainable development. It goes on to critique the SD models in use and describes how they can be misleading in the development of relevant indicators. Without a frame of reference for assessing the meaning of the generated indicators where there are complexities and uncertainties, the results can be difficult to interpret. Therefore, this chapter concludes with a framework for evaluating complex scientific evidence on environmental factors in disease causation.

Policy Integration Indicators

According to Article 6 of the EU Treaty, environmental protection requirements must be integrated into the definition and implementation of EU policies and activities. Thus, environmental policy integration (EPI) can be defined as inserting environmental requirements into other policies during their development and implementation (EEA 1999b, 1999c, 2005; CEC 2004). EPI is distinct from conventional environmental policymaking because it involves a continual process to ensure that environmental issues are reflected in all policymaking, which generally demands changes in political, organizational, and procedural activities. The aim is to secure coherent policies in all fields that can support environment and SD. Apart from demanding appropriate systems, structures, and processes to ensure that environmental considerations are taken into account, EPI should lead to real progress in terms of political commitment, policy change, and environmental improvement.

Why is there interest in EPI? It emerged because conventional environmental policy and legislation alone were insufficient to address the many driving forces and pressures exerted on the environment by key economic sectors such as energy, transport, and agriculture. Environmental concerns are insufficiently weighted in political, policy, and practical terms, leading to environmental concerns being traded off against economic concerns. Poor integration is caused by numerous factors, including a lack of high-level political commitment to environmental issues, diverging or conflicting policy objectives, and insufficiently coordinated administrations. There are many theories on the root causes of these problems, including the basic problem that organizations and their cultures are deeply entrenched and very slow to adapt to new demands and circumstances.

The European Commission's 5th Environmental Action Programme (5EAP), published in 1992, addressed integration of environment into key sectors, and in 1997–1998 increasing attention began to be paid to the critical role of key economic sectors in causing major environmental problems. This was reflected in the Cardiff Process on sectoral integration and in the EEA's "Europe's Environment: The Second Assessment" (EEA 1998). This raised the following question: How do we recognize progress and the related information gap? In order to fill this gap and to monitor progress toward sectoral integration, a number of criteria were proposed.

The criteria[1] (Table 9.1) were developed from the experience gained in applying them in particular to the Global Assessment of the 5EAP (EEA 1999b). Four sectors originally were covered at member state level: energy, transport, industry, and agriculture. Tourism was not included because it was not initially identified as a priority in the Cardiff Process.

These criteria are meant to steer assessments, information collection, and indicator development in order to be more effective for measuring integration, which is often overlooked and difficult to measure. The aim is to shed light on progress with integration in its different stages and manifestations by covering a wide range of facets of integration. This will lower reliance on end-of-pipe results arising from integration, which may take years to show up. Although these criteria were used by some organizations (e.g., CLM 1999), many of the criteria need further work to become operational.

After the initial focus in the EU in the 1990s on integrating environmental concerns into sectoral policies, increasing attention is now being given to policy coherence as a

Table 9.1. Some criteria for assessing environmental integration into economic sector activities.

A	Institutional Integration

1. Are environmental objectives (e.g., maintenance of natural capital and ecological services) identified as key sectoral objectives and as important as economic and social objectives) in a sector integration strategy?
2. Are synergies between economic, environmental, and social objectives maximized?
3. Are trade-offs between environmental, economic, and social objectives minimized and transparent?
4. Are environmental targets (e.g., for eco-efficiency) and timetables agreed? Are there adequate resources to achieve the targets within the timetables?
5. Is there effective horizontal integration between the sector, environment, and other key authorities (e.g., finance and planning)?
6. Is there effective vertical integration between the EU, national, regional, and local administrations, including adequate public and other stakeholder information and participation measures?

B	Market Integration

7. Have environmental costs and benefits been quantified by common methods?
8. Have environmental costs been internalized into market prices through market-based instruments?
9. Have revenues from these market-based instruments been directly recycled to maximize behavior change?
10. Have revenues from these market-based instruments been directly recycled to promote employment?
11. Have environmentally damaging subsidies and tax exemptions been withdrawn or refocused?
12. Have incentives been introduced that encourage environmental benefits?

C	Management Integration

13. Have environmental management systems been adopted?
14. Is there adequate strategic environmental assessment of policies, plans, and programs?
15. Is there adequate environmental impact assessment of projects before implementation?
16. Is there an effective green procurement (supply) program in public and private institutions?
17. Is there an effective product and service program that maximizes eco-efficiency (e.g., via demand-side management, eco-labeling, products to services)?
18. Are there effective environmental agreements that engage stakeholders in maximizing eco-efficiency?

(continued)

Table 9.1. Some criteria for assessing environmental integration into economic sector activities (*continued*).

D	Monitoring and Reporting Integration

19	Is there an adequate sector and environment reporting mechanism that tracks progress with these objectives, targets, and tools?
20	Is the effectiveness of the policies and tools for achieving integration evaluated and reported, and are the results applied?

Source: EEA (1999b).

whole. Coherence is a prominent feature of good governance (RMNO/EEAC 2003) and SD. Therefore, it is now the EU's SD strategy (European Commission 2001) and the EU governance agenda (European Commission 2001) that provide the broad framework for promoting the integration of economic, social, and environmental objectives in Europe. In practice this suggests a two-way integration, from environment into sectors and vice versa. However, EPI is specifically justified by the fact that environmental policy concerns have been persistently underemphasized in other policies. The more integrated and mutually reinforcing policies are in their formulation, the easier their effective (and cost-efficient) delivery should be. In the EU context, coherence at the political and policy levels eases the work of the institutions and subsequent (national, regional, or local) implementation efforts (Peters 1998; Wandén 2003). The burden on individual actors is also reduced if regulatory requirements are streamlined. Ultimately, policy coordination makes it more likely that multiple objectives will be met.

In this broader context, and in addition to the initial EEA EPI criteria, other attempts have been made to identify suitable ways to measure progress with integration. Prominent among these is the Organisation for Economic Co-operation and Development checklist on policy coherence and integration for SD (OECD 2002). This checklist contains five groups of questions, related to understanding, commitment and leadership, steering, stakeholder involvement, and knowledge and scientific input. Other approaches include national SD strategies and EU integration strategies (Persson 2002; Dalal-Clayton 2004; Fergusson et al. 2001).

The challenge still is to identify a small set of headline criteria and indicators that can be applied to assess progress at both the EU and the national levels, within different institutions, and relating to both cross-sectoral and sectoral efforts. Thus, building on past work, an evaluation framework for EPI was developed in 2003–2004 (Figure 9.1) from which a set of more concrete criteria were identified (Table 9.2). Presented as a checklist to ensure wide applicability, the criteria serve two main purposes: They provide a single framework for undertaking evaluations of EPI supporting consistency

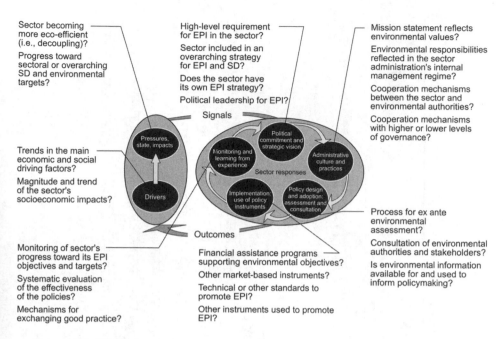

Figure 9.1. Virtuous cycle for EPI.

and shared learning between administrations and sectors, and they support understanding of how to promote integration.

Addressing the Context of Sustainability: Sustainable Development Models and the GEAR-SD Approach

The EEA role in the SD policy process lies mainly in ensuring that environmental concerns are addressed at an appropriate level in progress reports or when new policy proposals are being developed (sustainability impact assessment).

Assessing and reporting on progress with SD is a difficult and complex task. Current international SD reporting initiatives, such as the EU Spring Council reporting (using the "structural indicators") and ongoing work of the UN Commission on Sustainable Development (CSD), consist mainly of the bringing together of some key indicators developed for each one of the three SD pillars or spheres of interest (i.e., combining environmental indicators, social indicators, and economic indicators). The CSD also includes a fourth, institutional pillar addressing governance issues. However, SD will not be achieved simply by combinations of different sets of policy objectives because this would result in a weak compromise. Rather, reformulation and integra-

Table 9.2. A checklist of criteria for evaluating sectoral and cross-sectoral EPI.

CONTEXT FOR EPI	CROSS-SECTORAL	SECTOR-SPECIFIC
1. Trends in drivers, pressures, changes in state of the environment, impacts	1a. What are the main economic and social driving factors facing the administration?	1a. What are the trends in the sector's main economic and social driving factors?
	1b. What is the magnitude and trends of socioeconomic impacts?	1b. What is the magnitude and trend of the sector's socioeconomic impacts?
	1c. Is society becoming more eco-efficient, i.e. decoupling its economic activities and outputs from environmental pressures and impacts?	1c. Is the sector becoming more eco-efficient, i.e. decoupling its economic activities and outputs from environmental pressures and impacts?
	1d. Is progress being made towards key overarching SD/environmental targets and objectives?	1d. Is the sector contributing appropriately to key overarching SD/environmental targets and objectives?
		1e. Is the sector on track to reaching its own environmental targets and objectives?

EPI CATEGORIES	CROSS-SECTORAL	SECTOR SPECIFIC
2. Political commitment and strategic vision	2a. Is there a high level (i.e. constitutional/legal) requirement for EPI in general?	2a. Is there a high level (i.e. constitutional/legal) requirement for EPI in the sector?
	2b. Is there an overarching EPI or SD strategy, endorsed and reviewed by the prime minister or president?	2b. Is the sector included in an overarching strategy for EPI and/or for sustainable development?
		2c. Does the sector have its own EPI or sustainable development strategy?
	2c. Is there political leadership for EPI and/or sustainable development?	2d. Is there political leadership for EPI in the sector?
3. Administrative culture and practices	3a. Do the administration's regular planning, budgetary and audit exercises reflect EPI priorities?	3a. Does the sector administration's mission statement reflect environmental values?
	3b. Are environmental responsibilities reflected in the administration's internal management regime?	3b. Are environmental responsibilities reflected in the sector administration's internal management regime?

tion of policy objectives are needed to improve policy coherence so that optimal benefits can be gained from the synergistic effects of environmental, social, and economic policies. For sustainability assessments, this means that existing tools may no longer be adequate and that new impact assessment methods and indicators are needed to measure progress, especially at the synergistic interlinkages and overlaps between the traditionally separate areas of economic, social, and environmental policy. Furthermore, when assessments are designed to address sustainability, guidance is needed to identify the key interfaces on which to focus attention. This was the incentive behind the Guidelines for Environmental Assessment and Reporting in the Context of Sustainable Development (GEAR-SD).

The EEA founding regulation[2] requires the agency to report on the state and outlook of the environment, including the socioeconomic dimension, in the context of sustainable development. The limited progress made in developing and delivering truly useful SD-relevant information in a political decision-making context, as exemplified by the quality of the EU structural indicators, gives an immediate political focus to this work. The EEA needs to report on the environment in such a way that it provides useful information to policymakers to understand and respond to sustainability issues relevant to high-level decision makers. However, because of the breadth of sustainability concerns and wide interpretations of this concept, there are fundamental difficulties associated with identifying the relevant assessments and indicators needed to deliver this knowledge. For progress to occur, agreement is needed in a number of areas. This section examines our assumptions about SD embedded in the models of sustainability that we use to explain the concept and then presents GEAR-SD, which identifies main features that make sustainability operational in assessments and indicators.

The way we envisage sustainability must be examined because this will directly affect the features identified as important and the associated assessments and indicators needed. International consensus on the most suitable framework for describing SD is lacking. Nevertheless, some general requirements for applicable framework can be formulated. For example, within the EEA Expert Group on Guidelines and Reporting,[3] the following requirements have been raised:

• Sound conceptual foundation
• Ability to capture key information to measure sustainable development by selecting indicators
• Ability to clarify relationships between different indicators and policies
• Ability to integrate different dimensions of sustainable development

The model of sustainability that predominates thinking is composed of the social, economic, and environmental pillars. This is often visualized as a three-legged stool (Figure 9.2). There are many assumptions implicit in this model. Its main purpose is to register the need to consider all three domains to support sustainability. Beyond that, however, it contributes little and probably misleads greatly. In particular, it misses explicit

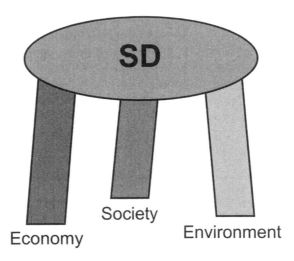

Figure 9.2. Three-legged stool model of sustainable development. The stool model emphasizes only the importance of the three pillars to support sustainable development but misses the all-important linkages (courtesy of the EEA).

representation of the all-important links between the pillars, where important synergies can be found and trade-offs are made. These are present in the model only implicitly in the need to keep the stool balanced to compensate for changes in one or the other pillar so that the stool does not fall over. A more explicit representation of this balancing act and the forces and trade-offs at play in such maneuvers would greatly improve the model and make transparent the hidden compensations in operation.

The three pillars sometimes are represented as overlapping circles (Figure 9.3). This model addresses the lack of linkages but offers no way of characterizing them. It promotes the notion that the nature of the three domains is the same and says nothing about the dependencies and dynamic interactions between domains. Furthermore, it does not illustrate the differences in problems within and between the different domains in regions and especially between developed and developing countries. These representations of SD are sometimes called the atomistic approach (EEA 2002).

Ironically, these models lead to a focus on addressing each pillar separately from the whole rather than a focus on the cooperation needed between the domains to produce the most efficient and effective sustainability outcomes. Furthermore, these models provide no insight on how to model the complex, reflexive interactions between domains. This leads to the false picture that each pillar can be organized and

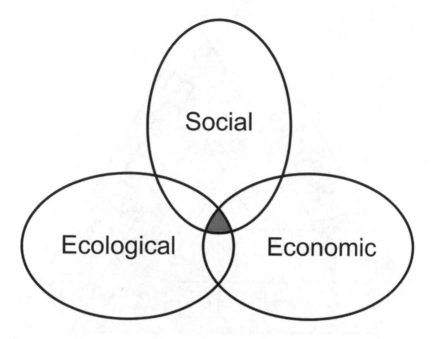

Figure 9.3. Sustainable development in three overlapping ellipses (Välimäki 2002).

measured independently of the others and that by adding them up, one can achieve SD (unconscious assumptions of independence and commutability, as seen in the EU structural indicators).

Within SD reporting, there is a strong emphasis on integrative or holistic reporting. The basic purpose of holistic reporting is to connect dimensions together (Figure 9.4). From the perspective of the holism–atomism debate, the basic question is whether it is reasonable to assume that sustainability is a property that can be found by simply incorporating the different dimensions together, or whether sustainability is more like an emerging property, not easily detected from the properties of different dimensions.

In contrast to these representations, the concentric ring model of SD (Figure 9.5) used in the EEA's "Turn of the Century" report (EEA 1999a) and the egg model of Prescott-Allen (2001) promotes an entirely different concept. It emphasizes the dependence of the socioeconomic system on the environment. It exemplifies the need to model both systems in order to understand the interactions and dependencies. It also visually encapsulates the concept of stocks of the socioeconomic and environmental systems so often forgotten in debates.

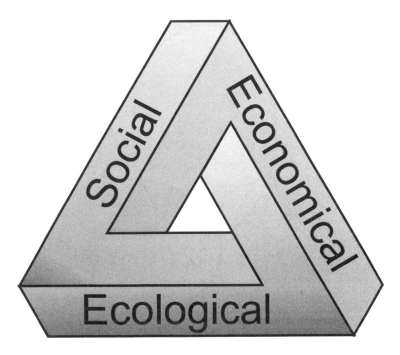

Figure 9.4. Never-ending triangle of sustainable development (Välimäki 2002).

The atomistic three-pillar model focuses not on cooperation but on strengthening the pillars separately. This can lead to false trade-offs being proposed, for example between social and environmental concerns against economic standards that are not commensurable in sustainability terms (e.g., pay for clean water for the whole world instead of reaching the Kyoto Protocol greenhouse gas emission targets). The overlapping circles model gives the impression that cooperation is needed only in the common areas; this suggests that only limited trade-offs are needed and puts no emphasis on looking for solutions in fundamental changes to whole systems. Finally, secondary (or system) benefits are difficult to identify and resolve in these discrete models.

The concentric ring and egg models instead emphasize symbiosis: The socioeconomic system is distinct but embedded in and dependent on the environment. From this flows integration and clearer trade-offs because the need for them to sustain the whole is apparent. Environment is not relegated to an optional extra ("if we try hard enough, perhaps we can stand on one or two legs only") but is identified as a system component, source, and sink.

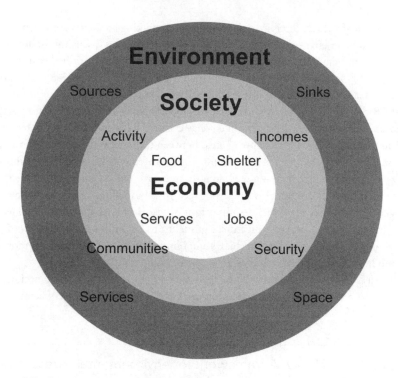

Figure 9.5. Concentric ring or egg model of sustainable development (EEA 1999a).

With these considerations in mind, it becomes clear that the SD models discussed here are too simple for guiding the identification of SD indicators. Indeed, once embedded in our thinking, they can explicitly or implicitly mislead us in the identification of important SD features. Crucial systemic and synergistic aspects of SD are particularly easy to overlook, and without them an oversimplified assessment of important characteristics can result.

To help guard against the pitfalls of inadequate models, some basic thinking was put into identifying underlying features of SD and what they mean for reporting on the environment. Emphasis was put on practical outcomes, which need to be made explicit in any analysis of environment and sustainability, regardless of which model is being used. The objective of going beyond the models in this way was to move the discussion away from trying to design an ideal framework of SD toward a practical means of identifying and checking that the agency was responding to its regulatory mandate and to assess the state, trends, and outlook of the environment in the context of SD.

As a first step, GEAR-SD is intended to stimulate thinking about what is meant by *sustainability* from an environmental point of view and to root this discussion in illustrative information and data. Eight SD key features (Box 9.1) have been identified that, from an environmental point of view, merit further analysis and development. These key features can be used as a checklist for testing the SD relevance of an assessment or indicator.

GEAR-SD does not address all SD-relevant aspects but focuses on those necessary to understand the SD context of environmental assessment. This domain is indicated in the diagram (Figure 9.6) as the overlapping areas between the environmental, economic, and social spheres and within the purely environmental sphere, which possesses some intrinsic aspects that demand SD thinking (e.g., long-term or irreversible environmental effects).

At the moment, GEAR-SD is simply a checklist, a guideline, and a tool: a checklist of key features to help tease out the important SD stories when conducting an assessment and to identify suitable indicators; a guideline to help identify SD-relevant issues to help compensate for unconscious biases and blind spots; and a tool and common language to help communicate SD issues.

The list is not complete and will be expanded and refined further. The checklist can be used to improve the reporting framework and can be useful for different actors at dif-

---

Box 9.1. GEAR-SD: A framework for environmental assessment and reporting in the context of SD.

- We want to provide future generations the same environmental potential as the current one (intergenerational equity).
- We want our economic growth to be less natural resource intensive and less polluting (decoupling).
- We want a better integration of sectoral and environmental policies (sector integration).
- We want to maintain and enhance the adaptive capacity of the environmental system (adaptability).
- We want to avoid irreversible and long-term environmental damage to ecosystems and human health (avoid irreversible damage).
- We want to avoid imposing unfair or high environmental costs on vulnerable population categories (distributional equity).
- We want the EU to assume responsibility for the environmental effects it has outside the EU geographic area (global responsibility).
- We want rules, processes, and practices to ensure the uptake of SD goals and implementation of cost-effective policies at all levels of governance (SD governance).

Source: EEA.

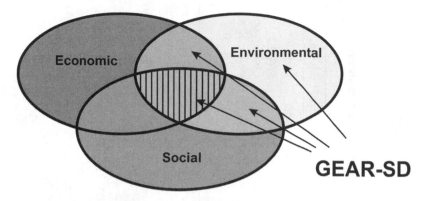

Figure 9.6. Scope of GEAR-SD (courtesy of the EEA).

ferent levels. Most important, it may help identify SD indicators at the critical SD interfaces. Similar analyses of the SD interfaces with the economic and social pillars, if applied, would greatly improve SD-relevant assessments of these domains and strengthen cross-sectoral, integrated thinking.

The Science–Policy Bridge: A Framework for Evaluating Complex Scientific Evidence on Environmental Factors in Disease Causation

In preparing a follow-up report to *Late Lessons from Early Warnings: The Precautionary Principle 1896–2000* (EEA 2001), the EEA has been developing a framework to assist with the practical application of the precautionary principle via common approaches to evidence evaluation at the science–policy interface. It has also been developing a simple analytical model for approaching such complex, multicausal phenomena as endocrine-disrupting substances, mediated diseases, and childhood asthma (EEA 2003).

The draft EEA framework in Table 9.3 uses just three strengths of evidence: weak (10–33% estimated probability), moderate (33–66% estimated probability), and strong (more than 66% probability), which are the same as the "low likelihood," "medium likelihood," and "likely" categories of the IPCC (Table 9.4). The draft framework also invites users to judge whether the overall evidence has become stronger or weaker over a relevant period of time between major evaluations of the evidence or since, say, 1992.

Preventive and precautionary actions must usually be taken on the basis of much less than scientific certainty and well before an understanding of the mechanisms of action

Table 9.3. An EEA framework for evaluating complex and conflicting scientific evidence on environment and disease.

Statement of Hypothesis*	Evaluation Factor				Overall Strength of Evidence	
	Association†	Plausibility†	Causality†	Mechanism of action?†	Overall weight of evidence (weak, moderate, or strong) ‡	Direction of evidence (stronger or weaker over last 5–10 years)
Outcome (e.g., childhood asthma)	Relevant exposure (e.g., indoor air pollution [NO$_x$])					

Note: This table is designed to be used to evaluate the scientific evidence for each hypothesis (at an appropriate level of detail, relevant to the expertise of those doing the evaluation) against the factors in the framework and judge the overall strength of evidence for the hypothesis and the direction in which the evidence is moving.

*From observations or theories.

†Commonly accepted among relevant scientists.

‡Commonly accepted by relevant scientists as "weak," "moderate," or "strong," categories based on Hill (1965), IPCC (2001), and IPCS/WHO (2002).

Quantitative Descriptor (probability bands based on IPCC 2001)*	Qualitative Descriptor	Illustrations
Very likely (90–99%)	• "Statistical significance" • "Beyond all reasonable doubt"	• Part of strong scientific "causation" evidence • Most criminal law; the Swedish chemical law, 1973 (for evidence of "safety" from manufacturers)
Likely (66–90%)	• "Reasonable certainty" • "Sufficient scientific evidence"	• Food Quality Protection Act, 1996 (U.S.) • World Trade Organisation SPS Agreement, Art. 2.2, 1995, to justify a trade restriction • International Agency for Research on Cancer (IARC) Category 1: "Probable Human Carcinogen"
Medium likelihood (33–66%)	• "Balance of evidence" • "Balance of probabilities" • "Limited evidence" • "Reasonable grounds for concern" • "Strong possibility" • "Scientific suspicion of risk"	• Intergovernmental Panel on Climate Change 1995 and 2001 • Much civil and some administrative law • IARC Category 2 B: "Possible Human Carcinogen" • European Commission on the Precautionary Principle 2000 • British Nuclear Fuels occupational radiation compensation scheme, 1984 (20–50% probabilities triggering different awards up to 50%, which triggers full compensation) • Swedish chemical law, 1973, for evidence required for regulators to take precautionary action on potential harm from substances
Low likelihood (10–33%)	• "Some evidence of carcinogenicity" • "Available pertinent information"	• IARC criterion for selecting substances for evaluation • WTO SPS Agreement, Art. 5.7, to justify a provisional trade restriction where "scientific information is insufficient"
Very unlikely (1–10%)	• "Low risk" • "Negligible and insignificant"	• Household fire insurance • Food Quality Protection Act, 1996 (U.S.)

100% probability
90%
50%
10%
0% probability

Increasing Probability

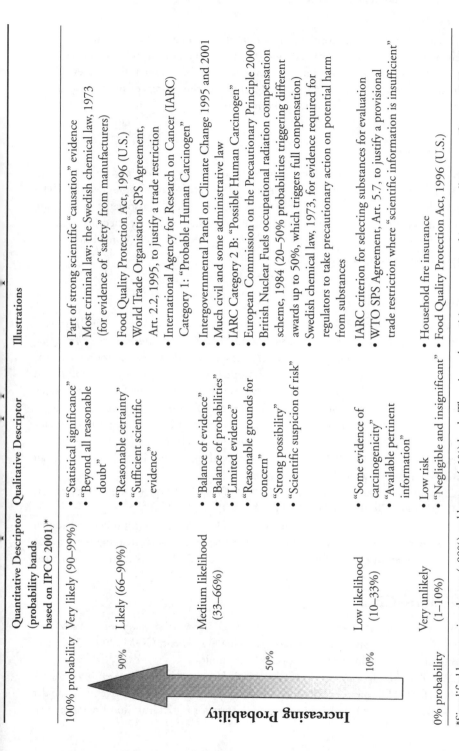

*Simplified by removing the top (>99%) and bottom (<1%) levels. There is rarely precision on contested issues to allocate specific numerical probabilities, and the boundaries between categories are in practice fuzzy: The quantitative descriptors illustrate broad categories of evidence based on has been achieved. The appropriate level of proof varies in each case, depending on the likely nature and scale of the hazards and the availability and feasibility of alternatives.

After further discussion and improvements, the EEA believes that this framework for evaluating scientific evidence will be a helpful tool in the process of producing consistent overviews of the existing states of knowledge.

Conclusion

Measuring sustainable development requires innovative techniques and indicators, rigorous underlying models, and frameworks for interpretation of complex evidence. The approaches and frameworks presented in this chapter and the associated critique are expected to contribute to improved assessment of sustainability by shedding light on new techniques, providing criticism of existing systems, and contributing new approaches to analyzing and interpreting results.

Notes

1. The criteria were initially introduced by the EEA in 1998–1999 in *Europe's Environment: The Second Assessment*, p. 284, and in *Europe's Environment at the Turn of the Century*, p. 20 (EEA 1998, 1999b). The criteria were based on key environmental programs such as the Rio Declaration; the European Commission's 5th Environmental Action Programme; the Pan-European Environmental Programme for Europe; policy papers produced to implement the EU Treaty provisions on integration, including the Commission of the European Communities Communication on Integration; conclusions of the Cardiff, Vienna, and Cologne summits; draft council papers on sectoral integration for the Helsinki Summit; and associated commentaries from the European Environmental Bureau and member states.

2. Council Regulation (EEC) no. 1210/90 of May 7, 1990, as amended by Council Regulation 933/1999 of April 29, 1999.

3. The EEA Expert Group on Guidelines and Reporting brought together national experts on the state of the environment and indicator reporting, meeting twice a year to discuss topics of mutual interest and to advise the EEA on its reporting activities.

Literature Cited

CEC. 2004. *Integrating environmental considerations into other policy areas: A stocktaking of the Cardiff process.* Commission Working Document, COM (2004) 394 final. Brussels: Commission of the European Communities.

CLM. 1999. *EU Agricultural policy after 2000: Has the environment been integrated?* A report for the European Environmental Bureau (CLM-446-1999). Utrecht: G. van der Bijl, Centrum voor Landbouw en Milieu (Centre for Agriculture and Environment).

Dalal-Clayton, B. 2004. *The EU strategy for sustainable development: Process and prospects.* Environmental Planning Issues no. 27. London: IIED.

EEA. 1998. *Europe's environment: The second assessment.* Copenhagen: European Environment Agency.

EEA. 1999a. *Europe's environment at the turn of the century.* Copenhagen: European Environment Agency.

EEA. 1999b. *Monitoring progress towards integration: A contribution to the "Global Assessment" of the Fifth Environmental Action Programme of the EU, 1992–1999.* An unpublished report from the European Environment Agency, Copenhagen, December.

EEA. 1999c. *Towards a Transport and Environment Reporting Mechanism (TERM) for the EU: Part I and II.* Technical Report no. 18. Copenhagen: European Environment Agency.

EEA. 2001. *Late lessons from early warnings: The precautionary principle 1896–2000.* Copenhagen: European Environment Agency.

EEA. 2002. *Sustainable development reporting: Frameworks and aggregated indicators.* Unpublished report of the EEA Expert Group on SoE Guidelines and Reporting. Copenhagen: European Environment Agency.

EEA. 2003. *An approach to multi-causality in environmental health.* EEA Background Paper no. 5 for the EU SCALE process. Copenhagen: European Environment Agency, October. Available at www.Brussels-conference.org.

EEA. 2005. *Environmental policy integration in Europe: State of play and an evaluation framework.* Technical Report no. 2/2005. Copenhagen: European Environment Agency.

European Commission. 2001. *A sustainable Europe for a better world: A European strategy for sustainable development.* COM(2001) 264 final. Brussels: European Commission.

European Commission. 2001. *White paper on European governance.* COM(2001)428. Brussels: European Commission.

Fergusson, M., C. Coffey, D. Wilkinson, B. Baldock, A. Farmer, R. A. Kraemer, and A. Mazurek. 2001. *The effectiveness of EU Council integration strategies and options for carrying forward the Cardiff process.* IEEP Report to the UK Department of Environment, Transport and the Regions. London: Institute for European Environment Policy.

Hill, A. B. 1965. Environment and disease: Association or causation? *Proceedings of the Royal Society of Medicine* 58:295–300.

IPCC. 2001. *Third assessment report on climate change: Summary for policymakers.* Intergovernmental Panel on Climate Change, WMO, UNEP.

IPCS/WHO. 2002. *Global assessment of the state of the science on endocrine disruptions.* International Programme on Chemical Safety and World Health Organization.

OECD. 2002. *Improving policy coherence and integration for sustainable development: A checklist.* Paris: Organisation for Economic Cooperation and Development. Available at www.oecd.org/dataoecd/60/1/1947305.pdf.

Persson, Å. 2002. *Environmental policy integration: A draft for comments.* Stockholm: Stockholm Environment Institute.

Peters, G. B. 1998. Managing horizontal government. *Public Administration* 76:295–311.

Prescott-Allen, R. 2001. *The Wellbeing of Nations: A country-by-country index of quality of life and the environment.* Washington, DC: Island Press.

RMNO/EEAC. 2003. *Environmental governance in Europe.* Background study no. V.02, edited by L. Meuleman, I. Niestroy, and C. Hey. The Hague: Advisory Council for Research on Spatial Planning, Nature and the Environment and European Environmental Advisory Councils.

Välimäki. 2002. *Sustainable development reporting: Aggregated indicators.* Background paper for discussion at EEA SoE Expert Group Workshop, December 16–17, Helsinki, Finland.

Wandén, S. 2003. *Society, systems and environmental objectives: A discussion of synergies, conflicts and ecological sustainability.* Naturvårdsverket Report 5272.

10

Integrated Assessment and Indicators

Arthur Lyon Dahl

Of all the potential uses of indicators of sustainability, integrated assessment is perhaps the most critical and also the most difficult because such assessments must bring together a wide variety of issues and topics. An assessment is by definition an evaluation, and indicators are one way of expressing the absolute or comparative value of something. In the context of sustainability, an assessment evaluates and draws conclusions about the state of and trends in some unit or component of society or the environment and its future perspectives. This component could be a local community, a corporation, an ecoregion, a nation, a continent, or the entire planet. This review focuses on international assessments, but the same principles apply at other levels. Various kinds of statistics, data sets, and indicators can serve as the basis for such assessments. An integrated assessment for sustainability involves a comprehensive consideration of the economic, social, environmental, institutional, and other relevant aspects of the entity, including the relationships between all these factors. In practice, our limited understanding of such complex human–environmental systems means that our assessments fall short of the ideal of full integration, and the issues may just be juxtaposed. There has been no comprehensive evaluation of the various attempts at integrated assessments, but the International Council for Science has proposed such a review (ICSU 2002).

This chapter explores the practice of and challenges in the use of indicators in integrated assessments, both to measure the states and trends in various components and, ideally, to indicate the behavior of the whole integrated system and its implications for the future. In this latter role, these would be true indicators of sustainability. The emphasis is on progress since the last review of the state of the art in sustainability indicators by the previous SCOPE project (Moldan et al. 1997). A very useful analysis and evaluation of recent efforts to produce more integrated indices has been prepared (in French) by Gadrey and Jany-Catrice (2003). It highlights the progress now being made

163

to produce indicators that begin to integrate over broad economic, social, and environmental areas.

At present, integration using indicators has followed two approaches: broad aggregations of indicators or indices that combine indicators across multiple sectors but do not analyze interactions and integration focusing on dynamic system behavior over time and the interrelationships between factors. Because of the difficulty of the latter, integrated assessments today have almost exclusively used the former, and it is those examples that are reviewed here. However, the challenges of climate change are pushing climate modelers to extend their computer models with an increasing number of environmental, social, and economic dimensions, which should accelerate future progress in integrated modeling. The end of this chapter includes some suggestions for future work on more dynamic and complete forms of integration to assess long-term human sustainability.

Scientific Validity of Definitions

Assuming that an integrated assessment is intended to report on sustainability, the most important and difficult definition is that of sustainability itself. Sustainability is not a goal to be achieved at some point in time but a characteristic of a dynamic human–environmental system able to maintain a functional productive state indefinitely (Dahl 1996, 1997a). Integrated indicators of sustainability therefore should measure the functional system processes that best represent its capacity to continue far into the future. Defining sustainability in terms of durability over time avoids the problem of specifying the characteristics of the system or entity to be maintained, which can be very subjective and specific, and where political, philosophical, and cultural differences can prevent any wide consensus. The optimal sustainability indicators are those that best show a scientifically verifiable trajectory of maintenance or improvement in system functions. Although the choice of indicators depends on the system in question, it is not their substance but their dynamic change over time that is important for measuring sustainability. Science cannot always validate the goals set for the system, but it can validate the ability of the indicators chosen to measure the trajectory toward those goals or the reduction in damaging factors threatening the system's sustainability.

Scientific approaches can also help us understand or model the complex operation of the system and thus ensure that the indicators selected reflect its most essential characteristics and are able to measure its sustainability within the limits of predictable system behavior.

Although sustainability assessment is needed most often in the context of sustainable development, and most integrated assessments specifically aim to do this, the concepts are not synonymous. The term *development* is often erroneously equated with growth, which by definition is not infinitely sustainable in a finite system. Sustainability requires the redefinition of development to mean improvements in human welfare and prosperity, including poverty reduction but respecting planetary limits, which may entail limited growth in some areas and perhaps reductions in consumption in others.

History and Existing Use

There are two fundamental starting points for linking integrated assessments and indicators based on two approaches to assessment. One uses expert opinion or consultation with stakeholders to produce an integrated assessment in text form and then develops indicators to explain, illustrate, and eventually complete or extend the results of the assessment. The second, more statistical approach is to assemble a set of indicators or statistics in some coherent statistical framework to produce a more numerical integrated assessment. Perhaps at some point in the future the two approaches will converge, but at present they involve different communities of natural and social scientists or statisticians.

Assessments with Indicators

Sampling of the principal integrated assessments at the international level illustrates the different ways indicators are being used today and the significant progress that has been made in the last decade. In most cases, the indicators are illustrative, providing numerical and graphical support to reinforce a text-based assessment. Generally such indicators are used only where good data are available, and many parts of the assessment may have little or no indicator support for this reason. Some assessments have been prepared by a one-off process producing a single report, and indicators for these are limited to the data available at the time. Other continuing assessment processes generate periodic reports. Where efforts are being made to build comprehensive and comparable global data sets as part of the assessment process, the number of indicators used in such reports is increasing.

The problem is that there are few compilers of globally consistent data sets or indicators, including the United Nations and its agencies (e.g., the Food and Agriculture Organization, United Nations Development Programme [UNDP], United Nations Environment Programme [UNEP], World Health Organization, World Meteorological Organization), the World Bank, the Organization for Economic Co-operation and Development (OECD) for its member countries, and a few national or nongovernment institutes (e.g., The Netherlands National Institute of Public Health and Environment, World Resources Institute).

A good illustration of the challenges of using indicators is the UNEP *Global Environment Outlook* (*GEO*) report series of integrated assessments. The first *GEO* report in 1997 (UNEP 1997) was largely qualitative in its assessments. Even illustrative data tables were limited to selected countries. The only use of a few indicators was in the scenarios giving some projections to 2050, which looked at regional changes in population, gross domestic product (GDP), primary energy consumption, energy intensity, agricultural production (maize), caloric intake, total water withdrawal, changes in land use and cover, and habitat loss. By *GEO 2000* (UNEP 2000), some indicators were given at the global and regional levels for the assessment of selected problems. In addition to the indicators used in *GEO 1*, these included cropland per capita, hunger, forest area, fishery production, carbon emissions, toxic waste, and urbanization. Many of these indicators were produced by one-off studies and did not present time series or trends. By *GEO 3* (UNEP 2002), the effort to develop the data necessary for globally consistent indicators for assessments began to show results. Nearly every page includes text and one or more indicator tables or graphics showing states or trends. However, the indicators used still measure specific problems or social or economic trends and do not attempt an integrated view of system behavior. The indicators are not really the tools for the assessment but illustrations with a function similar to photographs.

Other international assessments suffer from similar handicaps. The *Global Outlook 2000* (UN 1990) assembled chapters on various economic, social, and environmental trends illustrated with graphs and tables of selected indicators but with no integration across the sectors. Most global assessments follow the general model of the World Bank (2004) *World Development Report* with text-based assessments illustrated with a few indicators in graphs or tables, followed by tables of world development indicators by country. Such extensive data tables are useful for experts and have helped support many other assessment processes. They inspire confidence in the preceding assessment by emphasizing its quantitative scientific basis. However, they have little direct public impact, showing that too much numerical information without a framework to provide coherence and orientation has no meaning (Gadrey and Jany-Catrice 2003). The long history of economic indicators has allowed highly integrated indices such as the GDP to evolve, but there has been little effort to integrate beyond the economic sphere.

Even the assembly of such data tables suffers from serious problems of data gaps and inconsistencies, which make the production of indicators with sufficient consistency to permit integration a time-consuming and costly process even where it is possible. Few organizations can afford to do this, and once such data are made available, they are often endlessly and sometimes uncritically recycled from assessment to assessment.

The *World Resources* reports (UNDP, UNEP, World Bank, and World Resources Institute 2003), issued every 2 years, are among the most data- and indicator-rich global assessments, with analytical text and selected indicators combined with extensive data tables. Like the UNDP *Human Development* reports, each report develops a specific theme with data and indicators relevant to that theme. However, the data tables are relegated to the end of the report, and, if anything, the use of indicators has declined in recent years in favor of other forms of graphic communication and summary text. The UN Division for Sustainable Development prepared a *Critical Trends* report for the 5-year review of Agenda 21 (UN DPCSD 1997). Although it surveys the long-term trends in selected environmental and socioeconomic issues illustrated with appropriate indicators, it does not integrate them in any systematic way.

The report *Protecting Our Planet, Securing Our Future* (UNEP, NASA, and World Bank 1998) was a one-off attempt to identify and integrate the key scientific and policy links between major global environmental issues and between these issues and basic human needs. It uses a selection of indicators to show present environmental impacts and projected future trends, but again these are illustrative rather than the basis for integration.

Another approach is to build an assessment around important statistical trends, with a compiled index of several indicators as the central theme and attraction of the assessment, amplified by additional text, indicators, and data tables. The best example is UNDP's annual *Human Development Report* (UNDP 2004), which aims to get countries to focus on key issues of human development. The report makes headlines and attracts high-level political attention because it ranks countries with its Human Development Index (HDI). This simple index, combining only a few basic statistics (life expectancy, adult literacy, school enrollment, GDP per capita), was initially quite controversial but has had great impact. It is significant more as a communication tool to motivate countries to reexamine the impact of development on people rather than a truly integrated measure of sustainable development. It attracts people to read the report and to consider the other data tables and thematic analyses that amplify the basic message (Sen 1999). The annual thematic assessments provide an integrated view of key human development issues, but again the indicators are used just to support the text. They are illustrative rather than tools in themselves for integration.

These examples show a pattern of increasing use from scattered illustrations to an index as the flagship of the assessment, but the indicators still play only a supporting role rather than defining the behavior and sustainability of the human–environment system.

Assessments Based on Indicators

The second approach to integrated assessment has built on the long work of statisticians and economists to assemble integrated and coherent national economic accounts. Gadrey and Jany-Catrice (2003) have reviewed in detail the recent efforts to extend this work into indicators of wealth and development. This approach starts by compiling many different statistics and indicators into a comprehensive data set. The challenge of this approach is to identify a realistic and balanced set of indicators and to collect sufficient reliable data to avoid so much interpolation or estimation that the results are meaningless. As with the illustrative indicators in the text-based assessments, a compilation of indicators can demonstrate many facets of the problem but does not actually integrate them. Here the issues of selection and weighting become crucial, and there is no consensus on a scientifically valid solution. This approach is still at the stage of a better description of the present state of the economy and society and sometimes the reconstruction of past trends. Less work has been done on the potential to project such indicator-based assessments into the future to determine sustainability because this will require complex and conceptually challenging models. A few examples will illustrate the present state of the art.

Gadrey and Jany-Catrice (2003) cite the Index of Economic Well-Being developed by Osberg and Sharpe (2002) as the most methodologically sound of the integrated indices while combining both objective and subjective measures. It equally weights four components: consumption (market consumption per capita, government expenditure per capita, unpaid domestic work), wealth (physical capital per capita, R&D per capita, natural resources per capita, human capital and education, minus net exterior debt per capita, minus cost of environmental degradation), equality (poverty, Gini coefficient of inequality), and economic security (risk of unemployment, economic risk of illness, poverty risk in single-parent families, poverty risk of older adults). Some of the factors are only roughly estimated, but because the index measures change over time, the absolute values are less important than relative changes from year to year. The index is also insensitive to changes in weighting. The plots of this index and its components over time show that GDP per capita and well-being do not always correlate, and even between industrialized countries, the performance on the different components can vary widely. Although the focus of this index is economic, it includes social and environmental dimensions. It does not attempt to measure sustainability, but methodologically it shows what might be possible.

The World Economic Forum and Yale and Columbia Universities developed an Environmental Sustainability Index (ESI) as the basis for their report *Environmental Performance Measurement: The Global Report 2001–2002* (Esty and Cornelius 2002), recently updated in the *2005 Environmental Sustainability Index* (Esty et al. 2005) comparing the performance of 146 countries. The ESI is made up of twenty-one indicators and seventy-six variables. It is probably the environmental assessment that most directly uses indicators as the tool for its evaluation. However, the reliance on indicators did not reduce the subjective dimension of the assessment, which was simply reflected in the selection of indicators and the weighting method chosen. Widespread criticisms of the 2002 ESI led to significant modifications in the 2005 version, which also identified further improvements that would be desirable when the data permit. The index also aims only to provide an integrated measure of environmental sustainability and does not attempt to address economic or social sustainability.

The UN Commission on Sustainable Development work program on indicators has produced two compilations of methods for sustainable development indicators (UN DSD 1996, 2001) for use at the national level. These have conceptually attempted to provide the basis for integrating many dimensions of sustainable development as defined by governments in Agenda 21, but they have not actually been used to generate integrated assessments, leaving that responsibility to national governments. Governments have indicated that they did not want such indicators used to compare and assess their sustainable development at the international level out of fear that this might lead to conditionality in development assistance. However, the Commission on Sustainable Development (CSD) indicators is the only set benefiting from such high-level political acceptability through their trial by many governments and adoption by the CSD. The first trial set of 134 indicators was arranged in a driving force, state, response framework and grouped by chapters of Agenda 21 (UN DSD 1996). This could have provided the basis for integrating the indicators according to their roles in system sustainability, but the indicators were too few and disparate for such integration, and the framework served only to show how well key issues of sustainability were being covered. The second set of 58 core indicators (UN DSD 2001) aimed to show their policy relevance by clustering them by themes and subthemes. This strengthened their power to communicate but was less amenable to an integrated view of sustainability.

Examples

Building on the CSD and other work, the Consultative Group on Sustainable Development Indicators (CGSDI) (iisd.org/cgsdi/) has assembled a data set corresponding to the CSD indicators and developed an interesting tool, the Dashboard of Sustainability (esl.jrc.it/dc/index.htm), that provides an integrated presentation of such indicator sets. The CGSDI thought that integrating across economic, social, and environmental fields was conceptually difficult because there was no common denominator, but that economic indicators with monetary values, social indicators expressed per capita or in similar human terms, and environmental indicators based on scientific measurements could be integrated within those sectors and then cross-compared for a more complete view of sustainability. The result is not an assessment as such but a means by which each user can perform individual assessments. Because the Dashboard is a tool for an integrated view of any data set, it can be used to compare different indicator sets and to highlight and make transparent the assumptions and weightings, conscious or unconscious, behind each. It can therefore facilitate more open integrated assessments.

An interesting recent initiative to address sustainability more directly is the Environmental Vulnerability Index (EVI), developed by the South Pacific Applied Geoscience Commission (SOPAC) (Kaly et al. 2003; Pratt et al. 2004; SOPAC 2005). This uses fifty indicators to estimate the vulnerability of the environment to future shocks in 235 countries (www.vulnerabilityindex.net/). What is conceptually interesting about this index is its effort to relate the indicators to scientifically founded concepts or limits of what is sustainable rather than to simply give the range of countries from best to worst. The index is reported as a single dimensionless number, accompanied by several subindices and a country profile of the results for all indicators, showing where the specific problems lie. The index thus integrates and assesses all aspects of environmental vulnerability. Although there are still aspects that need refinement, the EVI approaches an integrated measure of environmental sustainability. It is intended to accompany another index of economic vulnerability also developed in the context of the 1994 Barbados Programme of Action for Small Island Developing States, which called for the development of a vulnerability index.

Nongovernment organizations have developed their own assessment approaches and reports in an effort to provide an alternative view to that of the official or dominant view of governments and economists. Some of these have pioneered integrated indices as the principal instrument for their assessments, supported only by short text commentary. A good example is the annual WWF *Living Planet Report* (WWF 2004). It includes a Living Planet Index averaged from indices of global terrestrial, freshwater, and marine species and a World Ecological Footprint compiled from cropland, grazing land, forest, fishing ground, and energy footprints. It also includes scenarios projecting key indicators into the future. *The Wellbeing of Nations* (Prescott-Allen 2001) is another example of an assessment of nations' environmental status and quality of life based on several highly aggregated indices. However, it would best be described as a status report rather than a sustainability assessment.

As these examples show, although indicators are becoming increasingly common in integrated assessments, they are still largely illustrative of specific factors or the comparative state of such factors and are far from reflecting or driving the integrated perspective itself or capturing the dynamic processes underlying sustainability. However, some recent initiatives are beginning to make progress in that direction.

Methodological Aspects

The use of indicators in integrated assessments faces the same challenges as with other uses of indicators: selection of appropriate indicators, data availability, comparisons between disparate topics and forms of measurement, weighting, and total and relative numbers of indicators selected (which often implies an inherent weighting). In addition, there is the challenge of integration itself: finding indicators that reflect the whole and not just the parts. In the present state of the art of integrated assessment, this question has not yet been resolved. One approach will be through complex computerized system models that mathematically reproduce the structure and dynamics of the system. As assessments come to be based on such models, as is now at least partly the case for climate change assessments, indicators can be derived from the models to reflect system resilience, susceptibility to perturbation, and ability to maintain basic functions and outputs over long time periods. Once these new indicators of system performance and sustainability have been validated by such models, they can be implemented with models driven by real data streams.

Some specific types of indicators have an integrating aspect useful for integrated assessments, such as indicators of material flows (Adriaanse et al. 1997), energy intensity, and decoupling of resource inputs from outputs. Indicators that show vectors of trends toward or away from a sustainable state or convergence with a target can also be helpful (Dahl 1997b).

One of the most difficult aspects to treat in a methodologically and scientifically rigorous way is the underlying assumptions guiding the assessment and therefore the selection of indicators. Different individuals, organizations, sectors of society, and cultural groups have their own worldviews, visions of the future, perspectives, and values. There is an inherent tendency to select indicators and make assessments that validate a preconceived view of the world or confirm inherent biases. Such assessments tend to be more popular and influential and receive acceptance in policy circles not because they are scientifically valid or right but because they say what people want to hear. An indicator set that reflects the views of corporate leaders in a materialistic, free enterprise economic system will be very different from one prepared by environmental groups or social activists in undeveloped countries.

The methodological challenge is first to make these different perspectives and biases transparent and then to separate the normative dimension of sustainability from the scientifically verifiable trends in that particular context. Integrated measures of the sustainability of a system for warfare or development assistance should be possible without moral judgments about the goals of the activity. After all, the integrated index of gross national product was first developed to measure the American war effort. Once indicators of system behavior and sustainability have been developed, it will be necessary to try to step outside the context of the various dominant worldviews and to judge sustainability with respect to planetary limits, at least for the factors that can be established scientifically. This scientific perspective on sustainable limits can then be reintegrated transparently with value judgments about the choices to be made to keep the human economic and social system within those limits.

Relevance to Sustainable Development

As the state of the art in integrating indicators progresses, synthetic indices combining many indicators will become increasingly relevant as the basis for assessing and communicating sustainability. At present, the assembly of increasingly comprehensive data sets of indicators covering the state of and trends in economic, social, and environmental factors relevant to sustainability provides a first approximation of where we are and where we are going. However, these data sets do not capture the interactions between factors and the broader dynamics of the system that are critical to sustainability.

One important issue is the distinction between development, as commonly understood, and sustainability. Development often is equated with growth, whether in wealth and economic activity, infrastructure, or institutions. However, where growth has pushed a society beyond sustainable limits, long-term sustainability may entail a reduction in certain economic activities, technologies, or resource uses and a simplification in lifestyles (Meadows et al. 1992). This entails a broader vision of human development that may combine higher levels of social integration, culture, science, and the arts with a more moderate approach to the material side of life. Care must be taken to select indicators of sustainability that capture all the dimensions of a rich and rewarding human society contributing to social and human sustainability, not just the material aspects of sustainability on this planet.

Policy Relevance and Legitimacy

Most recent integrated assessments give a high priority to policy relevance and ensure that issues of concern to policymakers are explicitly addressed. The use of extensive supporting data tables and indicators increases their legitimacy by demonstrating the objective foundations of their analyses. However, such data tables by themselves will have little direct impact on decision makers, who need simpler and more explicit indicators of sustainability to communicate the key messages. The HDI is a good example of a simple indicator that reaches policymakers and opens the door to a more detailed consideration of underlying causal factors. The HDI leverages much greater impact from the whole *Human Development Report* (Sen 1999). Integrated assessments should aim to have both detailed indicators of key problems and trends for specialists and technical advisors and one or more flagship indices that will attract the attention of policymakers and the media.

The real problem is that the best-integrated assessment based on substantial data is still not sufficient to convince the major actors in society, whether in government or the private sector, to look beyond their immediate short-term interests. Sustainability is inevitably a long-term issue. There are rarely problems that threaten our very survival tomorrow. It is hard to motivate people to make sacrifices to avoid crises that will affect only future generations. The development of some high-impact indicators of sustainability together with models and scenarios in support of integrated assessments should help to make society more responsive. Involvement of users and laypeople in the development of the indicators can also increase buy-in and relevance. Participatory approaches with wide stakeholder involvement are increasingly used to legitimate assessment processes.

Another problem with the policy relevance of assessments and their indicators is that their acceptance often depends on who produces them. People tend to have confidence in those who think like them and share their values, and reject assessments produced by those with opposing views. Businesspeople appreciate the indicators developed by the World Economic Forum (Esty and Cornelius 2002; Esty et al. 2005); conservationists prefer those of the WWF (2004). For some, the UNEP is suspect because it is environmental; for others the World Bank is suspect because it is the World Bank. This reinforces the need to build a more scientific basis for the legitimacy of indicator sets and assessments. Legitimacy and acceptance also depend partly on the track record over time. New indices often are controversial, but if they demonstrate their usefulness and impact over time, they increasingly come to be accepted.

Extent of Applicability

Integrated assessments will be in growing demand as the best way to provide policy guidance on the major directions for future society. They will be needed for a variety of institutions at different levels of governance, from local to global. As the principles for integrated indicators of sustainability are worked out, they should be applicable at a variety of levels and adaptable to different contexts. The techniques for indicating the sustainability of processes and trends, irrespective of the goals of the entity being assessed, should be of general use in many integrated assessment processes, whereas many other indicators will be case-specific.

Gaps in Knowledge and Research Needs

The major challenge is how to integrate indicators of many types across sectors to give an overall evaluation of sustainability. Improved data sets will be an essential prerequisite, but new integrated or linkage indicators are also needed. Just as the GDP measures the flow of money through an economy and thus gives an integrated measure of economic activity, new indicators are needed to measure such features as the flow of natural resources for human use as related to their rate of renewal, the changing balance in various forms of natural capital, the stability of social institutions and networks such as the family, the community and local associations, the vulnerability and resilience of the society, the flow of information, the links between different social entities and environmental processes, and other factors that are critical to sustainability.

Research is needed to explore new approaches to indicators using satellite remote sensing and other observing technologies. These techniques can overcome data gaps by providing uniform planetary coverage and regular time series. For assessments of global sustainability, observing systems should be able to generate indicators of the state of the biosphere, land use trends, the balance between human impacts and natural processes, the status of natural resources, and the extent of poverty in human communities. The Integrated Global Observing Strategy Partnership (www.igospartners.org) and the intergovernmental Group on Earth Observations (earthobservations.org) provide mechanisms to plan and coordinate such efforts.

Another research priority is to find indicators able to capture the less tangible dimensions of human society for integrated assessments. Indicators are needed for the effectiveness of governance, the adequacy of legislation, the flowering of arts and culture, access to science and technology, and other important dimensions of development. The sustainability of a society also depends to a great extent on the strength of its ethics, norms, values, and spirituality (IEF 2002). Although it may be difficult to find direct indicators of these aspects, there may be surrogate measures that can be used to assess their importance and evolution over time (Bahá'í International Community 1998). Until these fundamental but intangible dimensions of society have adequate indicators, they will be invisible for assessment purposes.

Another missing dimension in present sustainability measures is the sustainability of societies themselves from generation to generation. A community or society is sustainable only if it transmits its knowledge, experience, science, culture, wisdom, and values from old people to younger ones before they are lost. Education is a key part of this process, but families, communities, religious and cultural organizations, and the media are also important. With rapid social change, traditional forms of transmission may be disrupted, and significant parts of a society's heritage may be lost before their importance is appreciated. Similarly, new media and information technologies may have both positive and negative impacts on the transmission of knowledge and values. These open a society to the world but often convey values, lifestyles, behavior patterns, and desires for consumption at odds with both the local culture and the needs of sustainability, driving social change in directions with unanticipated consequences. Indicators therefore are needed that capture the effectiveness with which intergenerational information transfer is taking place and the directions in which it is pushing social and cultural evolution.

It may be helpful in identifying indicators of sustainability for society as a whole to undertake a historical analysis of the factors causing the unsustainability and collapse of past civilizations. There may be interactions between social, environmental, political, and cultural factors, or sequences of destabilizing processes, that will stand out better in such retrospective analyses than in any attempt to detect them today. Such analyses could provide a long-term perspective on critical dimensions of the sustainability of civilizations that is lacking in our own society. Indicators could then be developed to follow these dimensions in our own time.

Integrated assessments represent the most difficult challenge for indicators of sustainability because of their need to capture and integrate all aspects of the assessment. Some progress is being made in this direction (Gadrey and Jany-Catrice 2003), but there is still a long way to go before indicators can fully support the integrative aspect of these assessments.

Literature Cited

Adriaanse, A., S. Bringezu, A. Hammond, Y. Moriguchi, E. Rodenburg, D. Rogich, and H. Schütz. 1997. *Resource flows: The material basis of industrial economies.* Washington, DC: World Resources Institute.

Bahá'í International Community. 1998. *Valuing spirituality in development: Initial considerations regarding the creation of spiritually based indicators for development.* London: Bahá'í Publishing Trust.

Dahl, A. L. 1996. *The eco principle: Ecology and economics in symbiosis.* London: Zed Books; Oxford: George Ronald.

Dahl, A. L. 1997a. The big picture: Comprehensive approaches, Part One: Introduction. Pp. 69–83 in *Sustainability indicators: A report on the project on indicators of sus-*

tainable development, SCOPE 58, edited by B. Moldan, S. Billharz, and R. Matravers. Chichester, UK: Wiley.

Dahl, A. L. 1997b. From concept to indicator: Dimensions expressed as vectors, Box 2H. Pp. 125–127 in *Sustainability indicators: A report on the project on indicators of sustainable development,* SCOPE 58, edited by B. Moldan, S. Billharz, and R. Matravers. Chichester, UK: Wiley.

Esty, D. C., and P. K. Cornelius. 2002. *Environmental performance measurement: The global report 2001–2002.* New York: Oxford University Press.

Esty, D. C., M. A. Levy, T. Srebotnjak, and A. de Sherbinin. 2005. *2005 Environmental Sustainability Index: Benchmarking national environmental stewardship.* New Haven, CT: Yale Center for Environmental Law & Policy. Available at www.yale.edu/esi/.

Gadrey, J., and F. Jany-Catrice. 2003. *Les indicateurs de richesse et de développement: Un bilan international en vue d'une initiative française.* Rapport de recherche pour la DARES, March 2003. Paris: Ministère de l'emploi, de la cohèsion sociale et du logement. Available at http://www.travail.gouv.fr/IMG/pdf/rapport-indicateurs-richesse-developpement.pdf.

ICSU (International Council for Science). 2002. ICSU Series on Science for Sustainable Development no. 8: *Making science for sustainable development more policy relevant: New tools for analysis.* Paris: ICSU.

IEF (International Environment Forum). 2002. Dialogue on indicators for sustainability. Forum on science, technology and innovation for sustainable development. World Summit on Sustainable Development, Johannesburg, August 27. In *Report of the Conference,* available at www.bcca.org/ief/conf6.htm.

Kaly, U., C. Pratt, J. Mitchell, and R. Howorth. 2003. The demonstration Environmental Vulnerability Index (EVI). SOPAC Technical Report 356. Suva, Fiji: South Pacific Applied Geoscience Commission. Available at www.vulnerability index.net/.

Meadows, D. H., D. L. Meadows, and J. Randers. 1992. *Beyond the limits: Confronting global collapse, envisioning a sustainable future.* Post Mills, VT: Chelsea Green.

Moldan, B., S. Billharz, and R. Matravers (eds.). 1997. *Sustainability indicators: A report on the project on indicators of sustainable development,* SCOPE 58. Chichester, UK: Wiley.

Osberg, L., and A. Sharpe. 2002. The index of economic well-being: An overview. *Indicators: The Journal of Social Health* 1(2):24–62.

Pratt, C., U. Kaly, J. Mitchell, and R. Howorth. 2004. *The Environmental Vulnerability Index (EVI): Update & final steps to completion.* SOPAC Technical Report 369. Suva, Fiji: United Nations Environment Programme and South Pacific Applied Geosciences Commission.

Prescott-Allen, R. 2001. *The Wellbeing of Nations: A country-by-country index of quality of life and the environment.* International Development Research Centre, IUCN,

International Institute for Environment and Development, FAO, Map Maker Ltd., UNEP-WCMC. Washington, DC: Island Press.

Sen, A. 1999. Assessing human development. P. 23 in *Human development report 1999*, UNDP. New York: Oxford University Press.

SOPAC. 2005. *Environmental Vulnerability Index*. Suva, Fiji: South Pacific Applied Geoscience Commission. Available at www.sopac.org/evi.

UN. 1990. *Global outlook 2000: An economic, social and environmental perspective*. New York: United Nations Publications.

UNDP. 2004. *Human development report 2004. Cultural liberty in today's diverse world*. New York: Oxford University Press.

UN DPCSD. 1997. *Critical trends: Global change and sustainable development*. New York: United Nations.

UNDP, UNEP, World Bank, and World Resources Institute. 2003. *World resources 2002–2004*. Washington, DC: World Resources Institute.

UN DSD. 1996. *Indicators of sustainable development: Framework and methodologies*. New York: United Nations.

UN DSD. 2001. *Indicators of sustainable development: Guidelines and methodologies*, 2nd ed. New York: United Nations.

UNEP. 1997. *Global environment outlook*. Oxford: Oxford University Press.

UNEP. 2000. *Global environment outlook 2000*. London: Earthscan.

UNEP. 2002. *Global environment outlook 3*. London: Earthscan.

UNEP, NASA, World Bank. 1998. *Protecting our planet, securing our future: Linkages among global environmental issues and human needs*. Nairobi, Kenya: United Nations Environment Programme; Washington, DC: US National Aeronautics and Space Administration, The World Bank

World Bank. 2004. *World development report 2004: Making services work for poor people*. New York: Oxford University Press.

WWF. 2004. *Living planet report 2004*. Gland, Switzerland: WWF International.

11

Qualitative System Sustainability Index: A New Type of Sustainability Indicator

Jasper Grosskurth and Jan Rotmans

The Conceptual Challenge

In Chapter 10 Arthur Dahl challenges the scientific community to develop sustainability indicators that "measure the functional system processes that best represent its capacity to continue far into the future." According to Dahl, these indicators should "reflect the whole and not just the parts." Indicators should highlight problems rather than symptoms. We agree with Dahl's perception that existing sustainability indicators do not reflect the whole: "Increasingly comprehensive data sets of indicators covering the state of and trends in economic, social, and environmental factors relevant to sustainability . . . do not capture the interactions between factors and the broader dynamics of the system that are critical to sustainability." In our contribution, we present a concept for an indicator for the sustainability of systems that is designed to address Dahl's challenge.

We define an indicator, following Rotmans and de Vries (1997), as "a characteristic of the status and dynamic behaviour of the system concerned. Or equivalently: an indicator is a one-dimensional systems description, which may consist of a single variable or a set of variables." The characteristic of the system that we are most interested in is its ability to sustain itself in the long run in a desired state or on a desired trajectory. A system with that ability is sustainable.

In order to evaluate the sustainability of a system, we would optimally take into account time, scale, and domain. A measure of sustainability should represent changes in the system that are relevant in the long term of 25 to 50 years. It should reflect developments within the system and trade-offs to systems on other scale levels. It should cover the economic, ecological, and social aspects of sustainability.

The Practical Challenge

When we provided advice on regional sustainability to civil servants in strategic departments of subnational governance bodies, we found that they were quite capable of identifying economic, environmental, and social states and trends in their region, such as a low education profile in the population or a rapid loss of biodiversity.

The target audience could not map these fragmented parts as a system of stocks and flows. Consequently, they could not understand the complex dynamics resulting from these interactions. The target audience was even less able to design and test policy actions for their effects on the sustainability of their region. This results in inefficient actions and counterintuitive surprises. Forrester (1968) discusses the consequences for action resulting from an insufficient system understanding from a system dynamic perspective, Dörner (2003) from a psychological perspective, and Sterman (1994) from a learning perspective.

For our work, this meant that a selected list of sustainability indicators or an aggregated indicator of sustainability would not address the most urgent questions and would not help to make better decisions for a sustainable future. Instead of focusing on the states of the parts of the system, we needed to focus on the system structure itself.

Any indicator based on the system structure must use a representation of the real world in a system format. The representation and evaluation of a system are intrinsically subjective, normative, ambiguous, uncertain, and incomplete (Rotmans and de Vries 1997). In response, several authors have proposed public participation (Spangenberg and Bonniot 1998) or stakeholder participation (Jaeger et al. 1997). Participation at the very least is potentially enriching for the scientific analysis of sustainability while addressing issues of legitimacy and transparency of sustainability studies and facilitating communication processes. For our work this implied the challenge to develop methods or interfaces for integrated sustainability assessments that are suitable for stakeholder participation.

Where available, quantitative indicators proved useful in communicating the urgency of key issues in participatory settings. However, the exact same information caused confusion and frustration when the goal of the information was extended beyond raising awareness to include the development of policy options for improvement. The reason for this was the lack of information on the systemic causes of indicator values and thus a lack of information on what one could do to change the value of a given indicator without negatively affecting other indicators. The availability of quantitative information caused a neglect of issues for which quantitative indicators were not available. In response, we had to develop methods to better integrate qualitative and quantitative information.

Describing the System

With conceptual and practical challenges in mind, we developed and applied the SoCial, ENvironmental, and Economic model (SCENE) approach to map a region as a system (Grosskurth and Rotmans 2005). The three domains of sustainability provide the basic structure for SCENE. The approach is based on the participative and qualitative representation of stocks and flows in the format of an influence diagram.

Stocks describe core elements of a system that change slowly. In contrast to the system dynamic notion given to the terms *stock* and *flow*, SCENE stocks can be quite generic titles, such as *lifestyle* or *economic vitality*. These titles can be interpreted multidimensionally. In the SCENE approach, we generally take four dimensions of a stock into account: quantity, quality, function, and spatial dimension. This breaks with the legacy of system dynamic modeling, where generally only one dimension of a stock is taken into account (i.e., quantity). Flows are relationships between stocks. Flows can represent material flows, information flows, or other relationships that follow a cause–effect line. Some more recent system descriptions also include actors as an endogenous part of the system.

Stocks, flows, and actors are essential parts of the system. The description of the system is a conceptual model of the real world. In the past 5 years we have drafted such models, applying the SCENE approach at national, provincial, and urban scale levels. Similar descriptions of systems are quite common in sustainability studies but rare in development of sustainability indicators.

Proposing a System Indicator

If sustainable development is interpreted as the balanced long-term development of the three domains of sustainability, then the development of one part of the system toward a desirable state should not occur structurally at the cost of developments elsewhere in the system because this would compromise its continuity and functionality. In this section we propose the Qualitative System Sustainability Index (QSSI) as an indicator for the degree to which the system structure causes such compromises.

We illustrate the QSSI on the basis of a conceptual model that we drafted for a SCENE case study on the transition in Dutch river basin management between 1970 and 2000 (van der Brugge et al. 2004). The model in Figure 11.1 describes the system for the year 1970. We chose this example because it is the smallest model we have drafted in terms of stock (10) and flow (23) numbers. The system properties in terms of connectedness, the relative number of feedback cycles, and cluster formation are comparable with those of other systems we have drafted. We therefore consider this model to be representative for illustration purposes.

The matrix is read from left to right. Each cell in the matrix stands for a potential flow from the stock in the row toward stock in the column of that cell. A flow from flood risk to dams (if there is a threat of floods, more dams are being built) can be found

Cause \ Effect	Flood risk	Costs	Dams	Retention capacity	Space for water	Space for land	Agriculture	Buildings	Potential damage	Nature
Flood risk			1						1	
Costs										
Dams	-1	1			-1	1				-1
Retention capacity	-1									
Space for water					1					1
Space for land							1	1		
Agriculture		-1			-1				1	-1
Buildings		-1			-1				1	-1
Potential damage		1								
Nature					1		-1			

Figure 11.1. Conceptual model for a river basin management system.

in the third cell of the first row and is represented by the value 1. The reverse flow from dams to flood risk (if more dams are built, the flood risk decreases) is represented by the value −1 in the first cell of the third row. As in the notation of system dynamics, positive flows (1) reinforce the original signal, and negative flows (−1) dampen the original signal.

We will not explain the content of the conceptual model in detail here because it serves as an illustration only. However, in order to introduce the reader to the kind of causal thinking represented in this diagram, we will follow some of the causal chains. For example, the size of the retention capacity influences the risk of floods. The risk of floods is an important factor when the number and size of dams to be built are determined. These dams take up land that could be used for agriculture and buildings if there were no dams. Buildings and agriculture produce economic benefits that reduce opportunity costs. The flows connect stocks that are directly related in that there is no other stock in the system through which the flow described would take effect. In other words, the arrows represent only first-order relationships.

Because this description gives rise to different possible interpretations, the background material of each case study contains extensive documentation on the arguments and discussions raised during the drafting of the system, the interpretations considered, and the relevant scientific literature.

If we want to evaluate the sustainability of the system described in our model, we need to define sustainability by making the inherently normative choice of what is desirable. We proposed to let these judgments be formulated in a stakeholder process. This way, the normativity is made explicit rather than being buried implicitly in the framework underlying an indicator or index.

For the purpose of the QSSI this choice is made for every stock separately, independently of the effects of that stock on other stocks. The choice to attribute a low desired state to agriculture therefore cannot be based on the argument that agriculture reduces retention capacity or has a detrimental impact on nature. Because of this criterion of independence of other stocks, the discussion will not be concerned with the choices that should be made between mutually exclusive stocks.

We add the information on the desired direction of the stock to our model in the form of "+" and "–" (Figure 11.2). These symbols stand for the desired state of each stock in terms of direction independent of the other stocks in the system and also independent of the system consequences of that target. A "+" indicates that a high value for that stock is desired, a "–" indicates that a low level is desired (ceteris paribus).

Desired direction		–	–	–	+	+	+	+	+	–	+
		Flood risk	Costs	Dams	Retention capacity	Space for water	Space for land	Agriculture	Buildings	Potential damage	Nature
–	Flood risk				1					1	
–	Costs										
–	Dams	-1	1			-1		1			-1
+	Retention capacity	-1									
+	Space for water					1					1
+	Space for land							1	1		
+	Agriculture		-1			-1				1	-1
+	Buildings		-1			-1				1	-1
–	Potential damage	1									
+	Nature					1		-1			

Figure 11.2. Checking for inconsistencies in the system.

In order to evaluate the long-term continuity and functionality of the system, we tested the consistency of the desired directions and flows connecting the stocks. The consistency check for a flow is passed if the value of the flow ("+1" or "−1") is equal to the desired direction of the originating stock multiplied by the desired direction of the receiving stock of that flow. For example: −1 (flood risk) × −1 (dams) = 1 (the value of the flow from flood risk to dams found in the third cell of the first row). If these signs are not consistent, then the desired direction of a stock is not consistent with the flow driving it. For example, −1 (dams) × −1 (flood risk) ≠ −1 (the value of the flow from dams to flood risk).

An inconsistency implies that if a desired development for one stock is realized, another stock comes under some pressure to develop in an undesired direction. Thus we can pinpoint the system elements the development of which undermines the system elsewhere. The more of these inconsistencies we encounter, the more difficult it will be to steer the system onto what is normatively chosen to be a sustainable trajectory. In our example, 9 out of 23 flows are inconsistent with our sustainability goals. The inconsistencies are shaded in Figure 11.2.

These inconsistencies are not inconsistencies in the policy interventions. They are structural patterns within the system that make a consistent policy strategy for sustainability impossible. Weakening or removing these inconsistencies is a necessary but not sufficient condition for an integrated sustainability strategy. Each inconsistency in itself seems trivial: If more room is provided for nature, this comes at the cost of agriculture. The building of more dams comes at the cost of land for other purposes. The construction of buildings within the river's flood bed increases the potential damage. Summed up, they determine the difficulties of sustainable management of the river basin.

Calculating QSSI

The simplest version of a sustainability indicator related to the model in the previous section is calculated by dividing the number of inconsistencies by the total number of flows, resulting in an index between 0 and 1 (in our example, 9/23 = .39). A lower number implies a higher level of sustainability and vice versa. This indicator is independent of the current state of the stocks or related indicators. Similar to existing sustainability indicators, the QSSI consists of two layers of information: an index (i.e., a single number that summarizes the information contained in the underlying system) and the body of information that in this case consists of a model in matrix form and in more common approaches of lists of indicators. In both cases, a sound interpretation of the index is impossible without an insight into the underlying body of information. By the logic of the QSSI, the number of inconsistencies in the system can initially be decreased in three ways.

- We can try to find a way to make an inconsistent flow disappear (decoupling of stocks). In our example there is an inconsistency as buildings take up retention capacity. Both are desired, and if there was a possibility to build houses above the potential flood water level (e.g., on poles), then the negative flow between the two stocks would disappear. This would result in a QSSI of 8/22 = .36.
- The QSSI is reduced even more if we manage to turn the sign of the flow. If we could stimulate such agriculture that can be applied in retention areas and actually profit from occasional flooding of these areas, then the QSSI in our example would decrease from .39 to 8/23 = .35.
- A third way to decrease the QSSI is to add consistent flows. If we can find a way to give space to water in such a way that agricultural activity is stimulated at no cost for nature development (admittedly a far-fetched example), then we would add a consistent flow. The QSSI would then decrease from .39 to 9/24 = .38.

These interventions for increasing the ability of the system to sustain itself on a desired trajectory in the long run are all concerned with the structure of the system. By changing the system structure (adding, removing, or changing flows) in a desired manner we make the system more sustainable. Considering the enormous resources needed to actually change existing flows in the real world, the QSSI is likely to change very slowly. This reflects the fact that the indicator is concerned with the slowly changing system structure and its direction rather than short-term symptoms in the form of flows.

But not all inconsistencies can be solved this way. At some point no further system improvements are possible. We will have to make choices between different sets of inconsistencies. In our example one of these choices could be to weigh the importance of agriculture and the economic benefits of it against the importance of retention capacity and the potential damage done to the agriculture by floods and nature (all the stocks that agriculture is influencing). It is at this point that the desired direction of a stock dependent on the desired direction of other stocks is discussed. In the Dutch case, policymakers have made a very explicit choice and decided that agriculture should recede in order to provide room for nature (RIVM 1997).

At the cost of the foregone economic gain from agriculture, we resolved our conflicts with the retention capacity, the potential agricultural damage caused by floods, and nature. We thus add one inconsistency and remove three others. This results in a QSSI of 7/23 = .30. The QSSI is reduced through a change in our goals. This choice is not a choice to be made in a sustainability assessment but one that should be delegated to democratic processes.

The ultimate consequence of this is that we can achieve a higher degree of sustainability if we are willing to give up some of our goals and thus redefine what is desired. Traditional lists of indicators can help us make an informed choice about desired states and the relative urgency of different inconsistencies.

QSSI Extensions

We extended the QSSI to accommodate flows of different strength by assigning a value to the flows ranging from 1 = *weak* to 3 = *strong*. The weights of the flows can be deduced from stakeholder participation, expert judgments, and empirical evidence. Figure 11.3 illustrates this.

With weights assigned, the QSSI is calculated as the sum of the weights of the inconsistencies divided by the sum of weights of all flows. In our example, the QSSI is 18/47 = .38. Solving a strong inconsistent flow has a larger impact on the QSSI than removing a weak flow. The QSSI is .33 when a strong inconsistency is removed and .36 when a weak inconsistency is removed. With weights added, possible interventions can also be aimed at the strength of flows, giving room for issues such as eco-efficiency and technological progress that push the frontiers of sustainability without fundamentally changing the system.

In addition, weighted flows and inconsistencies provide arguments to reject inconsistencies based on their significance. If a stock is strongly consistent with one and weakly inconsistent with another stock, then the weak inconsistency is of low priority when it comes to action for sustainability unless the inconsistency itself is highly undesirable. The weighted QSSI implicitly takes this into account. It requires a high con-

Desired direction		Flood risk (−)	Costs (−)	Dams (−)	Retention capacity (+)	Space for water (+)	Space for land (+)	Agriculture (+)	Buildings (+)	Potential damage (−)	Nature (+)
−	Flood risk			3						3	
−	Costs										
−	Dams	-3	3			-3	2				-1
+	Retention capacity	-1									
+	Space for water				2						2
+	Space for land							2	2		
+	Agriculture		-1		-1					1	-2
+	Buildings		-2		-3					2	-2
−	Potential damage		2								
+	Nature				2			-2			

Figure 11.3. Influence matrix with weighted flows.

sistency (thus a high divisor) to compensate for a weak inconsistency, resulting in a low QSSI value.

With the addition of a set of standard stocks and a standard procedure to define flows for certain classes of systems (e.g., industrial sectors or regions) the QSSI could be extended in such a way that it allows comparisons between different systems.

Also, with methods of general network analysis it is possible to derive more system information, such as the character of stocks (active or reactive), and process additional information, such as the speed of change of stocks or the influence of different stakeholders. Any type of relevant information that can be translated into a range (e.g., from low to high), a vector of ranges, or an ordinary scale theoretically can be processed. This leaves plenty of room for refinements of the QSSI.

We based the QSSI on the SCENE approach. The QSSI is only one aspect of our broader aim to inspire methods for the qualitative understanding of systems and thus integrate quantitative and qualitative information (Grosskurth et al. 2004). Any other approach that comprehensively maps system components and the relationships between them can be used as a base. Therefore, the concept is not restricted to the three pillars of sustainability but can be adjusted to accommodate additional domains (e.g., institutional) or entirely different structures (e.g., the driving force, pressure, state, impact, response approach).

Conclusion

We propose the QSSI as a first example of an indicator that is not based primarily on the measurement of flows. The QSSI combines methods from soft and hard system thinking. The QSSI has roots in our earlier work on the SCENE model and work in progress on the method of qualitative system analysis. By presenting the method for this indicator at an early stage, we want to stimulate the discussion on how sustainability indicators might reflect the whole instead of the parts.

The question is whether the QSSI is a useful addition to the existing set of indicators. The QSSI emphasizes the system properties, is suitable for stakeholder participation, and is able to process more qualitative information than any other sustainability indicator. It focuses on the long-term dynamics of a system by addressing the system structure driving the parts rather than the short-term development of individual parts. The QSSI also integrates the three domains of sustainability through its roots in the SCENE approach. In decision making, the strength of the QSSI lies in its indication of interventions for more than cosmetic sustainability. The QSSI will register only such changes in the system that fundamentally change the ability of the system to sustain itself in the long run.

The models are formulated in such a way that any interested layperson can understand and contest them. The structure of the system therefore is open to discussion, and several versions of the system can be developed in parallel. These different representations of the same system make differences of opinion and some of the underlying uncertainties explicit.

Most important, the resulting priorities for sustainability action are easily understood and communicated and quite different from priorities derived from traditional indicator sets. Where a traditional set might point to a loss in biodiversity, the QSSI might point to the trade-off between the building of industrial areas and reserving space for nature (which in turn supports biodiversity as well). In other words, the QSSI points out the problem rather than the symptom.

These advantages come at a cost. A major drawback of the QSSI is its lack of comparability. The QSSI is highly context dependent because the normative judgments of the stakeholders are a crucial element of the QSSI. In addition, the process of deriving the QSSI is costly and requires the commitment of stakeholders. If that commitment is present, the issue of communication is lightened, but the QSSI itself remains difficult to communicate to a broad audience.

Currently, the QSSI is not more than a concept for a new type of sustainability indicator. Much research is still needed in order to develop a sound, solid, and robust sustainability indicator. Future research questions must include its relationship to existing indicators, the normativity of the index, its results' robustness, its susceptibility to different perspectives, and questions of scale and system boundaries, the role of stakeholders, the suitability of different frameworks to develop systems from which the QSSI can be calculated, and the potential use in the policy cycle. Potential extensions of the index could help to address some of these issues.

We are confident that the road we have taken will stimulate a fruitful discussion on alternative types of sustainability indicators and that we are progressing toward an indicator that resolves the challenge of reflecting the whole instead of the parts.

Literature Cited

Dörner, D. 2003. *Die Logik des Misslingens: Strategisches Denken in komplexen Situationen* [The logic of failure]. Hamburg: Rororo.

Forrester, J. W. 1968. *Principles of systems.* Cambridge, MA: Wright-Allen Press.

Grosskurth, J., N. Rijkens, and J. Rotmans. 2004. *Eindrapport Harmonisatie Prognoses POL* [Final report of the POL project on harmonizing forecasts]. Maastricht, the Netherlands: ICIS.

Grosskurth, J., and J. Rotmans. 2005. The SCENE model: Getting a grip on sustainable development in policy making. *Environment, Development and Sustainability* 7(1):133–149.

Jaeger, C., S. Shackley, É. Darier, and C. Waterton. 1997. *ULYSSES: Urban lifestyles, sustainability and integrated environmental assessment—Towards a polylogue on climate change and global modelling.* Darmstadt, Germany: Darmstadt Technical University.

RIVM. 1997. *National environmental outlook 4 1997–2020* [in Dutch]. Bilthoven, the Netherlands: Samsom H.D. Tjeenk Willink.

Rotmans, J., and H. J. M. de Vries. 1997. *Perspectives on global change: The TARGETS approach.* Cambridge: Cambridge University Press.

Spangenberg, J. H., and O. Bonniot. 1998. *Sustainability indicators: A compass on the road towards sustainability.* Wuppertal, Germany: Wuppertal Institute.

Sterman, J. 1994. Learning in and about complex systems. *System Dynamics Review* 10(2–3):291–330.

Van der Brugge, R., J. Rotmans, and D. A. Loorbach. 2004. The transition in Dutch water management. In *Information to support sustainable water management: From local to global levels,* edited by J. G. Timmerman, H. W. A. Behrens, F. Bernardini, D. Daler, P. Ross, K. J. M. Van Ruiten, and R. C. Ward, Proceedings of the Conference "Monitoring Tailor-Made 2004." Alphen aan den Rijn, the Netherlands: Alfabase.

PART IV
System and Sectoral Approaches
Arthur Lyon Dahl

The progress being made in assessing sustainability can be illustrated best through some examples of specific indicators and indices measuring key properties or processes in the human–environment system. These indicators help to define significant dimensions of sustainability more clearly, often in a way that is directly relevant to policy targets or that highlights areas for management action.

Chapter 12, "Indicators of Natural Resource Use and Consumption," shows the usefulness of material flow analysis to give a more systemic view of the operation of the economy. Instead of looking at the flow of money, as in gross domestic product (GDP), material flow analysis measures the physical movement of materials and the resulting mass balances as raw materials, products, and wastes are transferred from place to place. Because certain environmental impacts are strongly linked to natural resource extraction and waste disposal, this analysis and the indicators it generates are useful to assess the environmental impacts of economic activities. By looking at domestic extraction rates and trade balances in physical terms, it can illustrate the extent to which a country may be exploiting its natural resources unsustainably or exporting its environmental burden. Because most materials become wastes sooner or later, it can illustrate the waste potential created by certain activities and facilitate waste management or reduction. It can also account for materials accumulated in stocks and infrastructure. The right material flow indicators can measure efforts to decouple economic progress from material consumption, with increased resource productivity and recycling and the dematerialization of the economy.

In Chapter 13, indicators that measure the decoupling of environmental pressures from the relevant economic driving forces are further explained. By tracking temporal changes, these indicators aim to reduce environmental pressures and to encourage decoupling from specific environmental impacts. Although relative decoupling can signal a desirable trend, absolute decoupling is usually necessary to achieve sustainability. Because decoupling indicators tend to change over shorter time periods than environmental state indicators, they can be particularly useful for assessing policy effectiveness. By linking such indicators to decoupling targets, we can produce performance standards for products.

Taking this approach a step further, Chapter 14 proposes a Geobiosphere Load Index. This index combines material flow analysis, energy flow accounting, and a modified form of the ecological footprint concept, including indicators of input, output, and consumption, to assess human pressures on the environmental capacities of materials, energy, and land and thus on related ecosystem services. Such an index could be calculated per square kilometer, per capita, or per unit of GDP for resource efficiency. Although this is still work in progress, it provides a simple, easily understood measure of fundamental dimensions of sustainability.

As one example of the use of indicators in a specific field, the health sector demonstrates how indicators are playing an increasingly important role in communicating complex health and environment information. Chapter 15 reviews the use of health and environment indicators in support of sustainable development, emphasizing indicators of the linkages between environment and health at the national, sectoral, and community or neighborhood levels. Early attempts to simplify information for policymakers and the public included a driving force, pressure, state, exposure, health effect, action framework, but this was found to be too linear and oversimplified and thus potentially misleading. Core indicators with harmonized and rationalized methods are being developed to lessen reporting burdens and to facilitate between-country comparisons. The World Health Organization has developed indicators and targets to measure policy and program implementation and to support the World Health Report. Environmental health indicators are particularly useful in sustainable development planning and in supporting regional health information systems. A recent focus is on children's environmental health indicators in a more flexible model that takes into account multiple exposures. All such efforts face limitations in the available data, necessitating further development and harmonization of data collection and processing. The key future challenges for health and environment indicators include scale, capacity, data comparability, and reliability.

For another sectoral approach, Chapter 16 provides a comprehensive review of the status of biodiversity indicators, including the basic concepts, current developments, and future possibilities. Responding to the international target to reduce biodiversity loss and the need to measure the effectiveness of measures adopted under the Convention on Biological Diversity, recent indicator work has generated a certain number of trial indicators in a field suffering from inadequate data, great complexity, and poor understanding of the causal links between biodiversity and ecosystem services. The present conceptual framework combines three levels of organization (ecosystem, organism, and gene) with three aspects (composition, structure, and function). There are still problems in establishing a baseline for biodiversity loss, measuring richness and evenness, and finding proxies for immediate use where the information is inadequate. Although rapid progress is being made in studying biodiversity at the genetic level, it will be some time before indicators can be developed at this level, complementing the present species-based indicators.

Among the most common biodiversity indicators in use, species richness is the most widespread, but like endemism it is not good for measuring loss. Species extinction is an easy indicator to understand but difficult to prove in practice. Population abundance–based indicators in theory can give the necessary early warning of biodiversity loss, but they generally lack the time series data to do so. Area-based indicators, such as the area of ecosystems and the extent of fragmentation, can address abundance at the ecosystem level but suffer from methodological problems that produce inconsistent data and an absence of long-term time series. Chapter 16 describes other still embryonic approaches, including phylogenetic and evolutionary indicators, functional indicators, integrity indicators, and various composite indices. Where there is no information on biodiversity trends, indicators of pressures that are correlated with biodiversity loss can be used instead, as demonstrated in Chapter 17. For policy-relevant indicators, there is a mismatch between the available information primarily on species composition and the policy concern to conserve the functional attributes of ecosystems.

The chapter addresses data issues by recommending a more pragmatic approach, focusing on well-known groups and using qualitative and informal as well as quantitative information. The immediate need is for land cover and use, the distribution of plants and vertebrates, trends in the populations of key species, genetic diversity within domesticated species, and the impacts of land use on species. Significant gaps on which research should focus include the functional relationships between biodiversity and ecosystem services, predictors of the consequences of human activities, genetic relatedness and redundancy, maps of land use and species distributions, historical ecology to provide baselines, and the establishment of biodiversity observation and assessment systems.

One example of an aggregate process-oriented indicator highly relevant to sustainability is the human appropriation of net primary production and the pressure it exerts on natural biodiversity, as described in Chapter 17. This pressure indicator is defined as the alteration in primary productivity from human land use (the productivity prevented) plus the human extraction of net primary productivity (the biomass harvested), giving the difference between the potential vegetation and the part remaining after harvest. It serves as an aggregate indicator of human-induced changes in ecosystem processes and therefore of human domination of terrestrial ecosystems. Its significance for sustainability is shown by the estimate that 40 percent of global terrestrial primary productivity is appropriated for human uses. The indicator is linearly correlated with the natural state of landscapes and shows a high correlation with species richness, making it a good pressure indicator for biodiversity loss. It is also relevant for carbon stocks and flows in ecosystems and can be related to economic activities, consumption, and GDP. The human appropriation of net primary production is thus an excellent candidate for a set of sustainability indicators.

12

Indicators of Natural Resource Use and Consumption

Nina Eisenmenger, Marina Fischer-Kowalski, and Helga Weisz

The limits to growth debate in the 1970s was concerned with exponential growth rates of gross domestic product (GDP) and population, which were seen as major drivers of resource use and waste production (Meadows et al. 1972; Hardin 1968; Ehrlich and Holdren 1971). The main environmental concern was resource scarcity of nonrenewable raw materials and environmental and health damage through growing amounts of toxic wastes and emissions. In the 1980s, when earlier expectations about the exhaustion of natural resources did not materialize, environmental concerns focused on the output side of the social metabolism, particularly on pollution. Finally in the 1990s, the notion of sustainability became the leading environmental discourse supporting a conceptual shift (World Commission on Environment and Development 1987). The focus moved from the output side of the production system to an integrated understanding of the biophysical dimension of the economy (Munasinghe and McNeely 1995; Cleveland and Ruth 1997). One important idea emerging from the sustainability concept was that it is not the growth of the monetary economy (measured in GDP) but the growth of the physical economy that causes environmental burdens. With this, new policy targets aiming at a decoupling of the monetary and the physical economy became popular. Evidently, information on the state of the environment is not sufficient to support such policies. What is needed are environmental information systems that are conceptually linked to socioeconomic information systems, above all to the system of national accounts, and allow the compilation of pressure indictors (Eurostat 1999; UN et al. 2003). The emergence of material flow accounting (MFA) has to be seen in this context (Fischer-Kowalski and Hüttler 1998).

MFA dates back to the 1960s (Ayres and Kneese 1969; Gofman et al. 1974; Wolman 1965) and reappeared in the early 1990s (Steurer 1992; Bringezu 1993; Fischer-

Kowalski et al. 1994; Japan Environment Agency 1992), when the notion of sustainable development became the new leading paradigm of environmental policy. In the late 1990s the World Resources Institute coordinated the first comparative material flow studies (Adriaanse et al. 1997; Matthews et al. 2000). While several European countries started to include MFA reporting in their environmental statistics, the statistical office of the European Union undertook a concerted effort of harmonization, leading to a methodology guide (Eurostat 2001a). Eurostat and the European Environment Agency (EEA) also initiated the establishment of harmonized data compendia for the European countries (Eurostat 2002; ETC-WMF 2003; Weisz et al. 2005b).

Currently, processes are ongoing on the EU and Organization for Economic Cooperation and Development (OECD) levels that focus on the development of policies for sustainable resource use (Commission of the European Communities 2005; OECD 2004a). This in turn has enforced further methodological development and implementation of MFA.

One of the strengths of MFA is its systemic approach and the consistent application of the mass balance principle (i.e., material inputs equal material outputs minus stock increases). MFA provides a biophysical account of the level of national economies in analogy to economic accounting (GDP). (For an early conceptualization of the relation of MFAs and economic accounting, see Ayres and Kneese 1969.)

MFA keeps track of all materials that enter and leave the economy within 1 year. These flows comprise extracted or imported materials to be used within the national economy and all material released to the environment as wastes and emissions, exported to other economies, or added to societal stocks (Figure 12.1). The term *used* refers to acquiring value within the economic system (Eurostat 2001a).

In MFA the economy is usually treated as a black box (except for net additions to societal stocks, which are accounted for to close the mass balance equation). The boundary of the physical economy is defined in a fashion as compatible as possible with the system of national accounts (SNA) in order to facilitate integrated monetary and biophysical analysis (Eurostat 2001a). A detailed MFA database normally comprises flow data for several hundred different input materials. Material flow data are based on statistical data from either international or national statistics. For material categories where no statistical data are available, estimation methods were developed, such as for grazing and straw or for construction minerals (Eurostat 2002; Weisz et al. 2005b). MFA thus provides a comprehensive description of the physical economy in the form of a consistent database. From this, various highly aggregated national material flow indicators can be derived.

In MFA, the unit of measurement is metric tons. Arguably, the choice of one simple unit of measurement has disadvantages. In particular, mass units are insensitive to the quality of the materials (i.e., their specific environmental impact). We will return to this question later on. However, mass as an accounting unit has the clear advantage

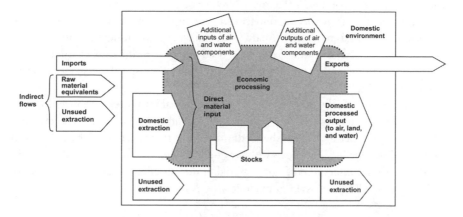

Figure 12.1. General scheme of material flow accounting (Matthews et al. 2000, modified according to Eurostat 2001a).

of being a common measure across countries and over time. Tons are not subject to fluctuating exchange rates and relative prices, as is money value, nor to competing expert opinions, as is the assessment of specific environmental impacts. The use of tons as a unit of measurement in MFA (and also in other physical accounting tools) is thus a consequence of its feasibility, transparency, and stability. It also reflects Herman Daly's (1973) argument that it is the scale of material throughput that exerts pressures on the environment (Fischer-Kowalski 1998).

Environmental Relevance of MFA and Implications for Policymakers

The introduction of the new framework of sustainable development in the Brundtland report (World Commission on Environment and Development 1987) focused policy interest on material management policies and accounts and has increased the importance of MFA. Before looking at MFA indicators in detail, we would like to look at how MFA can be linked to different policies.

In general, environmental impacts of material flows result from the specific impact (per ton) multiplied by the volume of the flow. Therefore, the environmental impact of a small flow of a hazardous substance may be of the same order of magnitude as that of a high-volume flow of a substance with low toxicity. Materials lying outside this range of similarly high environmental impact (outside the ellipse in Figure 12.2) are of lesser interest or do not exist (Steurer 1998). This graph is a rough illustration stressing that materials of little specific impact can pose significant environmental pressure because of their total mass used. Apart from that, the environmental impact as the result of both the total amount of resources used and the specific impact still can vary significantly and depends on other conditions, such as the specific ecosystem taken from or released to.

The socioeconomically used materials can be grouped into three categories (Steurer 1998): the high-volume or bulk flow group, such as sand and gravel, with very low environmental impact per mass unit; the medium-volume flows such as timber, steel, and cement; and the small-volume flows with very high environmental impact per mass unit. To address those different types of material flows, specific policy instruments are needed, which will be discussed in the following paragraphs.

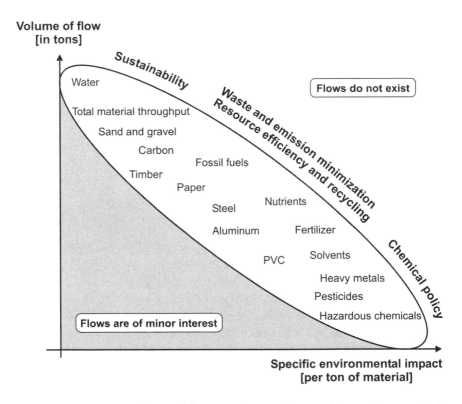

Figure 12.2. Grouping of material flows according to volume and impact (Steurer 1998).

Bulk Material Flows

The main sustainability problems resulting from extraction and use of bulk material flows such as sand and gravel but also biomass from grassland are loss of biodiversity, sealing of land area, disruption of ecosystems, and resource depletion caused by over-exploitation of resources (Fischer-Kowalski and Hüttler 1998). These problems are a result mainly of the amount of materials extracted and used, not of the composition of or substances included in these materials. Sustainability in relation to bulk material flows therefore asks for a reduction in the total material throughput or in the scale of the economy, as argued by Daly (1992). Policy instruments that address high-volume flows include taxes and voluntary agreements (Steurer 1998). For high-volume flows, more general sustainability considerations apply (Steurer 1998), and they focus on the overall reduction of the material throughput. The key interest in this area is dematerialization (i.e., decoupling of economic growth from material use, leading to an increase of material efficiency). In the 1990s these issues were most prominently addressed by Weizsäcker et al. (1995), who asked for a reduction of overall material use by a factor of four, and Schmidt-Bleek (1994), demanding a reduction by a factor of ten.

Medium-Volume Material Flows

Concerning medium-volume material flows such as iron ores and timber, policy focus is on material management, leading to an increase of material efficiency and waste minimization (Steurer 1998). Life cycle assessments are moving to the fore with recycling and reuse strategies and the reduction of wastes and emissions as actions toward sustainable development (Fischer-Kowalski and Hüttler 1998).

Small-Volume Flows with High Environmental Impact per Mass Unit

In this category, environmental problems arise from the environmental impact of specific substances with high toxicity for human health and the functioning of ecosystems. The control of these materials and substances therefore is politically important (Steurer 1998). In the discourse about environmental problems, this category was, and for decision makers still is, the predominant environmental issue. To reduce the use of hazardous substances Steurer (1998) proposes policies such as bans or phase-out protocols.

Analyses of environmental impact are based on information provided by substance flow analysis (SFA), which focuses on single substances such as nitrogen or phosphorus and their environmental impacts. MFA, on the other side, offers an overall accounting framework to which SFA can be linked and through which SFA is thereby linked to the whole economic process.

MFA Indicators and Their Relevance to Sustainability

A number of flow aggregates expressed in metric tons per year can be derived from the MFA framework for a country or a national economy (Table 12.1).[1]

By using different denominators, these (extensive) flow aggregates can be related to certain reference scales, resulting in a set of intensive variables that allow comparisons of socioeconomic systems of different scales in relation to relevant dimensions. The most important reference scales are population, indicating the material intensity of a society; size of territory, indicating the intensity of use of a given area; and size of the economy, resulting in an indicator of material efficiency (Weisz et al. 2005a). For some purposes, it is also useful to relate aggregate flows to one another (dimensionless indicators).

Table 12.1. Overview of material flow analysis indicators.

Indicator		Definition
DE	Domestic extraction	DE comprises the raw materials domestically extracted that enter the economy for further use in production or consumption processes (excluding water and air).
Im and Ex	Imports and exports	All materials and goods imported or exported with the weight they have at the time they cross the border.
DMI	Direct material input	DMI = DE + Im. DMI covers all materials entering the socioeconomic system for further processing or final consumption.
DMC	Direct material consumption	DMC = DE + Im − Ex. DMC measures all materials that are used within the observed economy for processing or final consumption.
PTB	Physical trade balance	PTB = Im − Ex. PTB is a measure of net imports or net exports.
DPO	Domestic processed output	DPO comprises all wastes and emissions released to the domestic environment.
NAS	Net addition to stocks	NAS = DMI − DPO. NAS measures the physical growth of an economy.
RME	Raw material equivalent	RME comprises all material extracted and used to produce traded goods (semimanufactured or finished).

Source: Eurostat (2001a).

Highly aggregated MFA indicators provide a rough overview of specific resource use patterns but cannot tell us anything about the environmental impacts. However, disaggregated MFA indicators on a level of detail that is "consistent and meaningful in terms of physical and chemical properties, economic use, and environmental pressure associated with the primary production of the materials" (Weisz et al. 2005a) provide enough information for an analysis of driving forces and a development toward a sustainable use of resources "without impairing the strength of MFA in providing an overall picture of the economy-wide material flows" (Weisz et al. 2005a).

Domestic Extraction (DE) of Raw Materials

The aggregate flow DE covers the annual amount of raw materials, apart from water and air, extracted from the national territory in order to be used as material factor inputs to economic processing. The term *used* refers to acquiring value within the economic system. "Inputs from the environment refer to the extraction or movement of natural materials on purpose and by humans or human controlled means of technology (i.e., involving labour). . . . Unused flows are materials that are extracted from the environment without the intention of using them" (Eurostat 2001a:17).

Domestic extraction consists of biomass,[2] fossil fuels, industrial minerals and gross ores, and construction minerals.[3] DE on the national level indicates resource depletion within a country's territory. This depletion may be temporary (as with agricultural biomass) or long term (as with the destruction of primary forest or the extraction of minerals and ores). This relates to intergenerational equity as a sustainability issue. On a global scale, DE reflects the overall scale of human activity on this planet. Comparing DE between countries indicates each country's share in the global raw material extraction. When differentiating by class of materials, DE relates to specific scarcities. DE of plant biomass, for example, can be related to global net primary production of biomass, DE of fossil fuels, minerals, and ores to known reserves.

DE per square kilometer indicates environmental pressure on the domestic natural environment in relation to its size and the intensity of its human use. The specific environmental pressure may consist of competitive pressure on other species and loss of biodiversity. Most fractions of DE somehow relate to area: Biomass is clearly area dependent; construction materials (always used in close proximity to their extraction) are used for sealing land and regulating water bodies, which competes with biotic uses of land; and fossil fuels enrich the atmospheric carbon cycle unless neutralized by the growth of green plants, which need area.

Imports and Exports of Materials and Goods

Within the MFA framework, import and export flows are accounted for by their weight at the time the material or product crosses the border of a nation state. In the country of origin, there usually occur additional material flows for production of the traded goods that are not part of the tonnage of imports and exports, called indirect flows (Eurostat 2001a) (Figure 12.1). If a country produces a commodity for domestic consumption internally, this typically generates more material flows within this country than importing the same commodity, and vice versa with exports. The global division of labor within the world economy enforces these processes of production chains spread over several countries. In MFA this creates a discrepancy between the materials accounted for as used within a country (domestic material consumption) and the overall amount of resources this socioeconomic system is based on. This will be discussed in the section on raw material equivalents (RMEs).

Relating biophysical imports and exports to DE shows the dependency of the observed socioeconomic system on trade. A high ratio of imports to DE expresses a high dependence on foreign natural resources and a tendency to outsource the environmental burdens of resource extraction and early stages of processing. The ratio of exports to DE indicates the role of a supplier of raw materials in the international division of labor. A high ratio of exports to DE describes a country that exploits its natural resources for final consumption elsewhere. Comparing the economic value of imported weights with that of exported weights appears to be an interesting measure for the economic value attributed to biophysical imports and exports. If the economic value of imported tons is much higher than the one of exported tons, a country loses natural resources without gaining real economic profits from exports. The other way around, a country with low economic values of imports and high values for biophysical exports gains high economic profits by trade without losing much natural resources.

Direct Material Input (DMI)

DMI comprises all materials that enter a socioeconomic production process. It encompasses DE and imports. In many studies, DMI used to be the main indicator for material use. Upon further analysis, however, it proved to be difficult to interpret as soon as international trade becomes a substantial component of material input. On one hand, a shift from producing domestically toward importing commodities reduces DMI in a country (because all wastes generated during extraction and production are externalized from its territory). This reduction is not real dematerialization but rather externalization of impact. On the other hand, countries that extract raw materials for export (particularly metal ores or highly processed biomass such as meat) tend to have very high DMI without a corresponding domestic material consumption level.

Relating DMI to GDP provides useful information on the material needs of the national economy. It is one of the most common indicators of the dematerialization of an economy. Mathematically, this indicator can also be formulated as material productivity of the economy (GDP/DMI).

Domestic Material Consumption (DMC)

DMC measures the annual amount of raw materials extracted from the domestic territory (DE) plus all imports minus all exports. The definition of DMC thus corresponds to apparent consumption, not final consumption. It is important to keep this definition in mind when interpreting DMC. From the point of view of final consumption, an imported commodity is functionally equivalent to a domestically produced commodity. In DMC, though, these functional equivalents lead to great differences. Similarly, producing a commodity for export is intuitively unrelated to the domestic consumption of materials, but according to the makeup of DMC, all wastes occurring in the course of this production are a component of domestic consumption. Thus, DMC can be better interpreted as domestic waste potential and refers to all materials used and consumed in both production and consumption processes (Weisz et al. 2005a). This national waste potential will either add to environmental pressure within the national territory (immediately or some time in the future) or create an international environmental pressure attributed to the responsibility of the country (in the case of CO_2 emissions).

Physical Trade Balance (PTB)

PTB equals physical imports minus physical exports. The physical trade balance thus is opposite to the monetary trade balance (exports minus imports) because in economies money and goods move in opposite directions (Eurostat 2001a). A physical trade surplus indicates a net import of materials, whereas a physical trade deficit indicates a net export.

Domestic Processed Output (DPO)

DPO includes all wastes and emissions from economic processing and final consumption. Accounting systems that calculate wastes and emissions existed before MFA was established, and usually these statistics provide good data for single materials or substances. What MFA can add to these existing databases is an overall accounting framework that includes all outflows to our domestic environment. Waste and emission policies can lead to a decrease of specific substances but simultaneously cause an increase of other wastes or emissions. Such trade-offs can become visible through an accounting framework provided by MFA that calculates all outflows to the natural environment.

Net Addition to Stocks (NAS)

NAS equals all materials added to stocks (i.e., with an expected commodity life span of more than a year) minus wastes and emissions from stocks. NAS represents the net growth rates of the material stocks of a society. NAS is highly relevant to sustainable development for the following three reasons:

- Material stocks comprise mainly built-up infrastructure such as roads. Increasing NAS therefore indicates an increase in land sealing.
- All material stocks bind future material inputs because the stocks have to be maintained and reproduced.
- In the future stocks will turn into wastes and emissions when they are disposed of. Therefore stocks have to be regarded as future waste flows and burdens to future generations.

Unfortunately, the present level of methodological sophistication in MFA hardly allows us to generate reliable indicators for NAS. To calculate NAS other than as a simple statistical difference between inputs and outputs, we need estimates for existing stocks and their life span so that annual wastes and emissions from stocks can be calculated, precise estimates for balancing items between inputs and outputs (additional air and water vapor; see Figure 12.1) so as to not statistically lump them together with stock changes, and above-average-quality data for the input of construction minerals.[4] Some efforts have been made to improve this situation (see Barbiero et al. 2003), and in light of the importance of this indicator for sustainability, further efforts seem well warranted.

Raw Material Equivalent (RME)

As explained earlier, it is not trivial to define, on the national level, indicators for natural resource use and consumption that would not be distorted by differences or changes in international trade. One way of consistently dealing with imports, exports, and domestic extraction on an equal footing has been proposed by Eurostat as RME.[5] RME equals the upstream requirements of used raw materials (i.e., used extraction) of the imported and exported products[6] (Eurostat 2001a). A proper calculation of RME requires an integration of material flow analyses and input–output analyses. Weisz (2006) provides a methodological suggestion for how this can be achieved. Similarly to what already exists in the field of energy requirement, international databases would have to be extended and adapted in a harmonized way to facilitate comparability of calculations.

RME of imports indicates the material extraction undertaken in other territories in order to produce and transport the commodities imported by the country observed. RME of exports expresses the part of DE and imports that is due to the provision of raw materials and commodities for export. Using RME, a new indicator called raw material domestic consumption (RMC) could be calculated (DE + RMEimp – RMEexp). Such an indicator definitely would measure the raw material requirements of domestic final consumption.

Likewise, an indicator called raw material trade balance (RTB), defined as RMEimp – RMEexp, could measure ecological terms of trade related to the use of materials standardized to a single system boundary definition (i.e., to the definition of used extraction). A positive RTB would indicate the extent to which the domestic final consumption of an economy relies on the imports of raw materials. A negative RTB, on the other hand, would indicate the amount of raw materials that have been extracted from the domestic environment but are used to satisfy final consumption elsewhere, a phenomenon analogous to carbon leakage.

Application and Policy Integration

In the last decades several national material flow accounts have been conducted, most of them for industrialized countries such as the United States (Adriaanse et al. 1997), Japan (Adriaanse et al. 1997), or the fifteen European Union member states (Weisz et al. 2005b). Also, some material flow accounts were conducted for developing countries, such as Chile (Giljum 2004), Brazil (Machado 2001), Venezuela (Castellano 2000), Philippines (Rapera 2004), Thailand (Weisz et al. 2005c), and Laos (Schandl et al. 2005). For a detailed list, see Weisz et al. (2005a, 2005b). Most of these studies provide data and detailed analysis on the material input side, including material exports. Fewer studies deal with material outputs (Matthews et al. 2000; Eurostat 2001b; ETC-WMF 2003). Because policies that specifically target the use of resources have been lacking, policy application of indicators for resource use has been very limited. In recent years, however, policies that aim at a sustainable use of resources have entered the political agenda in Japan, the EU, and the OECD.

Japan clearly is the international forerunner in sustainable resource policy. In 2003 the Japanese government enacted the Basic Law for Establishing a Sound Material-Cycle Society. It includes two laws on waste management and public cleansing and promotion of the use of recyclable resources (OECD 2004a). As the title indicates, the Japanese policy focuses on reduction of consumption of natural resources through the enforcement of recycling and reuse to reduce the environmental load. MFA is specifically used as an accounting framework from which indicators such as DMI, DPO, and material use efficiency can be derived. The Japanese government set three quantitative sustainability targets for the period 2000–2010 and focused on the containment of material flows:

- To increase resource productivity of the Japanese economy by 40 percent. With respect to indicators, this is defined as GDP/DMI.
- To increase the cyclical use rate, which is the amount of recycled materials per DMI, by 40 percent.
- To reduce wastes deposited (a problem particularly grave for a densely populated country such as Japan) by 50 percent. This will be monitored by conventional waste statistics.

Although many countries in the past decade sought to include dematerialization in their sustainability programs one way or another, we do not know of a country other than Japan where major policies were so clearly directed at using indicators from MFA for sustainability targets and monitoring. However, it has to be stressed that material productivity (GDP/DMI or GDP/DMC), like material intensity (DMI/GDP or DMC/GDP), describes a relative condition. Increasing material productivity (like decreasing material intensity) can be gained through either relative decoupling (i.e., material use grows slower than economic growth) or absolute decoupling (i.e., material use declines). Only the latter indicates a real reduction of materials used. Besides, because DMI and DMC cover only materials domestically used, dematerialization can easily be gained by shifting from producing to importing commodities.[7]

In the EU the final version of a strategy on the sustainable use of natural resources was launched by the European Commission in December 2005 (Commission of the European Communities 2005). The focus of the communication is on understanding and mapping the links between the use of resources and their environmental impacts, improving the knowledge base, and developing tools to monitor and report progress.

The European Commission states that the overarching goal of the strategy is to "reduce the negative environmental impacts generated by the use of natural resources in a growing economy" (Commission of the European Communities 2005:5). Contrary to the Japanese government, the EC did not specify quantitative targets, did not focus on a reduction of resource use, and did not suggest specific measures.

The essential elements of the EC formulation are that the strategy assumes that improving the state of the environment and facilitating economic growth are not competing goals and that the environmental impacts of resource use are stressed, not the use of resources themselves. Both assumptions are not unambiguously shared in the scientific community and pose quite a few challenges for the design of resource use indicators.

In particular, concentrating on the environmental impacts of resource use may be difficult to accomplish. Attempts have been made to weight material flows according to environmental impact, but they have not been fully satisfactory. A recent study carried out for the European Commission used Life Cycle Analysis (LCA) factors to weight material flows and to create a highly aggregated index for the environmental impacts of material flows (van der Voet et al. 2005). This study has been criticized for lack of standardization. Using exergy (i.e., workable energy; see Wall 1977) as a means to weight material flows according to their reactivity with the earth's crust, the ocean, or the air (Ayres et al. 2004) is more promising regarding standardized methods and conceptual clarity. However, because it is a very abstract concept from thermodynamics, the meaning and relevance of exergy may be difficult to communicate to a wider public. In a recent article we suggested using MFA indicators at a more disaggregated level in order to come up with the conflicting requirements of standardization, transparency, linkage to SNA, and environmental specificity (Weisz et al. 2005a).

Gaps in Knowledge and Further Research Needs

MFA offers an accounting framework, which provides an overall picture of the biophysical structure of an economy and thus allows the linking of socioeconomic processes to biophysical units and consequently to natural systems.

Although it is a young accounting tool, MFA has reached a high degree of methodological standardization and implementation into official statistics in the EU, Japan, and increasingly in the OECD. Additional research is needed. Several lines are already visible and will be followed in the coming years. These include the following:

- The application and improvement of calculations of NAS and RME. In developing these two indicators MFA will be opened toward discussions of the relationship between stocks and material flows and the global responsibility for material flows related to domestic final consumption.
- The disaggregation of MFA along economic sectors to improve the link of resource use to economic activities.
- The provision of physical make and use tables and in a standardized format.
- The establishment of links between the various physical accounting tools and MFA to foster integrated analysis (see Haberl et al. 2004; Krausmann et al. 2004).

Moreover, in a wider context of sustainability MFA must be linked to other issues such as transport and time use, sustainable consumption, material standard of living, and quality of life.

Notes

1. Typically, these flow aggregates are generated not just for one period but in time series. Calculating time series of flows contributes substantially to data reliability because both measurement errors and stochastic variations between periods come into view.

2. Livestock is considered part of the economic processing system, so livestock inputs, whether fed by humans or consumed directly, belong to DE, whereas livestock animals' body mass in the case of consumption is considered an internal transfer within the socioeconomic system. The body mass of wild animals, if used by a socioeconomic system, is accounted for as DE.

3. If these raw materials contain a certain amount of water or air, their weight is included and usually has to be balanced in the course of industrial transformation to wastes and emissions (see Figure 12.1).

4. Stocks consist mainly of built infrastructure. In most countries, the use of construction minerals is not very well documented (Weisz et al. 2005b).

5. For similar purposes, the Wuppertal Institute has introduced the indicator total material requirement (TMR). TMR includes mainly unused flows associated with domestically extracted and imported materials (e.g., overburden in mining or estimates of soil erosion) (Bringezu et al. 2003), whereas raw material equivalents are rarely accounted for in TMR (see Bringezu et al. 2001; Pedersen 2002; Eurostat 2001b). Despite many efforts to introduce more precision and reliability into the estimates, both conceptual consistency and measurement quality of TMR do not come close to the indicators discussed in this chapter.

6. The Eurostat methodological guide is not very specific about whether RME should include or exclude the weight of the imported and exported commodities. We suggest defining RME as the upstream requirements of used extraction, which means that the weight of the imported (or exported) goods is included. This definition is easier to apply methodologically, and it is also more meaningful from a conceptual point of view (see Weisz 2006).

7. Japan uses this indicator as one of the core indicators within its official sustainability program "Towards Establishing a Sound Material-Cycle Society." Being confronted with this program, Japanese industry argued that following government targets may mean a loss of industrial capacity and jobs in Japan (Y. Moriguchi, personal communication, 2003).

Literature Cited

Adriaanse, A., S. Bringezu, A. Hammond, Y. Moriguchi, E. Rodenburg, D. Rogich, and H. Schütz. 1997. *Resource flows: The material basis of industrial economies.* Washington, DC: World Resources Institute.

Ayres, R. U., L. W. Ayres, and B. Warr. 2004. Is the U.S. economy dematerializing? Main indicators and drivers. Pp. 57–93 in *Economics of industrial ecology. Materials, structural change, and spatial scales,* edited by J. C. J. M. van den Bergh and A. M. Janssen. Cambridge, MA: MIT Press.

Ayres, R. U., and A. V. Kneese. 1969. Production, consumption and externalities. *American Economic Review* 59(3):282–297.

Barbiero, G., S. Camponeschi, A. Femia, G. Greca, A. Tudini, and M. Vannozzi. 2003. *1980–1998 material-input–based indicators time series and 1997 material balances of the Italian economy.* Rome: ISTAT.

Bringezu, S. 1993. Towards increasing resource productivity: How to measure the total material consumption of regional or national economies. *Fresenius Environmental Bulletin* 2(8):437–442.

Bringezu, S., H. Schütz, and P. Bosch. 2001. *Total material requirement of the European Union.* Technical Part. Copenhagen: European Environmental Agency, 5–61.

Bringezu, S., H. Schütz, and S. Moll. 2003. Rationale for and interpretation of economy-wide materials flow analysis and derived indicators. *Journal of Industrial Ecology* 7(2):43–64.

Castellano, H. 2000. *Material flow analysis in Venezuela. National level.* Caracas: Centre for Environmental Studies; Universidad Central de Venezuela.

Cleveland, C. J., and M. Ruth. 1997. When, where and by how much do biophysical limits constrain the economic process? A survey of Nicholas Georgescu-Roegen's contribution to ecological economics. *Ecological Economics* 22(3):203–224.

Commission of the European Communities. 2005. Communication from the Commission to the Council, the European Parliament, the European Economic and

Social Committee and the Committee of the Regions. *Thematic strategy on the sustainable use of natural resources.* Brussels: COM(2005) 670 final.

Daly, H. E. 1973. *Toward a steady-state economy.* San Francisco: W.H. Freeman.

Daly, H. E. 1992. Allocation, distribution, and scale: Towards an economics that is efficient, just, and sustainable. *Ecological Economics* 6:185–193.

Ehrlich, P. R., and J. P. Holdren. 1971. Impact of population growth. *Science* 171:1212–1217.

ETC-WMF. 2003. *Zero study: Resource use in European countries. An estimate of materials and waste streams in the community, including imports and exports using the instrument of material flow analysis.* Copenhagen: European Topic Centre on Waste and Material Flows (ETC-WMF).

Eurostat. 1999. *Towards environmental pressure indicators for the EU.* Luxembourg: Office for Official Publications of the European Communities.

Eurostat. 2001a. *Economy-wide material flow accounts and derived indicators. A methodological guide.* Luxembourg: Office for Official Publications of the European Communities, 1–92.

Eurostat. 2001b. *Material use indicators for the European Union, 1980–1997. Economy-wide material flow accounts and balances and derived indicators of resource use.* Working Paper no. 2/2001/B/2. Luxembourg: European Commission/Eurostat.

Eurostat. 2002. *Material use in the European Union 1980–2000. Indicators and analysis.* Luxembourg: Eurostat, Office for Official Publications of the European Communities.

Fischer-Kowalski, M. 1998. Society's metabolism. The intellectual history of material flow analysis, Part I, 1860–1970. *Journal of Industrial Ecology* 2(1):61–78.

Fischer-Kowalski, M., H. Haberl, and H. Payer. 1994. A plethora of paradigms: Outlining and information system on physical exchanges between the economy and nature. Pp. 337–360 in *Industrial metabolism: Restructuring for sustainable development,* edited by R. U. Ayres and U. E. Simonis. Tokyo: United Nations University Press.

Fischer-Kowalski, M., and W. Hüttler. 1998. Society's metabolism. The intellectual history of material flow analysis, Part II: 1970–1998. *Journal of Industrial Ecology* 2(4):107–137.

Giljum, S. 2004. Trade, material flows and economic development in the South: The example of Chile. *Journal of Industrial Ecology* 8(1–2):241–261.

Gofman, K., M. Lemeschew, and N. Reimers. 1974. Die Ökonomie der Naturnutzung: Aufgaben einer neuen Wissenschaft [original in Russian]. *Nauka i shisn* 6.

Haberl, H., M. Fischer-Kowalski, F. Krausmann, H. Weisz, and V. Winiwarter. 2004. Progress towards sustainability? What the conceptual framework of material and energy flow accounting (MEFA) can offer. *Land Use Policy* 21(3):199–213.

Hardin, G. 1968. The tragedy of the commons. *Science* 162:1243–1248.

Japan Environment Agency. 1992. *Quality of the environment in Japan 1992.* Tokyo: Japan Environment Association.

Krausmann, F., H. Haberl, K.-H. Erb, and M. Wackernagel. 2004. Resource flows and land use in Austria 1950–2000: Using the MEFA framework to monitor society–nature interaction for sustainability. *Land Use Policy* 21(3):215–230.

Machado, J. A. 2001. *Material flow analysis in Brazil.* Unpublished.

Matthews, E., C. Amann, M. Fischer-Kowalski, S. Bringezu, W. Hüttler, R. Kleijn, Y. Moriguchi, C. Ottke, E. Rodenburg, D. Rogich, H. Schandl, H. Schütz, E. van der Voet, and H. Weisz. 2000. *The weight of nations: Material outflows from industrial economies.* Washington, DC: World Resources Institute.

Meadows, D. L., D. H. Meadows, and J. Randers. 1972. *The limits to growth.* New York: Universe Books.

Munasinghe, M., and J. A. McNeely. 1995. Key concepts and terminology of sustainable development. Pp. 19–56 in *Defining and measuring sustainability: The biogeophysical foundations,* edited by M. Munasinghe and W. Shearer. Washington, DC: United Nations University and World Bank.

OECD. 2004a. *Material flows and related indicators. Overview of material flow related activities in OECD countries and beyond.* Descriptive Sheets. Paris: OECD Working Group on Environmental Information and Outlooks.

OECD. 2004b. *Recommendation of the Council on Material Flows and Resource Productivity.* Paris: OECD.

Pedersen, O. G. 2002. *DMI and TMR for Denmark 1981, 1990, 1997. An assessment of the material requirements of the Danish economy.* Copenhagen: Statistics Denmark.

Rapera, C. L. 2004. *Southeast Asia in transition. The case of the Philippines 1981 to 2000.* SEAtrans Working Paper, unpublished.

Schandl, H., C. M. Grünbühel, S. Thongmanivong, B. Pathoumthong, and P. Inthapanya. 2005. *National and local material flow analysis for Lao PDR.* SEAtrans Working Paper, unpublished.

Schmidt-Bleek, F. 1994. *Wieviel Umwelt braucht der Mensch? MIPS: das Maß für ökologisches Wirtschaften.* Berlin: Birkhäuser.

Steurer, A. 1992. *Stoffstrombilanz Österreich 1988.* Social Ecology Working Paper no. 26. Vienna: IFF Social Ecology.

Steurer, A. 1998. *Material flow accounting: Frameworks and systems.* Subgroup meeting on "Statistics in Material Flows of Scarce and Harmful Substances" of the Working Group "Statistics of the Environment." Joint Eurostat/EFTA Group, meeting of January 26 and 27.

UN, EC, IMF, OECD, World Bank. 2003. *Handbook of national accounting: Integrated environmental and economic accounting 2003.* New York: United Nations.

van der Voet, E., L. van Oers, S. Moll, H. Schütz, S. Bringezu, S. M. De Bruyn, M. Sevenster, and G. Warringa. 2005. *Policy review on decoupling: Development of indicators to assess decoupling of economic development and environmental pressure in the EU-25 and AC-3 countries.* CML report 166. Leiden: Universitair Grafisch Bedrijf.

Wall, G. 1977. *Exergy: A useful concept within resource accounting.* Göteborg, Sweden: Institute of Theoretical Physics, Chalmers University of Technology and University of Göteborg.

Weisz, H. 2006. *Accounting for raw material equivalents of traded goods: A comparison of input–output approaches in physical, monetary, and mixed units.* Social Ecology Working Paper no. 87. Vienna: IFF-Social Ecology.

Weisz, H., F. Krausmann, C. Amann, N. Eisenmenger, K.-H. Erb, K. Hubacek, and M. Fischer-Kowalski. 2005a. The physical economy of the European Union: Cross-country comparison and determinants of material consumption. *Ecological Economics* 58 (4):676–698.

Weisz, H., F. Krausmann, N. Eisenmenger, C. Amann, and K. Hubacek. 2005b. *Development of material use in the European Union 1970–2001. Material composition, cross-country comparison, and material flow indicators.* Luxembourg: Eurostat, Office for Official Publications of the European Communities.

Weisz, H., F. Krausmann, and S. Sangkaman. 2005c. *Resource use in a transition economy. Material- and energy-flow analysis for Thailand 1970/1980–2000.* Social Ecology Working Paper no. 81. Vienna: IFF-Social Ecology.

Weizsäcker, E. U. v., A. B. Lovins, and H. L. Lovins. 1995. *Faktor Vier: Doppelter Wohlstand, halbierter Naturverbrauch. Der neue Bericht an den Club of Rome.* Munich: Droemer Knaur.

Wolman, A. 1965. The metabolism of cities. *Scientific American* 213(3):178–193.

World Commission on Environment and Development. 1987. *Our common future: The Brundtland report.* Oxford: Oxford University Press.

13

Indicators to Measure Decoupling of Environmental Pressure from Economic Growth

Kenneth Ruffing*

This chapter describes the conceptual basis for developing decoupling indicators, argues for their use in the policy cycle, reports on efforts to construct a large number of such indicators, primarily for Organisation for Economic Co-operation and Development (OECD) countries, and discusses the implications of the results.

Technically, the term *decoupling* refers to the relative growth rates of a pressure on the environment and of an economically relevant variable to which it is causally linked. For example, at the national level the growth rate of sulfur dioxide emissions may be compared with the growth rate of the gross domestic product (GDP); at a sectoral level, the growth rate of carbon dioxide emissions from energy use may be compared with the growth rate of total primary energy supply.

Decoupling indicators describe the relationship between environmental pressures and the relevant driving forces over the same period. These are also the first two components of the driving force, pressure, state, impact, response (DPSIR) framework widely used in environmental management. Indicators comprising variables belonging to other dimensions of the DPSIR framework (i.e., state, impact, or response) are not described in this chapter but are equally important. From a policy perspective, pressure indicators and the decoupling indicators derived from them are attractive because they are apt to change over shorter time periods than state indicators under the influence of, for example, environmental or economic policy. Reducing the absolute levels of environmental pressures to sustainable levels is the main goal of environmental policy, of

*The author is deputy director, Environment Directorate, Organisation for Economic Co-operation and Development (OECD). The views expressed here are his own and do not necessarily reflect the views of the OECD or its member countries.

course, but decoupling indicators are particularly suitable for monitoring changes over time in response to specific policy measures or to other factors and may be conveniently used as intermediate targets of environmental policy.

It should also be understood that the relationship between economic driving forces and environmental pressures usually is complex. Most driving forces have multiple environmental effects, and most environmental pressures are generated by multiple driving forces, which, in turn, are affected by societal responses. The DPSIR model will not reveal all such linkages, and so there is a need to use decoupling indicators within a more complete analytical framework.

The concept of decoupling is attractive for its simplicity. Graphs displaying a rising GDP juxtaposed with diminishing pollutant emissions or pollutant emissions rising faster than GDP convey a very clear message. However, graphs of synthetic decoupling indicators, such as one variable divided by another, often convey mixed messages. In a growing economy, relative decoupling will imply that environmental pressures are still rising. Likewise, if economic activity is falling, relative or even absolute decoupling may not imply a positive development for society as a whole.

Furthermore, the decoupling concept has no automatic link to the environment's capacity to sustain, absorb, or resist pressures of various kinds (deposition, discharges, or harvests). In the case of renewable natural resources, a meaningful interpretation of the relationship of environmental pressure to economic driving forces will require information about the intensity of use of the resource in question (i.e., of harvesting rate compared with the renewal rate). Some of this context information may be presented through a range of indicators (e.g., water abstraction as a share of available resources, forest harvesting versus annual growth increment). Therefore, decoupling indicators must be used together with targets expressed in absolute terms.

Decoupling indicators can also enrich the set of sustainable development indicators proposed in Chapter 3 of the OECD publication *Sustainable Development: Critical Issues* (OECD 2001b) because they help link selected environmental resource and outcome indicators with economic variables and indicators (such as GDP or sectoral outputs). Decoupling indicators help reveal prospects for longer-term developments that are essential for progressing toward sustainable development, at both the macro and the sectoral level. Decoupling indicators therefore are a useful tool to complement the extended national balance sheets integrating environmental and economic accounts and the development of underlying frameworks such as material flow accounts.

Scientific Validity of Definitions

Environmental variables in a decoupling indicator are usually expressed in physical units, and the economic variable (generally a socioeconomic driving force) is expressed either in monetary units at constant base year prices or in physical volumes. However, the notion of a driving force suggests that relevant variables may sometimes include others, such as population growth.[1]

Decoupling indicators are often expressed in terms of changes over time. Decoupling occurs when the growth rate of the environmentally relevant variable is less than that of its economic driving force (e.g., GDP)[2] over a given period. In most cases, however, absolute changes in environmental pressures are of fundamental concern, hence the importance of distinguishing between absolute and relative decoupling. If GDP displays positive growth, absolute decoupling is said to occur when the growth rate of the environmentally relevant variable is zero or negative (i.e., pressure on the environment is either stable or falling). Relative decoupling is said to occur when the growth rate of the environmental pressure variable is positive but less than the growth rate of GDP.[3]

The choice of decoupling indicator depends on the problem to be elucidated. Here, the term *decoupling indicator* is used to describe an indicator in which the particular environmental pressure appears as the numerator, and the driving force of interest appears as the denominator. Often, the numerator and the denominator are several steps removed from each other in the cause–effect chain of events. In some cases, it may be possible to decompose the main or primary indicator into two or more intermediate indicators, as follows:

Primary indicator = Intermediate indicator 1 × Intermediate indicator 2 × Intermediate indicator n.

For instance, in the case of emissions of air pollutants by the energy sector per unit of GDP, the following relationship can be written:

Emissions/GDP = Emissions/TPES × TPES/TFC × TFC/GDP

| | Depends on emission factors and fuel mix (TPES = total primary energy supply) | Depends on conversion efficiency and fuel mix (TFC = total final consumption of energy) | Depends on end use energy intensities, fuel mix, activity, and structure of the economy |

In other words, the decomposition of the relationship between emissions from the energy sector and economic growth allows us to distinguish between the effects of scale (or volume), sector composition, and technology. Each of these factors may in turn be influenced by policies and may be further decomposed (e.g., by fuel or by end use sector) (Figure 13.1).

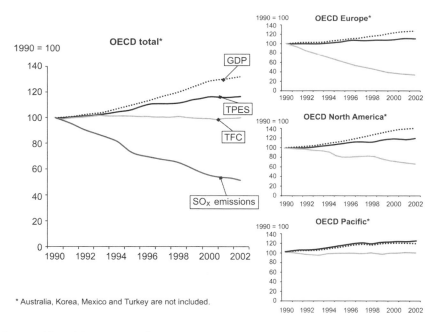

Figure 13.1. SO$_x$ emissions from energy use versus GDP, 1990–2002 (OECD 2005).

History and Existing Use

Many of the variables that figure in decoupling indicators also appear in the concepts of resource efficiency, resource intensity, and resource productivity. These synthetic measures may be calculated as ratios of averages, marginal quantities, or rates of change (to yield elasticities). For example, resource efficiency and resource intensity are calculated as ratios of resource use to economic value added, and resource productivity is the inverse ratio. Decoupling is usually conceived as an elasticity focusing on changes in volumes, whereas efficiency and intensity are more concerned with the actual values of these ratios. Which usage is chosen depends on the context and, often, on the audience being addressed. These terms appear to draw on the vocabulary of economics and business but do not always have the same meaning. For example, resource productivity is expressed in agricultural economics as the yield per hectare. As a marginal concept—the ratio of the addition to total value added to an additional unit of an input to production—this is simply marginal productivity and is widely used. The term *efficiency* is used quite differently. In economics, the term is used most often to refer to allocative efficiency, in which resources are deployed in such a way that the value added per unit of resource use is the same in all sectors. This type of efficiency (if adjusted to take into account environmental externalities) is a necessary but not sufficient condition for achieving environmentally sound use of natural resources—which also entails respecting the environmental sustainability criteria as defined in the *OECD Environmental Strategy* (OECD 2001a)—at the lowest economic cost to society. In the business community, a term widely promoted by the World Business Council on Sustainable Development is *eco-efficiency*, which refers to adding value while reducing the intensity of resource use, or achieving the improvements in resource efficiency that are also commercially profitable.

A recent report prepared for the Swedish Environmental Advisory Council (Azar et al. 2002) presented and analyzed various decoupling trends (CO_2, energy, transportation, materials, chemicals, biomass, SO_2, NO_x, chromium, copper, waste) in Sweden, the European Union, Japan, the United States, Brazil, China, and India over the past 10–40 years, using an approach similar to the one discussed here. In the report they conveniently cite a selected number of papers from the literature on concepts related to decoupling since 1966.

They conclude that a general decoupling of materials and energy from economic development is less interesting than decoupling of specific environmental impacts that cause concern (e.g., emissions of metals and persistent chemicals foreign to nature, CO_2, and acidifying substances) and that reducing the absolute levels of these emissions should be the main focus of policy.

A recent report by the OECD (2002), on which this chapter is largely based, develops and analyzes a set of thirty-one decoupling indicators covering a broad spectrum of environmental issues. Sixteen indicators relate to the decoupling of environmental pressures from total economic activity under the headings *climate change, air pollution, water quality, waste disposal, material use,* and *natural resources.* The remaining fifteen indicators focus on production and use in four specific sectors: energy, transport, agriculture, and manufacturing.

The evidence presented in the report shows that relative decoupling is widespread in OECD member countries. Absolute decoupling is also quite common, but for some environmental pressures little decoupling is occurring. The evidence also suggests that further decoupling is possible because absolute decoupling was recorded in at least one OECD country for all but two of the decoupling indicators examined at the national level.

The OECD Environmental Policy Committee uses these or similar decoupling indicators in its environmental performance reviews. Closely related indicators are also used in the sustainable development chapters of the economic reviews undertaken by the OECD Economic and Development Review Committee.

Although each of the indicator sets currently used by various international organizations was developed for a different purpose or from a different perspective, they are all based on the long-established data collection efforts of the OECD, International Energy Agency, United Nations, or Food and Agriculture Organization. Most of these indicators are also based on a common set of selection criteria for environmental indicators, such as those published by the OECD in 1993 (Table 13.1). These criteria are

Table 13.1. OECD environmental indicator selection criteria.

Policy relevance and utility for users	An environmental indicator should • Provide a representative picture of environmental conditions, pressures on the environment, or society's responses • Be simple, easy to interpret, and able to show trends over time • Be responsive to changes in the environment and related human activities • Provide a basis for international comparisons • Be either national in scope or applicable to regional environmental issues of national significance • Have a threshold or reference value against which to compare it so that users can assess the significance of the values associated with it
Analytical soundness	An environmental indicator should • Be theoretically well founded in technical and scientific terms • Be based on international standards and international consensus about its validity •Lend itself to being linked to economic models, forecasting, and information systems
Measurability	The data needed to support the indicator should be • Readily available or made available at a reasonable cost • Adequately documented and of known quality • Updated at regular intervals in accordance with reliable procedures

Note: These criteria describe the ideal indicator; not all of them will be met in practice.

also valid for decoupling indicators. In view of the difficulty of obtaining agreement on indicators, it is encouraging that many variables in the set of decoupling indicators found in the OECD report also show up in other indicator sets.

Methodological Aspects

Most decoupling indicators are country specific and do not usually address the cross-border flow of environmental externalities. However, material flow accounts and the Ecological Footprint methods do address this issue explicitly and could be used to construct decoupling indicators capable of tracking such trends. Often quoted in this respect are the greenhouse gas (GHG) emissions associated with a country's imports and exports (OECD 2003). A different type of example can be found in the fishery sector, when fisheries are not confined within national boundaries,[4] necessitating decoupling indicators developed at an appropriate level of aggregation.

The most direct manner of displaying decoupling between an environmental pressure and an economic driving force is to plot two indexed (e.g., 1980 = 100) time series on the same graph, as in Figure 13.1. From such a graph, it is immediately clear whether the economic driving force is growing or shrinking, whether decoupling (absolute or relative) is occurring, when it started, and whether it continues. This method is used by the OECD in displaying overall trends for all OECD countries and for the three main OECD regions (i.e., Europe, North America, and Pacific). Some of these qualities are lost if decoupling is presented as a single line (i.e., a time series of the ratio of environmental pressure to driving force), although the idea of improvement in efficiency or intensity is better communicated this way.

However, neither of these presentations lends itself well to numerical displays of decoupling trends for a large number of countries. To compare decoupling between countries, the ratio of the value of the decoupling indicator at the end and the start of a given time period may be defined as follows:

$$\text{Ratio} = (EP/DF)_{\text{end of period}} / (EP/DF)_{\text{start of period}}$$

where EP = environmental pressure and DF = driving force.

If the ratio is less than 1, decoupling has occurred during the period, although it does not indicate whether decoupling was absolute of relative. To avoid displaying (on a bar graph) small values when decoupling is significant, a decoupling factor is defined as follows:

$$\text{Decoupling factor} = 1 - \text{Decoupling ratio}.$$

The decoupling factor is zero or negative in the absence of decoupling and has a maximum value of 1 when environmental pressure reaches zero.[5] Decoupling factors for CO_2 in OECD countries for the interval 1990 to 1999 are displayed in Figure 13.2.

When several pollutants have similar effects, aggregation can reduce the information load. For example, to construct a single indicator accounting for the overall effect of all six GHGs on the climate system, conversion factors (based on the relative radiative force of the individual gases) are used to construct the decoupling indicator "GHG emissions from all sources per unit of GDP." Similarly straightforward procedures have been used elsewhere to construct indicators for acidification, toxic contamination, ozone depletion, or low-level ozone formation. Such aggregated indicators can often be linked to an appropriate driving force to obtain aggregated decoupling indicators.

In other cases, more complex aggregation procedures may be needed. One approach used to produce a weighted index of local air pollution is based on the health effects of

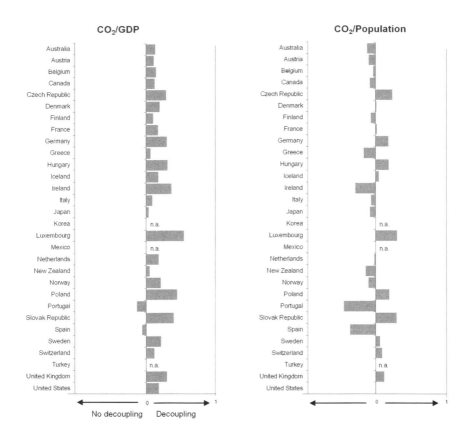

Figure 13.2. Decoupling factors for CO_2, 1990–2002 (OECD 2005).

exposure to specific levels of individual pollutants. A variety of other aggregated environmental indicators have been proposed in recent years, some of which are used for communication (e.g., various urban air pollution indices) or policymaking (e.g., the indicators associated with the Dutch National Environmental Policy Plan). Other indicators push the aggregation even further, using a common unit (such as tons) to aggregate measures across a variety of environmental and natural resource issues. Nevertheless, aggregated indicators are still far from being universally accepted.

Decoupling Indicators and Sustainable Development Targets

Because decoupling indicators relate environmental pressures to socioeconomic driving forces, their evolution over time must be evaluated against the absolute levels of environmental pressure or state variables. For many of these it is necessary to have explicit policy targets. For pollution pressures, most OECD countries have set both environmental quality standards (e.g., receiving water and ambient air quality standards) and effluent or emission limits. These targets are generally based on a scientific assessment of environmental quality objectives and associated levels of maximum environmental pressure. Some countries also apply benefit–cost analysis to ensure that the costs to society of ensuring compliance with the limits do not exceed the benefits of doing so. A number of countries also use other evaluation tools such as strategic environmental assessments or sustainability impact assessments. At the regional and international levels, there are several legally binding conventions and agreements setting targets (emission ceilings) for particular environmental pressures (e.g., sulfur dioxide, volatile organic substances).

For natural resources, uses can be viewed either as a driving force (expressed in value terms) or as a pressure on the environment (measured in physical units). For example, the use of nonrenewable resources (e.g., fossil fuels) is often treated as a driving force when the environmental variable of interest is a pollutant. Regarding renewable resources (e.g., water, fish, wood), their use is considered mostly a pressure changing in response to an economic driver, such as household consumption. For some resources, many countries have imposed constraints on natural resource use because of the environmental pressure they generate. For example, the abstraction of surface water sometimes is regulated in order to maintain adequate flows in rivers for the protection of aquatic ecosystems. However, these targets are less well developed than those for pollution pressures.

Pressure reduction targets at a national or international level have so far been expressed in absolute (e.g., tons of sulfur dioxide per annum) rather than relative terms (emissions per unit of GDP or resource use per unit of GDP). However, in some cases only relative decoupling targets may be politically feasible even where the long-term objective may be a reduction in the absolute levels of environmental pressure.[6] Applied to products, decoupling targets, usually called performance standards, are increasingly used (e.g., minimum energy performance standards for electrical appliances). Where possible, trends in decoupling can be compared with policy targets to show the "distance to target." However, even in the absence of defined thresholds, ceilings, or targets, decoupling indicators are useful to compare countries, to identify similarities and differences, and as a starting point to assess the potential for improved performance. They are particularly well suited for assessing the adequacy of the policy measures implemented to achieve decoupling.

Policy Relevance and Legitimacy

As mentioned earlier, trends in decoupling may be decomposed in a number of intermediate steps. These may include changes in the scale of the economy, consumption patterns, and economic structure, including the extent to which demand is satisfied by domestic production or by imports. Other mechanisms in the causal chain include the adoption of cleaner technology, the use of higher-quality inputs, and the post facto cleanup of pollution and treatment of waste.

Over time, these mechanisms will change for a variety of reasons. Many of them can be influenced, directly or indirectly, by sectoral and environmental policies. For example, consumer behavior can be changed through the promotion of ecolabels or the imposition of product taxes. Incentives can be provided to enterprises to undertake life cycle analyses of products. Cleaner production technologies can be promoted through measures that internalize environmental externalities and by favorable tax treatment of environmental research and development. Toxic additives in gasoline can be banned, and minimum energy performance standards can be imposed for cars or electrical appliances.

An example of the different role of these mechanisms is provided by emissions of sulfur dioxide. Indicators presented in the OECD report on decoupling (OECD 2002) show an absolute decoupling of sulfur dioxide emissions from energy production (e.g., as a result of regulations on and incentives for the use of low-sulfur fuels) in nearly all OECD countries. This decoupling partly reflects the reduction in energy use from GDP (e.g., through greater energy efficiency or shifting demand to less-energy-intensive goods and services). Another example of the role of these different mechanisms is provided by discharges of nitrogen; in the future, these could be decoupled from conventional agricultural production (aiming at less and better use of nitrogenous fertilizers) as demand shifts toward ecolabeled products or low-meat diets. Similarly, the negative environmental impacts of waste can by reduced by technologies that minimize the release of dioxins from incineration and the leaching of hazardous substances and methane from landfills; however, they can also be reduced by waste prevention policies designed to reduce the demand for waste disposal and its growth relative to GDP or total consumption.

Extent of Applicability

When such indicators can be assembled on a time series basis, mathematical functions can be fitted to the data and used to compute other parameters useful for policy analysis. Decoupling indicators can also be formulated at the product or enterprise level, as is being attempted at present by the Global Reporting Initiative.

Decoupling indicators are intended primarily to track, for a single country, temporal changes in the relationship between environmental pressures and economic driving forces. When decoupling indicators are used to compare environmental performance between countries, the national circumstances of each country must also be taken into account. These include factors such as country size, population density, natural resource endowments, energy profile, changes in economic structure, and stage of economic development. Moreover, the initial level of an environmental pressure and the choice of time period considered can affect the interpretation of the results because countries proceed according to different timetables.

Gaps in Knowledge and Research Needs

Despite efforts over the past few decades to improve coverage, gaps in statistical data remain pervasive; definitions often differ across countries (e.g., for waste management) and change over time. More work in estimating missing data points could extend the time periods and number of countries covered for some indicators. Even more important are the gaps resulting from some data not being collected at all. For example, of the thirty-one indicators considered by the OECD, for only ten of them were data available for at least twenty of the thirty member countries from at least 1990. Moreover, nine indicators were assessed as needing further work for a variety of reasons (e.g., concept, definition, measurement). To be sure, existing data were collected for a range of purposes other than for constructing decoupling indicators. This underscores the importance of reviewing the information needs associated with the decoupling perspective, determining what other information would be needed, and assessing whether it would be worthwhile to collect it.

Beyond statistical gaps are science gaps. Many ecological systems are still poorly understood. Scientists have pointed to the need for caution in setting sustainable limits to environmental pressures because ecological processes are nonlinear, and we know little about thresholds and trajectories. Certain pressures can continue to grow without apparent effect and then, after crossing some unsuspected threshold or ceiling, suddenly show dramatic discontinuities or even complete collapse (as has happened with some fisheries). Policymakers need to be aware of these gaps when using environmental indicators. This is particularly true for decoupling indicators, which can convey a positive message (i.e., relative decoupling) while in reality a country's ecosystems may be heading toward breakdown. More often than not, the complex nature of these thresholds cannot be shoehorned[7] into the format of a decoupling indicator. In these cases, caution is needed in interpreting decoupling indicators, or any other partial indicators, for that matter.

Notes

1. Population growth becomes relevant when demand for certain environmentally relevant goods or services become saturated at high levels of per capita income.

2. The term *decoupling* is not used when the environmental pressure variable increases at a higher rate than the economic driving force (a case of "supercoupling").

3. In the literature, the terms *strong* and *weak* are sometimes used as synonyms for *absolute* and *relative*, respectively.

4. Information about the intensity of resource use can be presented only on the basis

of discrete stocks of particular fish species. But where fish stocks are exploited by foreign fishing fleets, it becomes difficult to link such information to country-based decoupling indicators.

5. Note that the decoupling factor generally will not change linearly, even if both environmental pressure and driving force do.

6. Such targets (sometimes called dynamic) were used in voluntary agreements on CO_2 emission reductions in Germany.

7. Differences of scale are another reason why country-based decoupling indicators are not well suited to take account of ecosystem constraints. Neither environmental pressure nor ecological carrying capacity is evenly distributed across a country's surface area, and local ecosystem collapses are likely to occur long before nationally averaged pressures approach critical values.

Literature Cited

Azar, C., J. Holmberg, and S. Karlsson. 2002. *Decoupling: Past trends and prospects for the future.* Paper commissioned by the Swedish Environmental Advisory Council (Miljövårdsberedningen). Available at www.sou.gov.se/mvb/pdf/decoupling.pdf.

OECD. 2001a. *OECD environmental strategy for the first decade of the 21st century.* Paris: OECD.

OECD. 2001b. *Sustainable development: Critical issues.* Paris: OECD.

OECD. 2002. *Indicators to measure decoupling of environmental pressure from economic growth* (SG/SD(2002)1/FINAL). Paris: OECD.

OECD. 2003. *Carbon dioxide emissions embodied in international trade in goods.* STI Working Paper no. 2003/15 (DSTI/DOC[2003]15). Paris: OECD.

OECD. 2005. *Indicators to measure decoupling of environmental pressure from economic growth.* Paris: OCED.

14

Geobiosphere Load: Proposal for an Index

Bedřich Moldan, Tomáš Hák, Jan Kovanda, Miroslav Havránek, and Petra Kusková

Chapter 40 of Agenda 21 acknowledges that "commonly used indicators such as GNP and measurement of individual source or pollution flows do not provide adequate indications of sustainability" (paragraph 40.4)(UNCED 1992). The problem with attempts to monitor and evaluate progress toward sustainable development (paragraph 8.6) is not the lack of potential indicators but their multiplicity and their interdependence. Given the divergent views on indicators, the challenge after Rio was "to develop a concept of indicators of sustainable development in order to identify such indicators" (paragraph 40.6) and to reach consensus on a suitable set of indicators that can adequately reflect the wide range of concerns encompassed by sustainable development (UNCED 1992).

Now, 13 years later, we see that the challenge put forward by Agenda 21 is still not fulfilled, despite substantial progress in the concept of sustainable development. The development of indicators is still seen as a major topic in sustainable development projects and programs (OECD 2004).

Currently it is recognized that the fundamental elements of sustainable development are its three pillars: social, economic, and environmental (the importance of adequate institutions for sustainable development is also sometimes stressed). The latest, most authoritative event, the World Summit on Sustainable Development in Johannesburg in 2002, acknowledged this concept in its Plan of Implementation and motto: "People, Planet, Prosperity."

The three pillars are characterized by distinct sets of variables and indicators, some of which stress linkages between pillars (e.g., indicators of decoupling; OECD 2002). The economic and social dimensions have rather well developed and (which is critical) generally accepted indicators such as gross domestic product (GDP), unemployment rate, life expectancy, and literacy rate. On the social side, a novel aggregated indicator, the Human Development Index, introduced by UNDP in 1990, has been gradually gaining respect.

Historically, the idea of sustainable development originated among environmentalists (IUCN 1980). The development of environmental indicators has been under way for about 30 years, since the OECD introduced its core set of environmental indicators. There is a lot of progress in both methods and actual use of indicators, as we can see in the universal acceptance of the driving force, pressure, state, impact, response (DPSIR) framework and the development of various linkage indicators (e.g., the decoupling ones). As we see it, there are two important problems to be solved: finding the correct mix of indicators for synthesis (few indicators but less precise information) and analysis (many indicators but risk to overinform stakeholders), which is fundamentally linked to the usage situation of the assessment; and defining the relationship to sustainability more clearly.

We regard the challenge of capturing the issues of sustainability as the most important one. We assume that anthropogenic activity exerts its impact on the environment (carrying capacity). The problem with the carrying capacity concept is that it is probably not possible to quantify it in principle. We know only that carrying capacity is limited and that we cannot systematically deplete all the earth's natural assets (Daly 1990; Costanza et al. 1997). Unfortunately, knowledge on carrying capacity is still limited. On the other hand, we do know how the carrying capacity is affected by anthropogenic activities. With the DPSIR framework in mind, we have chosen to focus on pressures for our basic approach. The pressure indicators best describe the fundamental stresses human activities place on environment. The same conclusion was achieved by Eurostat's project resulting in the Pressure Index (Eurostat 2001b). Also, pressures are selected for decoupling indicators to characterize the environmental "evils" (OECD 2002). Pressures caused by humankind, which have already been and will remain a major environmental force for future millennia, even gave a name to the present period: Anthropocene (Crutzen 2002).

Methodology

We call the pressures in question a geobiosphere load (GBLoad). Despite the fact that the term *load* might imply "impact," we concentrate merely on pressures exerted on the environment by social and economic developments (e.g., resource extraction, resource transformation into products and services, and subsequent emissions). To put it clearly, the pressure from a ton of coal equals the pressure from a ton of biomass.

Referring to our previous research (Moldan and Billharz 1997; Hák 2002), we focus on the geobiosphere load in three categories: material and energy flows and land requirements. Material flow analysis, energy flow accounting, and ecological footprints are useful points of departure for the development of the specific indicators in these three categories.

The GBLoad index is calculated in three forms (subindices related to either area, population, or GDP) in a transparent way following a straightforward formula. By proposing a single index based on three clearly defined indicators only, we fulfill the first of the fundamental prerequisites: a small number of individual indicators. Formally, our proposed GBLoad resembles the UNDP's Human Development Index based on three fundamental, rather obvious components of dignified human lives: health, education, and income. These three items are characterized by comprehensible indicators that are then joined by a simple and transparent mathematical formula. Our index is constructed in a similar fashion and is based on three indicators that, in our opinion, capture the most important factors of environmental sustainability.

Energy, materials, and land can be regarded as the essential components and prerequisites of nature's services (Daily 1997). The idea of ecosystem services is well established and is being developed as a fundamental concept by the ongoing Millennium Ecosystem Assessment program (MA 2003). Provision of energy and materials basically equals the provisioning services of ecosystems (e.g., food, fiber, energy resources, biochemicals, or freshwater). Land, in relation to other environmental media, is a prerequisite for all kinds of ecosystem services (beyond the provisions mentioned, land provides further supporting services such as primary production, regulating services such as climate regulation, and cultural services such as recreation).

The material component of the GBLoad index is based on data and indicators of economy-wide material flow analysis (MFA). MFA was developed in the 1990s with the cooperation of many research institutes and organizations, including the World Resources Institute; the Wuppertal Institute for Climate, Environment and Energy; the Institute for Interdisciplinary Studies of Austrian Universities (Department of Social Ecology), and Eurostat. In 2001, these methods were standardized in a methodological guide (Eurostat 2001a). The aim of the method is to quantify the physical exchange between the national economy, the environment, and foreign economies on the basis of total material mass flowing across the boundaries of the national economy. These flows consist of material inputs to and material outputs from the national economy. Material inputs are all mined raw materials and consumed biomass. Material outputs are air and water emissions, solid waste, and so-called dissipative use of products, such as are fertilizers, pesticides, and winter filling. The difference between inputs and outputs is the quantity of materials accumulated in the economic system in the form of construction, transport infrastructure, durable products, and so on (net addition to stock [NAS]). It is also important to include so-called unused extractions or hidden flows. Unused extractions are material flows that have taken place as the result of resource extraction but do not directly enter the economic system. Examples include biomass left in forests after logging, overburden from extraction of raw materials (as in open cast coal mining), earth movements resulting from the building of infrastructure, and dredged deposits from rivers. Foreign trade and related indirect flows (such as overburden in coal mined abroad and subsequently imported) also play an important role in the analysis because they also represent an important flow of material across the boundary of the economic system.

MFA provides an important database to infer series of environmental pressure indicators. The most commonly used material flow indicators can be divided into several groups:

Input indicators: Direct material input (DMI) equals domestic used extraction (excavated raw material, harvested biomass) plus imports; total material requirement (TMR) includes domestic used and unused extractions, imports, and their indirect flows.

Output indicators: Domestic processed output (DPO) comprises emissions to air, landfilled wastes from industrial processes and households, the material load in wastewater, and dissipative uses and losses of products; total domestic output (TDO) includes DPO and unused domestic extractions.

Consumption indicators: Domestic material consumption (DMC) is calculated as DMI minus exports; total material consumption (TMC) is TMR minus exports and their indirect flows; NAS measures the physical growth rate of the economy. Each year new materials are added to economic stocks, such as new buildings and durable goods, and old materials are removed from this stock and become wastes.

All these robust indicators have been developed into a fixed methodological framework, and they are characterized by a transparent method of data aggregation. Because all components of MFA indicators (e.g., extracted minerals, mined fossil fuels, or harvested biomass in case of DMI) are measured in tons, there is no limitation with regard to aggregation of data in different physical units (this applies for direct and used flows; indirect and unused flows usually are calculated by means of conversion coefficients). MFA indicators meet common policy relevance criteria. They relate directly to human pressure on the environment. They are designed to cover all material flows, so they are representative and comprehensive. Moreover, they are comparable because they are constructed on the basis of a standardized method. For these reasons, material flow indicators currently appear more often in official results of many organizations such as Eurostat, the European Environment Agency, and UN agencies (Eurostat 2001a; EEA 2004; UN 2001), even though some of MFA-related issues have not yet been addressed (e.g., linkage of pressures expressed by material flow indicators to specific impacts). The developing research in this field focuses on the fact that some enormous flows are not necessarily very harmful (e.g., overburdens from mines), whereas smaller highly toxic ones can be much more damaging to humans and nature (Steurer 1996; Van der Voet et al. 2004).

One of the aforementioned indicators will be selected or a new indicator (a combination of the existing ones) will be developed for the GBLoad index. One must keep in mind that all of these indicators are highly correlated. At present, the TMC indicator seems to best fulfill the requirements of a suitable indicator for the GBLoad concept; it is a subject of our current research.

Energy flow accounting (EFA) aims at establishing a complete balance of energy inputs, internal transformations, and energy outputs of a society, or of a defined socio-economic unit. On a macroeconomic level, EFA uses a similar concept as MFA (Haberl 2001). Its aim is to assess all inputs and outputs of a socioeconomic system in energy units (joules). EFA uses existing notions and methods of conventional energy balance as far as possible in order to trace energy flows through an economy and obtain indicators for the amount of energy a society is able to harness for its purposes (Krausmann and Haberl 2002; Schandl et al. 2004). Contrary to the conventional energy balances (IEA 2000; UN 2000), EFA includes inputs of energy-rich materials not directly used for energy conversion (e.g., wood for furniture, construction, or the paper industry) and also includes inputs of domestic animal and human work. These have a crucial impact on energy balances in preindustrial societies or small localities, where human and animal work can be significant components. EFA also includes inputs of nonmaterial energy carriers such as wind power, hydropower, heat, and electricity. EFA provides an important database for the derivation of a number of energy indicators. These quality-adjusted measures of energy flows are very useful in understanding the biophysical inputs needed for economic growth (Ayres et al. 2002). As with material flow indicators, it is possible to use EFA to assess changes or trends of crucial importance for the sustainability of national systems: intensity of energy use, energy consumption patterns, or energy use of regions.

EFA provides conceptually similar environmental pressure indicators as MFA does. The most important energy flow indicators can be grouped as follows:

Input indicators: Direct energy input (DEI) is the total amount of energy entering the socioeconomic component (either by domestic extraction or by imports); total primary energy input (TPEI) is defined as direct energy input and hidden flows (HFs), which can be classified as domestic hidden flows (DHFs) and imported hidden flows (IHFs).

Output indicators: Useful energy (UE) means total energy benefit, connected with the end use of energy. This is counted as final energy use (FEU) multiplied by energy efficiency of end-use devices; FEU is a commonly used indicator (not only in EFA) and is the energy sold to end users. Against widely used FEU, EFA also counts biomass as an energy-rich material that results in higher values.

Consumption indicators: Domestic energy consumption (DEC) is calculated as DEI minus exports; total energy consumption (TEC) is TPEI minus exports and their hidden flows; NAS measures the growth rate of the economy. Each year some energy-rich materials are added to economic stocks (e.g., new wooden buildings, energy carriers, and food in tins), and old materials are removed from this stock (e.g., eaten, burned) and become wastes.

Each of these aggregated indicators is compiled through a transparent method of data aggregation. At the end, all components of EFA indicators (e.g., electricity, fuels, or harvested biomass) are expressed in the same energy unit (joules). However, before the aggregation, some conversion factors must be used. EFA indicators also meet other criteria, such as being meaningful (EFA indicators relate directly to human pressure on the environment), representative (they are designed to cover all anthropogenic energy flows), and comparable (they are constructed on the basis of a standardized methodology).

Similarly to material flow indicators, the selection of the most appropriate indicator in this case is not finished. Analogously to TMC, the best candidate for the suitable indicator in the GBLoad context seems to be TEC. Again, research activities to resolve this question are under way. In general, the EFA method is less developed than its MFA counterpart, but research in this field is extensive and includes our own.

Together with energy and material flows, land and land area requirements are the third important category of resource input for economic activities. There is no doubt about the importance of land use for ecological processes. Land provides the spatial context (i.e., source function) for and bears the impacts of (i.e., sink function) human activities. Land-based trade balances illustrate the nondomestic land areas appropriated for the production of goods and services abroad (i.e., imports) and the domestic land area needed to produce the goods and services exported to the rest of the world (Hubacek and Giljum 2003). Several approaches or concepts try to assess human use of land, linking land use and socioeconomic data. The most popular ones include the Ecological Footprint (EF), which assesses land needs for individual consumption of goods and services (Wackernagel and Rees 1996). The EF concept is not based on actual land use or land cover data, and its results are hypothetical area units. EF will be taken as a point of departure (as a proxy for the "land requirements" indicator to provide results for pilot calculations). We have recently done research for a suitable indicator and its adaptation to provide the "land (area) requirement subindex" for GBLoad. We are considering tying the land pressure indicator to the degree of "naturalness." It will combine several variables (e.g., land cover, fragmentation, biodiversity) and will use approaches for evaluating anthropogenic influence on productive land and sea area. There are several examples to be modified and used: an assessment of land types based on their "naturalness" (Míchal 1994), the concept of human appropriation of the net primary production (Vitousek et al. 1986; Krausmann 2001; Haberl et al. 2004), and the classification developed by Daly (1996) for the hierarchization of capital stocks.

Although some degree of subjectivity is inherent in the GBLoad construction (selection and definition of the main realms of human-induced pressure on the environment), there are advantages that outweigh it: All three component indicators (subindices) of the GBLoad are expressed in clearly defined physical units, and they are not subject to any weighting or assumptions affecting the mathematical calculation. Such steps in the index method always jeopardize the credibility of results. We simply count tons, joules, and hectares or any other units of matter, energy, and surface area.

The resulting index, called GBLoad, is constructed as an average of the individual indicators or subindices (material flow, energy flow, and pressure on land resources). A similar approach has been used for some other indices, such as the Human Development Index, the Environmental Sustainability Index (ESI), and the Index of Environmental Friendliness (Puolama et al. 1996; Global Leaders for Tomorrow Environment Task Force 2002; UNDP 2003). The GBLoad index is constructed as an average of the three cumulative standard normal distributions of z scores of subindices, which were first related to a chosen reference scale (area, population, or GDP). A similar approach is used for calculation of ESI. The construction of the subindices from variables follows standard statistical methods and is shown in Box 14.1.

Box 14.1. OECD envionmental indicator selection criteria.

In the first step we choose the reference scales for the individual indicators/sub-indices: area, population size and economic performance (GDP) to make them comparable among nations. Then, we create z-scores from all indicators' components

$$Z_{tmc} = \frac{TMC_x - \mu_{tmc}}{\sigma_{tmc}}, \ Z_{ef} = \frac{EF_x - \mu_{ef}}{\sigma_{ef}}, \ Z_{tpes} = \frac{TPES_x - \mu_{tpes}}{\sigma_{tpes}}$$

Where z-value ($Z_{tmc, ef, tpes}$) expresses the divergence of the TMC (EF, TPES) of country x from the most probable result μ (mean of entire set of tmc or ef or $tpes$) as a number of standard deviations. The larger the value of z, the less probable the experimental result is due to chance. The probability for component (tmc, ef, $tpes$) and country (x) can be calculated from the cumulative standard normal distribution:

$$P_{tmc(x)} = \int_{-\infty}^{z} \frac{1}{\sqrt{2\pi}} e^{\frac{-u^2}{2}} \partial u$$

which gives the probability P that an experimental result with a z value less than or equal to that observed is due to chance. Final calculation is simple averaging of three component scores and transforming them for better visualization into 0-100 scale:

$$GBL_x = \frac{P_{tmc(x)} + P_{ef(x)} + P_{tpes(x)}}{3} \times 100$$

The final Geobiosphere Load Index is a dimensionless number that is appropriate for ranking individual countries or other units.

Results and Discussion

GBLoad was calculated in three forms relating subindices to either GDP, population, or area. Each form brings different information: $GBLoad_{GDP}$ points at the economy's efficiency of resource transformation into economic outputs, $GBLoad_{POP}$ captures the environmental justice aspects (equity and equal resource sharing) because all people should have equal rights to consume natural resources, and the results related to the area seem the most suitable from an environmental point of view because that index ($GBLoad_{AREA}$) shows pressure exerted on the geobiosphere, which should not exceed the carrying capacity of a given area.

Using the $GBLoad_{AREA}$ form, the best results (i.e., lowest GBLoad values) are achieved by countries with low population density (Figure 14.1). They differ in consumption: Brazil and Venezuela have low consumption per capita (expressed by material, energy, and area needs). On the other hand, Finland, Norway, and Sweden have high per capita consumption. The worst results (i.e., the highest GBLoad values) are achieved by countries with both high per capita consumption and population density (the Netherlands, Belgium and Luxembourg). The highest values for $GBLoad_{POP}$ are attained by countries with high consumption per capita but low population density (Figure 14.2). Concerning this group of countries, the way in which results are expressed has great impact. Although these countries achieved the lowest GBLoad values from the viewpoint of pressure on the environment, they are the black sheep from the viewpoint of equality of resource use. Quite a different situation applies to the Czech Republic, where GBLoad equaled 50 in both cases. Greece and Spain are also rather balanced, whereas Hungary achieved very different results for each GBLoad form, as did the Scandinavian countries, even though the absolute values of the indices are much lower in Hungary's case. As regards the last form, $GBLoad_{GDP}$, the lowest values (best results) were achieved by the economically developed European Union (EU) countries, Japan, and Norway, and countries such as Venezuela and Brazil placed much worse (Figure 14.3). Significantly worse results were also achieved by the countries that recently joined the EU (Czech Republic, Hungary). The authors did not investigate the reasons further, but the obvious reason for these disparities is the differences in the use of modern technologies and in labor productivity. Results of all three types of GBLoad presentation are shown in Figure 14.4.

Obviously, the core of our proposal is the three indicators (subindexes), selected as fundamental components of the entire human pressure on the environment. The GBLoad certainly does not capture all environmental problems caused by human activity (e.g., all the harmful effects of transport, dispersal of chemicals, and direct influence on climate and biodiversity). However, given the high correlation between selected indicators and such

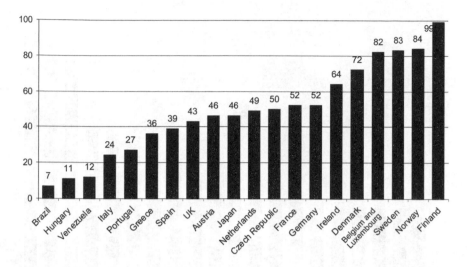

Figure 14.1. GBLoad$_{AREA}$, 2000.

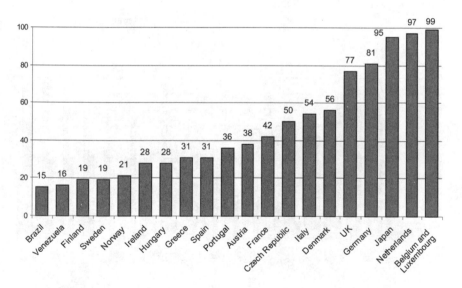

Figure 14.2. GBLoad$_{POP}$, 2000.

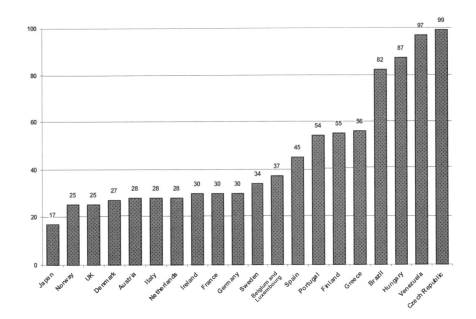

Figure 14.3. GBLoad$_{GDP}$, 2000.

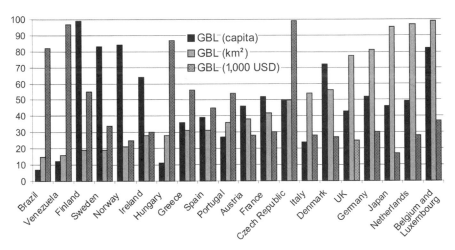

Figure 14.4. GBLoad$_{POP}$, GBLoad$_{AREA}$, and GBLoad$_{GDP}$ combined results, 2000 (GDP in exchange rate, at constant prices 1995).

phenomena as production and consumption of chemicals (correlated with material indicators), greenhouse gas emissions (correlation with energy indicators), and loss of habitats (correlation with land indicators), these factors are captured to some extent. We are conducting research on these and other correlations and relationships. We believe that the overall human-induced environmental pressure is captured in a fairly comprehensive way.

The proposed materials, energy, and land indicators are readily understandable by a broad public because no special environmental education is needed. Both their absolute values and the rank of entities being compared (townships, regions, countries) could be highly policy relevant when used for naming and faming or shaming. Local or national targets can be set. The analysis of time series may be revealing. The selected indicators can be complemented by a plethora of derived indicators (e.g., normalized by surface area or capturing the so-called decoupling phenomenon, i.e., links to GDP or other economically or socially critical variables).

GBLoad includes a challenge that will be addressed in future research: a possible double-counting (or even triple-counting) of some variables into the index. As an example we can take the pressure exerted by biomass harvesting: It is currently counted by material flow indicators in the form of matter of the biomass, by energy flow indicators in the form of biomass energetic content, and by Ecological Footprint in the form of the area needed to grow the biomass. Apart from this possible conceptual shortcoming, this multiple counting can be misleading for international comparison because a country that is more dependent on nonrenewable resources might have a lower GBLoad. This distortion occurs because biomass is counted three times, whereas nonrenewables are counted just once (e.g., minerals by material flow indicators) or twice (e.g., fossil fuels by material and energy flow indicators). The multiple counting might be justified on the basis of linking flows with impacts: The more impacts the particular flow would have, the greater weight would be given to it. However, the pressure impact analysis using material, energy, or land use data has not yet provided sufficient foundation.

Literature Cited

Ayres, R. U., L. W. Ayres, and B. Warr. 2002. *Exergy, power and work in the US economy, 1900–1998.* 2002/52/EPS/CMER. Fontainebleau, France: INSEAD, Centre for the Management of Environmental Resources.

Costanza, R., R. d'Arge, R. de Groot, S. Farber, M. Grasso, B. Hannon, K. Limburg, S. Naeem, R. V. O'Neill, J. Paruelo, R. G. Raskin, P. Sutton, and M. van der Belt. 1997. The value of the world's ecosystem services and natural capital. *Nature* 387:253–260.

Crutzen, P. J. 2002. Geology of mankind. *Nature* 415:23.

Daly, H. E. 1990. Towards some operational principles of sustainable development. *Ecological Economics* 2:16.

Daly, H. E. 1996. *Beyond growth: The economics of sustainable development.* Boston: Beacon.

Daily, G. C. (ed.). 1997. *Nature's services: Societal dependence on natural system.* Washington, DC: Island Press.

EEA. 2004. *EEA signals.* Copenhagen: European Environmental Agency.

Eurostat. 2001a. *Economy-wide material flow accounts and derived indicators. A methodological guide.* Luxembourg: Eurostat.

Eurostat. 2001b. *Towards environmental pressure indicators for the EU,* 2nd ed. Luxembourg: Eurostat.

Global Leaders for Tomorrow Environment Task Force. 2002. *Environmental Sustainability Index. An initiative of the Global Leaders for Tomorrow Environment Task Force, World Economic Forum.* Annual Meeting, 2002. Available at www.ciesin.org/indicators/ESI/.

Haberl, H. 2001. The energetic metabolism of societies. Part I: Accounting concepts. *Journal of Industrial Ecology* 5(1):11–33.

Haberl, H., M. Wackernagel, F. Krausmann, K.-H. Erb, and C. Monfreda. 2004. Ecological footprints and human appropriation of net primary production: A comparison. *Land Use Policy* 21:279–288.

Hák, T. 2002. *Material flows and environmental pressure.* Ph.D. thesis (in Czech), Institute for Environmental Studies, Charles University in Prague, Czech Republic.

Hubacek, K., and S. Giljum. 2003. Applying physical input–output analysis to estimate land appropriation (ecological footprints) of international trade activities. *Ecological Economics* 44:137–151.

IEA. 2000. *Energy balances of OECD countries 1997–1998.* Paris: International Energy Agency, Organization for Economic Co-operation and Development.

IUCN. 1980. *The world conservation strategy: Living resource conservation for sustainable development.* Gland, Switzerland: IUCN/UNEP/WWF.

Krausmann, F. 2001. Land use and industrial modernization, an empirical analysis of human influence on the functioning of ecosystems in Austria 1830–1995. *Land Use Policy* 18:17–26.

Krausmann, F., and H. Haberl. 2002. The process of industrialization from the perspective of energy metabolism: Socioeconomic energy flows in Austria 1830–1995. *Ecological Economics* 41:177–201.

MA. 2003. *Ecosystem and human well-being.* A report of the conceptual framework working group of the Millenium Ecosystem Assessment. Washington, DC: Island Press.

Míchal, I. 1994. *Ecological stability* [in Czech]. Ministry of the Environment of the Czech Republic, Prague.

Moldan, B., and S. Billharz (eds.). 1997. *Sustainability indicators. Report of the project on indicators of sustainable development.* SCOPE 58. Chichester, UK: Wiley.

OECD. 2002. *Indicators to measure decoupling of environmental pressures from economic growth.* SG/SD(2002)1. Paris: OECD.

OECD. 2004. *Measuring sustainable development: Integrated economic, environmental and social frameworks.* Paris: OECD.

Puolama, M., M. Kaplas, and T. Reinikainen. 1996. *Statistics Finland: Index of environmental friendliness. A methodological study.* Statistics Finland, Eurostat. Helsinki: SVT.

Schandl, H., C. G. Grünbühel, H. Haberl, and O. Weisz. 2004. *Handbook of physical accounting: Measuring bio-physical dimensions of socio-economic activities MFA–EFA–HANPP.* Social Ecology Working Paper 73. Vienna: IFF Social Ecology.

Steurer, A. (ed.). 1996. Material flow accounting and analysis: Where to go at a European level. Pp. 217–221 in *Statistics Sweden: Third meeting of the London Group on Natural Resource and Environmental Accounting,* Proceedings Volume, Stockholm, Sweden.

UN. 2000. *Energy statistics yearbook 1998.* New York: United Nations Statistics Division.

UN. 2001. *Indicators of sustainable development: Guidelines and methodologies.* New York: United Nations.

UNCED. 1992. *Rio declaration on environment and development.* Report of the United Nations Conference on Environment and Development. Distr. General, August 12, A/CONF.151/26 (Vol. I).

UNDP. 2003. *2003 Human development report.* New York: Oxford University Press.

Van der Voet, E., L. van Oers, and I. Nikolic. 2004. *Dematerialisation: Not just a matter of weight.* CML Report 160, Section Substances & Products. Leiden, the Netherlands: Centre of Environmental Science, Leiden University.

Vitousek, P. M., P. R. Ehrlich, A. H. Ehrlich, and P. A. Matson. 1986. Human appropriation of the products of photosynthesis. *BioScience* 36:368–373.

Wackernagel, M., and W. Rees. 1996. *Our ecological footprint. Reducing human impact on the earth.* Gabriola Island, BC: New Society Publishers.

15

Sustainable Development and the Use of Health and Environment Indicators

Yasmin von Schirnding

From international meetings held since Rio 1992, it has become evident that health issues are an increasingly important item on the broad environment and development agenda and that environmental issues are receiving more prominence on the public health agenda (von Schirnding 1998). Agenda 21 and the Johannesburg Plan of Implementation, negotiated in Johannesburg, placed much emphasis on the importance of health and included a chapter dedicated to human health. Health was also singled out as one of five priority issues in Johannesburg, along with water, energy, agriculture, and biodiversity (the WEHAB initiative).

The spectrum of health, environment, and development hazards has changed over the millennia of human existence. Yet despite impressive health gains, in many instances the health gaps between and within countries are widening. Sub-Saharan Africa, the world's poorest region, still has average life expectancies far below those of the wealthiest countries. Underlying much of this unequal burden of disease is the fact that environmental factors are a major contributor to sickness and death throughout the world, especially in the poorest regions (World Resources Institute, UNEP, UNDP, and the World Bank 1998).

Sustainable development cannot be achieved where there is debilitating illness, nor can good health be sustained when poor environments prevail (von Schirnding 2001, 2002b). Age-old public health hazards such as inadequate and unsafe food and water, microbiological contamination of the environment, and poor sanitation and environmental hygiene are still prevalent. In addition, new environment and development problems have emerged, some of which appear to threaten the entire ecosystem. The level of economic development and the policy choices of individual countries are important factors determining the nature of the problems faced and the ways in which they are addressed (von Schirnding 2002a).

Health concerns associated with air and water pollution, water supply and sanitation, waste disposal, or chemicals and food may be particularly relevant at the local or micro level (e.g., lead in household dust or environmental tobacco smoke) or may be important at the regional or global level (e.g., depletion of the ozone layer, global climate change, long-range transport of air pollution, or marine pollution). The problems to be dealt with are often simultaneously global and local. Global economic activities, escalation of travel and trade, and the changing use of technology all have significant implications for health and the environment.

Poverty remains the number one killer, with the poor bearing a disproportionate share of the global burden of ill health. Even in rich countries, the poor suffer worse health than do the better-off. Poor children are particularly affected: In the poorest regions of the world, one in five children dies before his or her first birthday, mostly from environment-related diseases such as acute respiratory infections, diarrhea, and malaria. Children are more heavily and frequently exposed to threats to their health in the environment because of their behavior (e.g., hand-to-mouth activity), because they are closer to the ground and more exposed to high concentrations of pollutants, and because they have a higher intake of harmful substances relative to their body weight than adults do. They are also more vulnerable to the ill effects on health because of their immature and changing stages of biological development. They often are particularly prone to preventable injuries and accidents.

Although the many hazards present in the environment today may have various effects on human health, it is difficult to quantify the risks attributable to these hazards with any degree of confidence. There are often difficulties in assessing people's exposures to environmental risk factors (which may vary widely in concentration from place to place and over time), and people's susceptibilities vary according to many factors. In the case of environmental pollution, the links to health are often uncertain and masked by other effects, such as social deprivation and lifestyle. Usually, large-scale, sophisticated epidemiological studies are needed to quantify health effects and to take account of other (nonenvironmental) factors that might influence the health outcomes. In addition, in many regions of the world the infrastructure for monitoring and health surveillance is poorly developed, so that the numbers of people at risk are largely unknown.

The Role of Indicators

Although health, environment, and development problems differ in various regions of the world, as do priorities in respect to their management, in all situations decision makers and the public at large need ready access to accurate information on health hazards associated with the links between development and the environment.

In the health and environment area, as in other areas, information is needed to monitor and assess trends, identify and prioritize problems, develop and evaluate policies and plans, guide research and development, set standards and guidelines, monitor progress, and inform the public. It is important that this data be conveyed in a readily comprehensible way but with due regard to the complexities and uncertainties inherent in the data. This is often a limiting characteristic of health and environment data.

Although there often is an abundance of available data and information (of variable quality) from monitoring and surveillance programs, this information may not always be policy relevant for decision makers. The information may thus be of limited use for informing the public and decision makers of key health and environmental problems and their causes or of possible management actions needed.

Indicators can play an important role in turning health and environment data into relevant information for decision makers and the public. Most important, they can help simplify a complex array of information with respect to the health–environment–development nexus. This way, they provide a synthesis view of existing conditions and trends that provides information for decision making in the public health sector.

Building on commonly accepted definitions of indicators, Briggs et al. (1996) define an environmental health indicator as "an expression of the link between environment and health, targeted at an issue of specific policy or management concern and presented in a form which facilitates interpretation for effective decision-making." Embodied in this definition is the concept of a link between a factor in the environment and a health outcome. An environmental or health outcome indicator can thus be regarded as an indicator of a health–environment relationship if there is some connection between the health indicator and the environment or between the environmental indicator and health.

Although composite measures are often used in health (e.g., a composite measure of the burden of disease based on the concept of disability-adjusted life years combines the years of healthy life lost due to premature death, disability, or disease; Murray and Lopez 1996), simple descriptive indicators are often very useful. They can be used to obtain baseline information on which to formulate subsequent policy options and plans and assess trends.

At all levels (global, regional, local), indicators that describe the overall state (quality) of the environment and highlight factors influencing environmental quality and potential impacts on and links with human health can be useful. They can provide an overview, a snapshot of a situation, or a profile of environment and health conditions, thereby exhibiting links and trends. They are visual depictions of data that, once combined, may reveal something about the assumed link between various factors. In this regard, the indicator framework described in this chapter may be applicable. However, indicators normally should not be used for the purpose of establishing causal links between factors in the environment potentially affecting human health. For this, sophisticated epidemiological studies in various settings and under differing conditions are usually needed and cannot be substituted by indicators.

Health and Environment Indicator Frameworks

Decision makers can use an indicator framework to obtain a better picture of the links between the complexity of factors in the environment–development process that might potentially influence human health. One such indicator framework is an adaptation of the pressure, state, response (PSR) framework developed by the OECD, which in turn is based on earlier work done by the Canadian government.

The PSR framework has been particularly useful in representing the way in which pollution affects the environment, for example by looking at the various pressures exerted on the environment, which affect the state (quality) of the environment and consequently demand a response for dealing with the situation. However, this framework has been criticized for being linear and unidirectional, and various adaptations have been proposed.

Driving Forces- Pressures- State- Exposures- Effects- and Actions (DPSEEA) Framework

From the perspective of human health impacts, both exposures and the resulting human health effects must be represented. These aspects have been taken into account in a further adaptation of the framework for health purposes, the DPSEEA framework (WHO 1995; Briggs et al. 1996). It is a descriptive representation of the way in which various driving forces generate pressures that affect the state of the environment and, ultimately, human health, through the various exposure pathways by which people come into contact with the environment.

People may become directly exposed to potential hazards in the environment when coming into direct contact with these media through breathing, drinking, or eating, for example. A variety of health effects may result, ranging from minor, subclinical effects (i.e., effects that may not manifest in overt symptoms) through to illness and sometimes death, depending on the intrinsic harmfulness of the pollutant, the severity and intensity of exposure, and the susceptibility of the person exposed (e.g., the elderly, the young, and the sick may be more susceptible than others).

Various actions can be implemented at different points in the framework and may take different forms. They might involve the policy development, standard setting, technical control measures, health education measures, and treatment of people with diseases.

Although the DPSEEA framework, like the PSR framework on which it is based, represents the various components in a linear fashion in order to more clearly articulate the connections between factors influencing health and the environment, in reality the situation is much more complex, with various interactions occurring at different levels between various components. The different components of the DPSEEA framework are given in Figure 15.1. The framework can be applied to information gathering and indicator development at the national level, the sectoral level, or the community or neighborhood level (Hammond et al. 1995; WHO 1997).

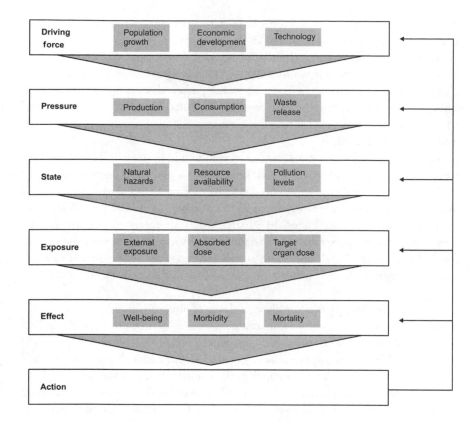

Figure 15.1. Components of the DPSEEA framework.

Core Indicators

There has been a lot of debate about and interest in the concept of a set of core indicators that can be used on a global basis to examine overall trends in environment and health conditions worldwide. Opponents of such a concept have argued that environment and health problems and priorities for their management differ significantly in various regions of the world, as do monitoring and analytical capabilities and resource availability, making it problematic to establish a core set of indicators that have universal applicability. Problems in standardizing definitions and difficulties in ensuring quality control procedures on a worldwide basis are other complicating factors.

On the other hand, most countries, regardless of their level of development or of other sociopolitical or cultural realities, deal with certain problems that are of universal significance. In the environmental domain these might include air quality, water and sanitation, food safety, waste disposal, and toxic substances. Although the specific dimensions of these problems will differ from country to country and within countries, sets of universally applicable indicators could be valuable in terms of improving shared knowledge on factors affecting the state of the global environment and their effects. Common sets of indicators have other obvious benefits. They enable aggregation at various levels (e.g., local, country, regional, global). They also provide momentum to countries for achieving uniform and rigorous standards. There may also be national reporting requirements under international treaties that may necessitate standardized indicators internationally.

The identification of a limited number of common indicators, based on those currently accepted and widely used by countries, is thus a potentially important tool for the harmonization and rationalization of indicators. Establishing agreement on such a limited set will significantly lessen the data reporting burden on countries. Where user needs are similar, indicators should be harmonized. Efforts by government departments, agencies, nongovernment organizations, civil society, and donor communities should be coordinated and should aim at strengthening data collection and management. Existing data should be drawn on as much as possible, with due recognition to its limitations.

Although standard, internationally agreed sets of indicators fulfill a major international role for between-country comparisons, nations may need other specific indicators to enable them to develop and evaluate national policies and plans. Any core set of indicators will have to be augmented in view of particular national, regional, and local policy concerns. Some indicators naturally will be more relevant at a national or global level (e.g., climate change), whereas others will be more locally relevant (e.g., drainage problems, problems with solid waste). Many issues (e.g., ambient air pollution) may necessitate management over different levels of government. At the national level, indicators to inform the setting of policies and standards may be fundamental, whereas at the local level indicators regarding service delivery and implementation of policies may be key. Information on these indicators could be collected and obtained at different geographic levels of resolution, for example at the local, national, and global levels.

The roles and responsibilities in respect to various environment and health management functions at different levels of government, the degree of decentralization of powers and functions, and other factors such as data availability and quality influence the extent to which it makes sense to examine data at different levels for international comparison purposes. Regardless of the level at which the data are aggregated and examined, however, most information normally will need to be collected in at the lowest level of resolution as is practicable and feasible.

Obtaining relevant data at country level remains a significant problem (particularly in poor countries), although most countries have some sort of health information system, even if fairly rudimentary. Nevertheless, there are often discrepancies in diagnosis, notification, and reporting and differences in referral procedures and misclassification of diseases. Differences in environmental sampling and measurement methods often affect results, and the data sometimes are unrepresentative. For all these reasons, procedures for checking accuracy, consistency, and comparability should be introduced. Uncertainty, when it exists, should be communicated effectively to the public and decision makers who must act on the basis of this information.

Development of Health Indicators

The World Health Organization (WHO) has been involved in efforts to develop indicators over a number of years. Efforts were most intensive in the mid- to late 1990s. Although not all indicator sets developed during this period are still in use, they may be of interest to those developing health indicators for other purposes. For example, indicators (and targets) were developed to assess WHO's Health-for-All (HFA) policy (WHO 1996b).

The purpose of the HFA indicators was to guide member states in the evaluation of their national strategies for HFA and to follow up on the implementation. HFA indicators dealt with trends in policy development, socioeconomic development, health and environment, health resources, health systems, health services, and health status. The framework used was based mainly on health services, health status, health determinants, and health resources. Various regions were also involved in the efforts to develop HFA indicators, as were individual countries (van der Water and van Herten 1996). In the late 1990s a new set of targets, incorporating indicators, was developed for the renewed HFA policy (WHO 1998).

Global indicators have also been used for reporting purposes in the World Health Report of WHO (2000), and health and health-related indicators have been used extensively in various regions (WHO/PAHO 1997). Over the years WHO has also developed various program indicators to monitor the health of infants and young children, the health of women, and the health of the general population. The indicators have been categorized according to whether they are outcome related (concerned with health status or death) or process related (concerned with health care delivery and management) or whether they are determinants (e.g., behavioral factors or environment and development factors that influence health outcomes). The indicators were intended to be used by public health administrators and health program and service managers (WHO 1996a).

Much work has also been done on indicators for environmental health (WHO 1995; Corvalan et al. 1997). *Linkage Methods for Environment and Health Analysis* (Briggs et al. 1996) deals with methods for linking health and environmental data and the application of indicators to quantify and monitor environmental health conditions. Field studies were carried out to obtain information on aspects of environmental health status and particular environmental health problems in the study areas (WHO 1995). No uniform set of environmental health indicators has been recommended by WHO, but suites of indicators that can be selected from for various purposes have been compiled, as have updated methodology sheets for constructing selected indicators (von Schirnding 2002a; Briggs 1999).

Work has been done at WHO on the development and use of health and environment indicators in the broader context of sustainable development, which emphasizes intersectoral planning processes and the way in which indicators have been used in elements of the planning cycle. Indicators are highlighted by media (e.g., air, water) and by sector (housing, agriculture, transport), and case studies of application at the national and local levels are presented (von Schirnding 2002a).

In addition, regions have been active in developing indicators for use in their country contexts. One such example is the European region, which has been developing a suite of environmental health indicators and, in particular, has focused on the application of these indicators in four topic areas: air pollution, noise, transport accidents, and water and sanitation. A pilot study has demonstrated the usefulness of indicators for assessment and reporting while also demonstrating the limitations of routinely collected data (WHO 2004).

The pilot study is part of the process of developing an environment and health information system by the WHO Regional Office for Europe, in collaboration with a number of countries, the European Environment Agency, and the European Commission. The process involves the selection of policy-relevant issues and the development of indicator methods as well as feasibility and pilot testing and the selection of core sets of indicators in thematic areas (Box 15.1).

Initial results from the pilot study show that indicators are powerful communication tools for policymakers, experts, and the general public. When fed into the policymaking process, they can evaluate and demonstrate the effectiveness of environment and health policies and facilitate the setting of priorities among competing policies.

Key lessons learned include the following:

- It is important for core sets of indicators to be chosen to minimize the additional burden of collecting and reporting data. Where indicators use existing sources, additional costs are not necessarily incurred.
- Indicators can shed light on environmental risk factors and health effects, their determinants, and the actions taken, thereby highlighting the potential impact of environment and health policy on the health of the population.
- Indicators have been able to document several examples of good practice. Across Europe, examples range from the banning of coal sales in Dublin, which led to reductions in air pollution and mortality; ecological taxation in Germany, which reduced exposure to PM10; noise reduction policy in the Netherlands, which reduced exposure to road noise despite a doubling of traffic volume; and the Bathing Water Directive, which resulted in significant improvements in recreational water quality in the EU between 1992 and 2002.
- The indicators have also helped provide a uniform approach to tracking progress in environment and health status, by monitoring time trends in individual countries or in a group of countries, and have also facilitated intercountry comparisons.

Box 15.1. Indicators and associated DPSEEA links.

Air pollution	Passenger transport demand by mode of transport	Driving force
	Road transport fuel consumption	Driving force
	Emissions of air pollutants	Pressure
	Exposure to ambient air pollutants (urban)	Exposure
	Years of life expectancy lost in 1 year	Effect
Noise	Population annoyance from noise	Effect
	Application of regulations, restrictions, and noise abatement measures	Action
Transport accidents	Mortality from transport accidents	Effect
	Road accident injuries	Effect
Water and sanitation	Urban wastewater treatment	Pressure
	Drinking water exceedances of microbiological guidelines	State
	Microbiological quality of recreational waters	State
	Access to piped, regulated drinking water	Exposure
	Outbreaks of waterborne diseases	Effect

Source: WHO (2004).

In the future, countries should be able to select indicators based on policy needs, feasibility, and scientific rationale and will be able to combine this information with other evidence to describe the potential for interventions and improvements in public health practices, including surveillance programs.

However, results to date also indicate that the level of comparability of indicators across Europe is limited, often because of deficiencies in surveillance and reporting methods in some countries. There is a need for progressive development and harmonization of data collection and processing (WHO 2004).

However, the DPSEEA model, on which the European pilot study was based, has been acknowledged as being an oversimplification of reality that, when read too literally, can seriously mislead (Briggs 2003). With its emphasis on anthropogenic causes, it is most relevant for hazards such as pollution but is less effective for other environmental health hazards and may neglect the complexity of environment and health associations and the multiplicity of risk factors and health effects involved. Many other models and frameworks exist, based on knowledge of the epidemiology of health–environment interactions and the causal pathways and complexities involved.

Work is under way to develop indicators to improve children's environmental health, with the launch of an initiative at the World Summit on Sustainable Development (WSSD). Indicators for children's environmental health are being pilot tested in various parts of the world (Briggs 2003). A model has been developed that is more flexible than the DPSEEA model and allows the consideration of multiple exposures and effects. It emphasizes that exposures in different settings can lead to many different health effects and that these are affected by contextual conditions such as social, economic, and demographic factors. Actions can be targeted at the exposures, the health effects, or the underlying contexts.

Although this might seem self-evident to environmental health practitioners, the model is perhaps of particular use to those with limited understanding or appreciation of the nature of health–environment associations and interactions, especially to non–health experts working outside the health sector. In reality, of course, exposures normally are not limited to one setting, such as home, school, or neighborhood, but rather transcend different settings and must be looked at in their totality. Thus, none of the existing models or frameworks are all-embracing, and they all have limitations (Briggs 2003).

In general, experience with the development and application of health and environment indicators to date confirms their potential usefulness in monitoring environmental health conditions, tracking progress, and informing the development and evaluation of policy. In 2000 a group of researchers, practitioners, and health professionals met in Canada to discuss the challenges in environmental health monitoring and surveillance and to discuss the possibility of developing consensus on some key issues. It was agreed that there was a need to develop environmental health indicators for rural as well as urban conditions and to expand the work done on environmental health indicators to encompass the social, economic, and political environments in addition to that of the physical environment (Furgal and Gosselin 2002).

Because the relationships between health and the environment are so complex, it is often difficult to know what to measure in any particular context when monitoring the status of environmental compartments, human health, and the relationship between them (Furgal and Gosselin 2002). Measuring all factors in the relationship chain would be too time-and resource-consuming, and it would necessitate the identification and monitoring of the health status of particular at-risk groups, such as children, who are most vulnerable. There is also a need for greater understanding of the processes of collecting, interpreting, and drawing conclusions from indicators for effective use in decision-making processes.

Capacities differ between jurisdictions, countries, and continents, calling for greater cooperation, coordination, and commitment between governments and agencies to take advantage of the benefits of new information technologies. This is of particular importance in relation to issues of global relevance. Key challenges for the future, as highlighted in the Canadian meeting on environmental health indicators, include issues of scale, capacity, data comparability, and reliability.

Literature Cited

Briggs, D. 1999. *Environmental health indicators: Framework and methodologies.* Geneva, Switzerland: WHO.

Briggs, D. 2003. *Making a difference: Indicators to improve children's environmental health.* Geneva, Switzerland: WHO.

Briggs, D., C. Corvalan, and M. Nurminen. 1996. *Linkage methods for environment and health analysis.* Geneva, Switzerland: UNEP/US EPA/WHO.

Corvalan, C., T. Kjellstrom, and D. Briggs. 1997. Health and environment indicators in relation to sustainable development. Pp. 274–283 in *Sustainability indicators,* SCOPE 58, edited by B. Moldan and S. Billharz. Chichester, UK: Wiley.

Furgal, C., and P. Gosselin. 2002. Challenges and directions for environmental public health indicators and surveillance. *Canadian Journal of Public Health* 93(Suppl. 1) S5–S8.

Hammond, A., A. Adriaanse, E. Rodenburg, D. Bryant, and R. Woodward. 1995. *Environmental indicators: A systematic approach to measuring and reporting on environmental policy performance in the context of sustainable development.* Washington, DC: World Resources Institute.

Murray, C., and A. Lopez. 1996. *The global burden of disease: A comprehensive assessment of mortality and disability from diseases, injuries and risk factors in 1990 and projected to 2020.* Cambridge, MA: Harvard University Press.

van der Water, H., and L. Van Herten. 1996. *Bull's eye or Achilles' heel: WHO's European Health for All targets evaluated in the Netherlands.* Leiden: TNO Prevention and Health.

von Schirnding, Y. 1998. Addressing health and environment concerns in sustainable development, with special reference to participatory planning initiatives such as Healthy Cities. *Ecosystem Health* 3(4):220–228.

von Schirnding, Y. 2001. Health in the context of Agenda 21 and sustainable development: Meeting the challenges of the 21st century. *Sustainable Development International* 3:171–174.

von Schirnding, Y. 2002a. *Health and environment in sustainable development planning: The role of indicators.* Geneva, Switzerland: WHO.

von Schirnding, Y. 2002b. Health and sustainable development: Can we rise to the challenge? *Lancet* 360:632–637.

WHO. 1995. Health and environment analysis and indicators for decision-making. *World Health Statistics Quarterly* 48(2).

WHO. 1996a. *Catalogue of health indicators: A selection of important health indicators recommended by WHO programmes.* Geneva, Switzerland: WHO.

WHO. 1996b. *Evaluating the implementation of the strategy for Health for All by the year 2000. Common framework: 3rd evaluation.* Geneva, Switzerland: WHO.

WHO. 1997. *Health and environment in sustainable development.* Geneva, Switzerland: WHO.

WHO. 1998. *Health-for-All in the 21st century.* Geneva, Switzerland: WHO.

WHO. 2000. *World health report.* Geneva, Switzerland: WHO.

WHO. 2004. *Environmental health indicators for Europe: A pilot indicator-based report.* Background document for Fourth Ministerial Conference on Environment and Health. EUR/04/5046267/BD/4. Copenhagen: WHO.

WHO/PAHO. 1997. *Basic indicators.* Washington, DC: PAHO.

World Resources Institute, UNEP, UNDP, and the World Bank. 1998. *1998–1999 world resources: A guide to the global environment. Environmental change and human health.* Oxford: Oxford University Press.

16

Biodiversity Indicators

Reinette (Oonsie) Biggs, Robert J. Scholes,
Ben J. E. ten Brink, and David Vačkář

Human actions compromise the information content of the biosphere, contained in its biological diversity, or biodiversity. The fundamental logic of biodiversity conservation is that the variety of living things matters for a range of utilitarian (human-centered) and intrinsic reasons. Variation is the raw material of evolution and the source of novel and useful biological products, forming the basis for activities such as plant and animal breeding and the development of pharmaceutical products. Biodiversity is also important for ecosystem functioning (MA 2005). An ecosystem composed of very similar organisms will react differently to imposed stresses than one composed of dissimilar organisms, although general predictive rules remain elusive. Biodiversity has aesthetic appeal in all cultures and underlies many recreation and tourism activities. Many people share a moral imperative to conserve a representative sample of the full range of biological variation. In short, it is widely agreed that more diversity is better than less, especially in the context of natural or self-regenerating ecosystems, but the critical limits are unknown.

Many biodiversity indicators have been proposed (Delbaere 2002a; Reid et al. 1993; CBD 2003c, 2003e), but as yet none have been universally accepted and applied (Royal Society 2003). The adoption by the World Summit on Sustainable Development and the Convention on Biological Diversity (CBD) of a target to reduce the rate of biodiversity loss by the year 2010 (CBD 2003a, 2003d) has accentuated the need to achieve convergence on this issue. As a result, the international discussion on indicators has been accelerated and preliminary agreement on five trial indicators reached within the CBD (2003b).

The difficulties in establishing operational indicators of biodiversity stem from three main sources: the inadequacy of much of the data, the loss of information that occurs when a complex and multidimensional concept is reduced to a one-dimensional indicator, and our rudimentary understanding of the causal links between biodiversity and ecosystem function. This chapter presents the basic concepts important to monitoring biodiversity, provides a broad overview of current developments in biodiversity indicators, and outlines future possibilities in this field. We focus on indicators at the national to global levels in order to support policy decisions related to sustainable development.

Theoretical Basis for Biodiversity Definitions

A scientific definition of biodiversity might be "the complexity of living systems at all organization levels." Many definitions are valid in the context of their specific use, and no simple definition can cover all aspects: natural versus human-altered diversity, evenness versus richness, the various spatial (α, β, and γ biodiversity) and temporal dimensions (phylogenetic biodiversity), and the biological incompatibilities of increasing diversity at all organizational levels simultaneously. Because of the broad definition, it is very difficult to derive verifiable targets and measurements of biodiversity at this time. Which biodiversity should be conserved at the expense of which other biodiversity?

The conceptual framework attributed to Noss (1990) is a useful way to organize the various interrelated facets of biodiversity (Figure 16.1). It proposes that biodiversity is expressed at three main levels of organization (ecosystem, species, and gene) and in three aspects (compositional, structural, and functional). Thus, focusing for instance at the species level, a sample composed of several different species is more diverse than one with fewer species. Even with only one species, a sample that includes both big and small individuals, perhaps organized into clumps, is structurally more diverse than one in which all the individuals are the same size, organized in a perfect grid. If several species were present but all did exactly the same thing, functionally they would be less diverse than a sample that included species that had very different roles (e.g., a plant, herbivore, carnivore, and decomposer).

The same three aspects of biodiversity can be defined at the supraorganism scale (the ecosystem) and the suborganism scale (the gene). A landscape that is a mixture of forest, grassland, and cropland is compositionally, structurally, and functionally more diverse than one that is forest only. A cloned crop in which each individual is identical to every other is less diverse than a traditional landrace that contains genetic variation. On land, ecosystems are mapped almost entirely based on the distribution of easily identified plant structural formations, such as forests or grasslands. Structure often is closely linked to function; the biggest potential divergence is between composition and function. For instance, the microclimate in a forest is more closely related to the relative proportion of tall, medium, and short plants than to the variety of species present.

The fundamental evolutionary process generating biodiversity is mutation and its stabilization within populations by speciation. This is a slow and still poorly understood process. Ultimate biodiversity loss occurs by extinction (of specific genes, species, or ecosystems), which occurs at an unsteady and only roughly quantified rate, even in the absence of human threats. It is nevertheless apparent that the earth is in a period of net biodiversity loss (Leakey and Lewin 1995).

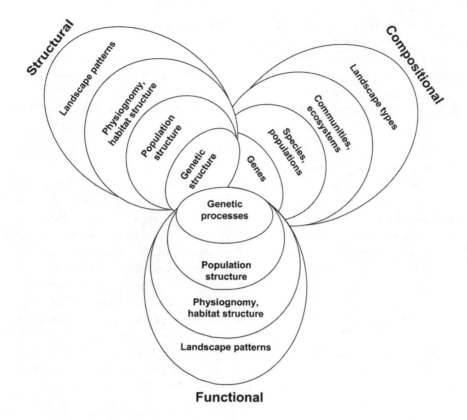

Figure 16.1. Compositional, structural, and functional attributes of biodiversity at four levels of organization (Noss 1990).

However, biodiversity loss entails much more than extinction and occurs at all levels of organization (CBD 2003a, 2003b, 2003d). At the ecosystem level it consists of a reduction in the extent, condition, or productivity of ecosystems; at the species level it consists of a decline in the abundance (ultimately extinction), distribution, or sustainable use of populations; and at the gene level it consists of loss of gene-level diversity within populations (genetic erosion). Anthropogenically driven biodiversity loss manifests as a decrease in abundance of many species and an increase in a few, leading to homogeneity at multiple scales.

Absolute biodiversity loss has to be measured relative to some baseline state, whose choice raises both political and practical concerns. A baseline state is not needed in order to determine relative biodiversity loss, but at least two measurements in time are needed. Four baselines have been proposed for use by the CBD (2003e):

Before any human interference (which is both illogical and unfeasible for many long-
    inhabited parts of the world)

Before major interference by industrial society (recommended as most appropriate)

From the time the CBD entered into force (i.e., 1993; results in a bias toward devel-
    oped countries)

Extinction threat according to the World Conservation Union (IUCN) Red List cate-
    gories

    Other possible baselines include viable population levels, species richness, a specific reference year, maximum sustainable yield levels, or a defined desired state.

    Richness and evenness are another set of ideas widely used in biodiversity observation. They have their origin in information theory (Pielou 1969). Information diversity indices combine the number of classes with the proportions in each particular class. Richness is a count of how many different varieties (classes) exist in a given sample. Evenness measures the degree of dominance among the different varieties. Many ecological diversity indices have been proposed (Magurran 2004). Interpretation of these indices is complex, and a range of approaches have been suggested (French 1994; Iszák and Papp 2000). Such indicators do not distinguish between native, introduced, and cultivated species, which is often desired by policymakers. A native species, as opposed to an introduced species, is one that occurs naturally in a particular area. Cultivated species usually are specifically bred for purposes of cultivation. A minimum description of diversity includes both richness and evenness concepts, in the same way that an analysis of poverty usually includes measures of both absolute income and income inequality.

α, β, and γ *Biodiversity*

Alpha (α) biodiversity is the number of types in a homogeneous patch, usually a very small area. Gamma (γ) biodiversity measures diversity at a more regional scale. Beta (β) biodiversity measures the species turnover between adjacent ecosystems. Alpha biodiversity can grow because of introduced species, whereas β and γ biodiversity decline because the systems share an increasing set of common species, the so-called homogenization process.

States, Drivers, and Proxies

As in all indicator systems, it is useful to distinguish between indicators of the state of biodiversity and indicators of the things causing changes in the state (drivers, e.g., harvest pressure, pollution, and fragmentation) and responses to the state (e.g., the area under formal protection). Because of data deficiencies, we are often forced to use proxies and surrogates as indicators of state and drivers. Exhaustive surveys in highly diverse regions cost a great deal of time and money. Therefore, simpler surrogates for mapping diversity, such as higher-taxon richness or the richness of indicator taxon groups, have been proposed and tested (Williams et al. 1998), with partial success. Bioindicator approaches measure the abundance of sensitive species (sentinels or detectors) or genetic markers in assessing ecosystem health but are not measures of biodiversity per se.

Biodiversity Indicators That Have Been Proposed

Several hundred indicators to measure the status of biodiversity have been proposed or are under development. There are indicators for all levels of biodiversity (genetic, species, and ecosystem) and all aspects of biodiversity (composition, structure, and function), but the amount of effort invested in these different areas has been uneven.

Gene-Level Indicators

At the subspecies level, diversity can be measured by richness counts of the number of different varieties within a species (e.g., the number of landraces of one crop) or the number of alleles present in a metapopulation. These measures should be accompanied by evenness measures of the relative dominance of the varieties.

Genetic biodiversity can be estimated by measuring appropriate molecular markers. Most studies on genetic variation within species are based on random markers. Current work is investigating the use of markers targeted at genes exhibiting ecologically relevant activity (Tienderen et al. 2002). Indices of relatedness, whether based on cladistics (Clark and Warwick 1998) or genetic distance (Nei 1972; Takezaki and Nei 1996), have potential for measuring gene-level diversity both between and within species. They have the advantage of not requiring the actual genes to be exhaustively inventoried. Examples are the measures that have been used to estimate soil microbial biodiversity (Pankhurst et al. 1997) when the species are unknown (and the species concept may not even apply).

Despite very rapid developments in this field, it is our assessment that robust and widely applied gene-level diversity indicators are still a decade away. Until then, the species, imperfect as it is, is the most robust scientific basis for constructing most indicators.

Species-Based Indicators

A conceptual problem that faces all species-based indicators is whether all species matter equally. If not, on what basis is the weighting assigned? Where a subset of species is used for practical reasons, can they be assumed to represent the unsampled species?

Species richness (i.e., the number of recognized and recorded species within a given set of taxonomic groups and within a given bounded area) is the simplest and most widely used biodiversity index. The data are derived from collection labels in museums or field observations and may be extrapolated over unsampled locations to create inferred species distribution maps, using either explicit techniques such as habitat modeling using GLIM or GARP (Crawley 1993; Stockwell 1994) or expert judgment. The principal problems with species richness as a biodiversity indicator are as follow:

It is critically dependent on the quality and completeness of collection and classification. Changes in richness often are simply a result of taxonomic revisions or observations rather than changes in the underlying biodiversity. Some stability has been achieved in richness numbers for well-known taxa such as birds, mammals, and some plant groups (Govaerts 2001; Bramwell 2002; Scotland and Wortley 2003) in well-studied areas, but for most of the biological realm, less than half of the postulated extant species have been formally described (WCMC 2000). In some parts of the world, even the plant and vertebrate observations are extremely incomplete (Prance et al. 2002).

Richness is strongly spatially dependent. Typically, the species richness in an ecological region rises to an asymptote in relation to the area sampled (the species–area curve). The shape of the curve varies between ecological regions and taxa and can itself be used as a biodiversity index (Cowling et al. 1989). Comparing species richness between locations without taking this relationship into account is misleading. Special procedures, such as rarefaction (Hurlbert 1971), correct for biases related to unequal sampling area.

Species richness is a very insensitive indicator of biodiversity loss. It provides no indication of changes in the abundance of component species in a community or of changes in their phylogenetic and functional diversity traits. A decrease in richness occurs only through extinction. Extinction is the loss throughout the world of a species or variety, whereas extirpation is the loss of a species or variety in a portion of its range.

Species richness per se does not distinguish between native and introduced species. In disturbed areas total species richness may increase because of introduced species, while populations of native species are reduced but not entirely extirpated. In this case species richness may provide perverse signals. It is therefore recommended (CGER 2000) that counts of introduced species be kept separate from those of native species. Maximizing richness by introducing alien species is a perverse objective.

Extinction, or the risk of it, has been widely used as a measure of biodiversity loss (e.g., IUCN 2002). It has the advantages of being extremely easily grasped and compelling, but as a biodiversity management indicator it has several drawbacks:

It is surprisingly difficult to prove that a species is extinct. How do you know that it is not just hiding?

If a species is not known to science, its extinction is invisible.

By the time a species has gone extinct (and been shown to be so), it is far too late to do anything about it. For this reason, the IUCN has created a variety of lev-

els of threat, which are indicators of the last phase of the lengthy process
of biodiversity loss.
Species extinction is a natural process, balancing, in the long term, the process of speciation. What rate of extinction is clearly too high?

Endemism is a widely applied refinement on species richness. This is the number of
species found only in a specified area and nowhere else on Earth (in the wild). Endemic
richness, at the species or higher taxonomic level and sometimes in conjunction with
an assessment of threat, is often used as an indicator of biodiversity hotspots (Reid et
al. 1993; Williams et al. 1998; Myers et al. 2000). Obviously, the loss of an endemic
species within its entire range is more critical than the local loss of an otherwise widespread species. However, endemism has the same problems as species richness.

COMPLEMENTARITY MEASURES

A sophisticated set of indices have been developed by conservation biologists for the purpose of optimizing the design of networks of protected areas (Margules and Pressey
2000). They are based on the assumption that a system of protected areas should represent the biodiversity in each region and separate it from processes that threaten its existence. Complementarity measures thus are based on quantitative targets that specify
how much of the landscape is needed to conserve a representative set of sample species
or habitats. Additionally, they may include considerations of threat or vulnerability.

PHYLOGENETIC AND EVOLUTIONARY INDICATORS

The fundamental unit of biodiversity is an evolutionary one (Faith 2002). Santini and
Angulo (2001) propose an index for the estimation of evolutionary potential, which
they define as the potential of a member of the genealogical hierarchy to persist for ecologically significant periods of time. It is the balance between diversification and extinction of the evolutionary unit considered. Phylogenetic diversity measures indicate the
amount of branch length or evolutionary history spanned by a set of taxa (Faith and
Williams 2006).

Population Abundance–Based Indicators

These indicators are based on trends in the abundance (i.e., number, biomass, or cover)
of individuals in a target population, which may be the total global population for that
species or some well-defined part of it. Abundance-based indicators provide the
advance warning of impending loss that richness-based indicators do not because the
underlying information is continuous rather than binary (present or absent). Such
indicators therefore are more sensitive and useful for policy purposes than richness-based
indicators, particularly for setting policy targets.

Hughes et al. (1997) estimate that there are on average about 200 separate populations per species. Because of the amount of effort needed to get reliable, repeated abundance estimates of wild species from just one population, such censuses are limited to

a tiny fraction of all known species. They are often either charismatic species, such as elephants, whales, and migrating waterfowl, or useful species, such as fish, timber, and domestic animal stocks. The Global Conservation Organization (WWF) marine ecosystem indicators are an attempt to apply population measures with a rapid, standardized sampling protocol (Wilkinson 2000). The Living Planet Index (Loh 2002) and species trend indices (e.g., Gregory et al. 2002) are a population approach based on a small set of species selected to represent the major groups. The Natural Capital Index and the Biodiversity Intactness Index are in principle both population-based measures.

Changes in species populations may be correlated with changes in diversity at the genetic level. The smaller the population, the smaller its genetic variation, particularly if losses are concentrated around the edges of the species' area of distribution.

Not all species have to be measured to indicate the overall process of biodiversity loss. A representative sample would be sufficient, and a deliberately selected set of indicators may be even more sensitive if the selection was well founded. Population-based indicators are the focus for much of the current activity in biodiversity indicator development.

Ecosystem-Level Area-Based Indicators

Area-based indicators can be considered as abundance-based measures at the ecosystem level, corresponding to population measures at the species level. They typically express the area (km^2) at a particular time of a defined ecosystem. The area may be expressed as a fraction of some reference state, such as the supposed original (potential) area. A well-known example is the *Global Forest Resources Assessment* (FAO 2001). Others are assessments of coral reefs (Wilkinson 2000) and mangroves (FAO 2003). Biodiversity, at all levels, needs an area in which to exist. When half the habitat of a given species is lost, the abundance of that species is roughly halved. This makes area an easily understood and easily measured indicator.

Fragmentation indicators are theoretically based on concepts of island biogeography, which predict that as fragments are isolated from a larger mass, they will lose species. They vary from the simple (e.g., mean fragment size, perimeter length to bounded area ratio) to the sophisticated (fractal-based indices). Road network density has been used as a proxy for fragmentation.

Area-based indicators are built on readily observable structural features, such as the cover by trees or coral reefs, and thus are typically at the highest level of ecosystem classification, the biome. In principle they are straightforward and uncontroversial, but in practice it is hard to ensure uniformity of application of the definition. As a result, the variation between different sources of information (compare FAO 2001 to Achard et al. 2002, for instance) usually is much higher than the temporal trend, at least in the short term. However, the relative temporal trend usually is consistent in direction, if not magnitude. Only a few time series of trends in biome extent exist at the global level (Jenkins et al. 2003).

It is not clear what the equivalent boundaries for oceanic ecosystems are because their edges would be expected to be somewhat more variable in time. The large marine ecosystems (Sherman et al. 1990) have fixed spatial definitions and therefore are not useful as indicators by themselves (although the fraction degraded within them would be useful). The marine biogeochemical provinces (Longhurst 1998) are mapped largely using remote sensing of sea surface temperature and chlorophyll content and are highly dynamic over a period of weeks, making them too variable to be useful in the short term. Marine ecosystems therefore may lend themselves better to indicators at the species or community (e.g., seagrass, coral reefs) level.

Functional Indicators

Indicators of functional biodiversity are least developed. There are many theories predicting a positive relationship between compositional biodiversity and functional attributes, but the generality and form of the link remain contested. It is unclear whether the link rests on biodiversity in general or on the presence of specific functional groups, or niche complementarity (Loreau et al. 2002). The debate on the functional link between biodiversity and ecosystem net primary productivity is converging on a conclusion that a weak positive link exists, which tends to level off at modest levels of plant diversity (on the order of ten species) (Naeem 2002; Kinzig et al. 2001).

For policy purposes it may be pragmatic to view function as synonymous with the capacity of ecosystems to supply ecosystem services, briefly defined as the benefits people obtain from ecosystems (MA 2003). For example, the capacities to supply water, fish, timber, food, or carbon storage services can be regarded as indicators of ecosystem function. Data on services often exist, but their link to biodiversity remains obscure. In many cases, an equivalent quantum of service could be supplied by a less diverse ecosystem, though perhaps with less resilience.

The Millennium Ecosystem Assessment (MA 2003) makes a useful distinction between biodiversity as an ecosystem service in its own right and biodiversity as an underlying condition necessary for the delivery of many other services. It is easy to confuse natural resources with biodiversity. For example, almost all the food we eat has its origin in a plant or animal, which in turn are components of biodiversity. But it is generally not the diversity of that source that is uppermost in our mind when we think of the nutritional service. Similar nutrition could be, and increasingly is, derived from a very small variety of sources. Nevertheless, there is a key role for biodiversity in food supply: It provides the source of variation needed to adapt the crop to a changing environment and to reduce the risk of catastrophic failure in the event of a widespread stress (e.g., drought or disease). The case for a close connection between biodiversity and human well-being is more direct for the regulating ecosystem services, such as control of pests and diseases.

Although monetary measures of natural capital (Costanza et al. 1997; De Groot 1992), which directly measure biodiversity function in monetary terms, are similarly of limited utility as biodiversity proxies, their ability to link to social and economic drivers and human use is a valuable attribute.

Non human-centered measures of ecosystem function usually boil down to direct or indirect measures of biogeochemical cycling, such as net primary productivity and evapotranspiration. The greenness of the land or ocean surface and its spatial and temporal variability are proxies for this type of measure that lend themselves to remote sensing.

At the species level, species can be reclassified into functional types or guilds: groups of species that share, to some degree, attributes such as basic physiology, reproductive strategy, trophic position, and response to stress or disturbance (Smith and Schugart 1996). Richness and evenness measures can then be applied to functional types rather than species. The problem is that there is no standard for the definition of functional types, which can be endlessly subdivided until species or even lower basic units are reached.

Integrity Indicators

Ecological integrity is the capacity to maintain a balanced, adaptive community of species having species composition, diversity, and functional organization comparable to those of natural habitats in the particular region. An example of an indicator is the Index of Biotic Integrity (Karr 2002), which scores a range of aquatic community parameters against defined standards. Terrestrial versions have also been proposed (Andreasen et al. 2001). Some integrity indices are based on the levels of connectedness and self-organization in ecosystems (Kutsch et al. 2001). Ecological health concepts focus on assessment of the functional aspects of an ecosystem, without necessarily referencing them to a natural state.

Inferential and Composite Indicators

There is a great temptation to reduce complexity by combining a range of indices into a single measurement. The Human Influence Index (Sanderson et al. 2002) is an example. Strictly speaking, doing this by summation or averaging makes mathematical sense only if the individual indices have the same units and measurement scales. Many social and economic indicators (such as the Human Development Index, UNDP 2003) are of such a multifactorial nature. Sometimes the problem of incommensurability is addressed by normalization of all the numbers to some benchmark or weighting of the various components (in which case the weightings are implicitly unit conversion factors). Axis scores in principal component analysis (or similar techniques) fall into this category. Multiplicative indices may make more mathematical sense but should be based on a sound conceptual model, or else they will make no ecological sense.

Given the complexity and variety of biodiversity in and between ecosystems, composite indicators probably are one of the few ways in which information can be made digestible and understandable for politicians and the public (CBD 2003a, 2003c). However, such indicators must be transparent about the way the various factors are transformed and combined. It must be possible to delve down into the individual components that make up the index and disaggregate them spatially.

Biodiversity Indicators in Use

Only a fraction of the hundreds of biodiversity indicators that have been proposed have been implemented on a regular basis (CBD 2003c; Delbaere 2002b). The most commonly used indicators are species richness, number of threatened and extinct species, number of endemic species, trends in abundance of particular species, areal loss of ecosystems, and the percentage of area protected and its derivatives. Some of these indicators are insensitive, provide perverse messages, or are not an indicator of state but of response. Biodiversity indicators currently in use and proposed under the CBD are listed in Tables 16.1 and 16.2.

Many biodiversity measures had their origins in the taxonomic disciplines, which continue to have a strong influence through their emphasis on creating complete and

Table 16.1. List of single indicators (single variable related to a reference value) that are in use.

Type	Indicator (not exhaustive)	Question
State and Trend		
Ecosystem	Area of ecosystem type (e.g., forest, agriculture, built-up land)	How much natural area remains, how much is agricultural, and how much is built up?
	Hotspots (areas of high endemism experiencing high human impact)	Which ecosystems with high diversity of endemic species are threatened?
Species	Trends in representative species, particular taxonomic groups, exploited species, endemic species, migratory species, Red List species	What is the quality of the remaining natural area and agricultural area? What are the trends at species level?
	Percentage of threatened and extinct species in particular groups	Which species are threatened?
Structure	Trends in structure variables (e.g., canopy cover, ratio of dead to living wood, percentage area vital coral reefs)	What are the trends of ecosystem structures?

(*continued*)

Table 16.1. List of single indicators (single variable related to a reference value) that are in use (*continued*).

Type	Indicator (not exhaustive)	Question
Genes	Number and share of livestock breeds and agricultural plant varieties Number of endangered varieties of livestock breeds and agricultural crops Share of major varieties in total production for individual crops	Which genetic resources are threatened?
II. Pressures and Threats		
Physical	Annual conversion of self-generating area as a percentage of remaining area Change in mean temperature and precipitation Road density Damming and canalization of rivers Fire	What are the size of the pressure and its trend?
Chemical	Acid deposition P or N deposition Exceeding soil, water, or air standards for particular pollutants	
Biological	Total number of invasive species Total amount harvested per species	
Indirect	Human population density, gross national product	What factors influence the direct pressures?
III. Use		
Provisioning	Total amount harvested per species or species group	What is the use? Is it sustainable?
	Per capita wood consumption	How many people depend on the system?
Regulating	Carbon stored in forests	What is the contribution to gross domestic product?
Cultural	Total revenue from ecotourism	
IV. Response		
Legislation	Total number of protected species as a percentage of particular groups	

(*continued*)

Table 16.1. List of single indicators (single variable related to a reference value) that are in use (*continued*).

Type	Indicator (not exhaustive)	Question
	Percentage protected area by IUCN category	
Targets	National Biodiversity Strategy and Action Plan objectives met	
Expenditure	Expenditure of abatement and nature management measures (US$)	
Management	Number of protected areas with management plan	
	Number of threatened and invasive species with a management plan	
	Effectiveness of protection measures in protected areas	
V. Capacity		
Personnel	Nature research capacity, in number of people	
	Conservation policy capacity, in number of people	
	Nature site management in number of people	
Legislation	Number of physical and chemical standards	
Monitoring	Number of physical, chemical, and biological variables measured	
	Local site support groups and number of volunteer monitors	

Source: CDB (2003c).

Table 16.2. Composite indicators that are currently in use.

Group	Indicator	Source
General state	Natural Capital Index	ten Brink (2000)
	Wilderness	Conservation International
	Living Planet Index	Loh (2002)
	Last of the Wild	Sanderson et al. (2002)
	Biodiversity Intactness Index	Scholes and Biggs (2005)
Trends of components	Species Assemblage Trend Indices (e.g., Bird Headline Indicator, Living Planet Index)	Gregory et al. (2002), Loh (2002)
Threat	Red List indicators on species groups	IUCN (2002)
	Hotspots	Myers et al. (2000)
	Human Footprint	Sanderson et al. (2002)
Pressures	Total Pressure Index	UNEP (2002)
	Habitat–species matrix (agricultural practices)	
Uses	Sustainability of total use	
Responses	Effectiveness of environmental measures	
	Effectiveness of area protection	
	Effectiveness of site management	

Source: CBD (2003c).

correct lists of species, hence efforts such as the Global Biodiversity Information Facility and Species 2000 (www.gbif.org/GBIF_org/what_is_gbif and www.sp2000.org). The rational structure such initiatives bring is welcome, but taxonomic completeness, if achievable at all, is still many decades down the road. Given the indications that the current rate of extinction is abnormally high, urgency is paramount, and robust and scientifically defensible shortcuts are needed. Absolute taxonomic completeness is not necessary to measure biodiversity loss. Sampling a limited number of well-described species can provide sufficient information to guide interventions.

The main conservation advocacy groups have relied on the perceived threat of extinction, captured in Red Data Lists, because these do not require complete species inventories. Flagship species, those with high public appeal, have attracted a disproportionate amount of attention. Over the past decade the policy focus has shifted toward an ecosystem-based approach, which is intended to be more holistic. Rather than focusing on individual species, there is growing emphasis on the protection of hotspots containing multiple endemic species in threatened locations. There is also a move toward protection of large untransformed areas at the ecosystem scale, based on indirect measures or expert judgment.

Another recent trend is toward the use of abundance-based indices rather than species richness indices. Abundance-based indices have circumvented data limitations by focusing on a small number of well-studied species (Gregory et al. 2002; Loh 2002), using models to supplement inadequate data series (ten Brink 2000), or using expert judgment in place of population censuses (Scholes and Biggs 2005).

In October 2003, the CBD accepted a working paper that proposes eighteen indicators for application at national scale (CBD 2003d). They include measures at all three levels (ecosystem, species, genetic) and all aspects (composition, structure, and function) but are not intended to be comprehensive or integrated. They are the result of a consultation process that began around the time of the Global Biodiversity Assessment (1995).

Because of the lack of data on biodiversity trends, there is a tendency to use indicators of pressure (drivers) instead. The Geobiosphere Load (Moldan 1997) and Ecological Footprint (Wackernagel et al. 2002) are indicators of pressure. Some pressures, such as population density and household dynamics, are easily quantified (Liu et al. 2003). It is generally recognized that human-dominated landscapes are species poor or invaded by alien species (Araújo 2003). Human appropriation of photosynthesis products (Vitousek et al. 1986; Rojstaczer et al. 2001) reduces the energy available to support wild populations and ecosystems.

Data Issues

The key issue in the applicability of indicators is access to reliable and consistent information, particularly when we are attempting to apply indices at continental or global scales. The quality of knowledge varies greatly across biological groups (e.g., from very good for birds to very poor for soil microbes). This reflects not their relative ecological importance but their charisma and ease of study. This unevenness of observation has geographic consequences: The tropics and the oceans are less well inventoried than the temperate landmasses because they have a greater biodiversity and a shorter record of scientific study. To an extent, broadening the information sources to include traditional or indigenous knowledge can help alleviate the problem for the more prominent groups, but it is unreasonable to expect indigenous people to have knowledge about subjects that may not have seemed necessary or even visible to them.

The pragmatic solution is to confine biodiversity indicators to taxonomic groups or topics that are well known (i.e., plants and vertebrates) and to make use of both qualitative (including informal) and quantitative information sources, at least until some parity in knowledge is achieved. In well-studied ecosystem types, biodiversity loss can be measured based on all well-known taxonomic groups, provided that they are included in an unbiased fashion.

The unwillingness of much of the traditional scientific community in the biodiversity domain to be pragmatic rather than perfectly rigorous is a significant impediment to addressing the urgent problem of biodiversity loss. Most indicators rely on very similar sets of input information, so the lack of consensus regarding the form of the indicator should not be an excuse to limit the collection and refinement of fundamental, spatially and temporally explicit data on

• Land cover and use and the spatial pattern of marine resource use
• The distribution of species, especially of plants and vertebrates
• Trends in the population size of key species
• The genetic diversity of domesticated species
• The impact of different land use activities on different species

A lesson in pragmatism can be drawn from the UN Climate Change Convention. When the Intergovernmental Panel on Climate Change was charged with developing performance indicators for greenhouse gas emissions, which can come from hundreds of activities and tens of thousands of individual sources, it chose activity-based approaches rather than full, source-by-source accounting. This approach establishes a statistical relationship between the intensity of an activity (e.g., agriculture) and its impact (in this case, on biodiversity). It calculates the score for a given geographic region by multiplying the area exposed to each activity by the impact factor. The Natural Capital Index and Biodiversity Impact Indices work in this way.

The increasing ease and decreasing cost of collecting genetic information may alter the emphasis currently placed on species-level information, especially for large groups of organisms whose taxonomy is poorly developed. For instance, in calculating endemism scores, the number of closely related species may be less important than the total genetic diversity in the system. In domesticated organisms, subject to intense breeding, the species concept is inapplicable in any case, and genetic methods are already widely applied. At present it is not practical to get full genetic profiles for all organisms. Community profiling is a method of choice for groups such as soil microbes, where a bulk sample is easily obtained.

Biodiversity Indicators for Policy Purposes

Several partially overlapping lists of desirable attributes of biodiversity indicators for policy purposes exist (CBD 2003a, 2003c, 2003d). An integrated list of criteria would include the following:

• Relevant to biodiversity policymaking
• Simple and easily understood
• Broadly accepted
• Scientifically credible
• Quantitative
• Normative (allowing comparison with a baseline situation and a policy target)

- Measurable in a sufficiently accurate way at an affordable cost
- Responsive to changes at policy-relevant time and space scales
- Usable for scenarios of future projections
- Allows aggregation and disaggregation between ecosystem, national, and international scales
- Usable in various composite indicators and for different purposes

To date, the bulk of the biodiversity research and media attention has been on species composition, whereas much of the policy-level justification for biodiversity conservation rests implicitly on ecosystem-level, functional attributes. This results in a mismatch between the type of information available and that needed for policy. Although all stakeholders acknowledge the variety of scales and aspects of biodiversity, in practice biodiversity indicators have focused on the species level and on a single compositional measure: species richness. Although the ecosystem approach is widely espoused, in reality measures of ecosystem diversity have seldom gone beyond statements about the areal extent of prominent ecosystem types, such as forests.

There is no single all-purpose, universally good indicator in the field of biodiversity. The challenge is to find a small set of complementary indicators (because it is apparent that one indicator will not suffice) that is simultaneously easy to grasp, widely applicable, and sensitive. The suitability of a particular indicator depends on the purpose for which it is used. Even for a specified purpose, there are generally two or three indicators that could satisfy the criteria equally well. On the other hand, once the purpose has been defined, it is possible to eliminate entire indicator categories as inappropriate. Thereafter, practical considerations may further reduce the suitable candidate indicators to two or three options. Some indicators, such as hybrid indicators that are arbitrarily weighted summations of mixtures of states, pressures, and responses, should be avoided.

Various types of biodiversity indices are applicable at different stages in the policy process. Indices that identify priority areas for conservation action, such as The Last of the Wild, Hotspots, or complementarity indices, are aimed at the planning phase and are usually calculated only once. Performance monitoring tools, repeated on a regular basis, are operational phase indicators used as early warning measures and for evaluating current and future policies.

The authors of this chapter have been involved in the development of two closely related biodiversity indices, which reflect our convictions about what type of measures are likely to meet the criteria listed in this chapter and thereby fulfill policy needs while remaining scientifically legitimate. They are the Natural Capital Index (NCI; CBD 1997a, 1997b; ten Brink 2000; CBD 2003a, 2003c) and the Biodiversity Intactness Index (BII; Scholes and Biggs 2004, 2005). In principle, both measure the deviation of abundance of a broad spectrum of species from some reference state. The NCI uses actual population estimates of a limited number of species, supplemented where necessary by statistical population models. Its disadvantage is that such comprehensive data are available only in well-studied, low-biodiversity areas. The BII is more applicable in species-rich but data-poor regions.

It uses a panel of experts to assess the impacts of various types of human actions on abundance within functional groups of species. Its disadvantage is the uncertainty associated with such judgments. The numbers produced by both indices are easy to understand and explain, and they integrate both species-level information (richness, abundance) and ecosystem-level information (area extent of ecosystems, overlaid by area extent of human activities). They are structural and compositional but can be adapted for functional views as well when applied to functional types.

Gaps in Knowledge and Research Needs

This chapter has attempted to assess biodiversity concepts and indicators relevant to monitoring progress in sustainable development at national to global scales. Although certain aspects of biodiversity have been well researched, key gaps remain in the information needed for policy purposes. These include better information on

- Functional relationships between biodiversity and ecosystem services, and especially the presence of thresholds where these exist
- Genetic relatedness and redundancy within and between species
- Robust predictors, for all major ecosystem types in the differing parts of the world, of the consequences of major human activities, such as extensive and intensive agriculture, harvesting, settlement, and industrial pollution, on the various categories of biodiversity
- Consistent and reliable maps of land use (and ocean use) and species distributions at regional and global scales
- Historical ecology (in order to understand processes and construct baselines)
- Biodiversity observation and assessment systems for supplying consistent, long-term data for indicator construction

Literature Cited

Achard, F., H. D. Eva, H. J. Stibig, P. Mayaux, J. Gallego, T. Richards, and J. Malingreau. 2002. Determination of deforestation rates of the world's humid tropical forests. *Science* 297:999–1002.

Andreasen, J. K., R. O'Neill, R. Noss, and N. C. Slosser. 2001. Considerations for the development of a terrestrial index of biological integrity. *Ecological Indicators* 1:21–35.

Araújo, M. B. 2003. The coincidence of people and biodiversity in Europe. *Global Ecology and Biogeography* 12:5–12.

Bramwell, D. 2002. How many plant species are there? *Plant Talk* 28:32–33.

CBD. 1997a. *Recommendations for a core set of indicators of biological diversity.* UNEP/CBD/SBSTTA/3/9. Montreal, Canada: Convention on Biological Diversity.

CBD. 1997b. *Recommendations for a core set of indicators on biological diversity.* Background paper prepared by the Liaison Group on Indicators of Biological Diversity. UNEP/CBD/SBSTTA/3/INF/13. Montreal, Canada: Convention on Biological Diversity.

CBD. 2003a. *Consideration of the results of the meeting on "2010: The Global Biodiversity Challenge".* UNEP/CBD/SBSTTA/9/inf/9. Montreal, Canada: Convention on Biological Diversity.

CBD. 2003b. *Integration of outcome-oriented targets into the programmes of work of the convention, taking into account the 2010 biodiversity target, the Global Strategy for Plant Conservation, and relevant targets set by the World Summit on Sustainable Development.* Decision IX/13. Montreal, Canada: Convention on Biological Diversity.

CBD. 2003c. *Monitoring and indicators: Designing national-level monitoring programmes and indicators.* UNEP/CBD/SBSTTA/9/10. Montreal, Canada: Convention on Biological Diversity.

CBD. 2003d. *Proposed indicators relevant to the 2010 target.* UNEP/CBD/SBSTTA/9/inf/26. Montreal, Canada: Convention on Biological Diversity.

CDB. 2003e. *Report of the expert meeting on indicators of biological diversity including indicators for rapid assessment of inland water ecosystems.* UNEP/CBD/SBSTTA/9/INF/7. Montreal, Canada: Convention on Biological Diversity.

CGER. 2000. *Ecological indicators for the nation.* Washington, DC: National Academy Press.

Clark, K. R., and R. M. Warwick. 1998. A taxonomic distinctness index and its statistical properties. *Journal of Applied Ecology* 35:523–531.

Constanza, R., R. D'Arge, R. De Groot, S. Farber, M. Grasso, B. Hannon, K. Limburg, S. Naeem, R. V. O'Neill, J. Paruelo, R. G. Raskin, P. Sutton, and M. van den Belt. 1997. The value of the world's ecosystem services and natural capital. *Nature* 387:253–260.

Cowling, R. M., G. E. Gibbs Russell, M. T. Hoffman, and C. Hilton-Taylor. 1989. Patterns of plant species diversity in southern Africa. Pp. 19–50 in *Biotic diversity in southern Africa: Concepts and conservation,* edited by B. J. Huntley. Cape Town, South Africa: Oxford University Press.

Crawley, M. J. 1993. *GLIM for ecologists.* London: Blackwell Scientific.

De Groot, R. S. 1992. *Functions of nature. Evaluation of nature in environmental planning, management and decision making.* Amsterdam: Wolters-Noordhoff.

Delbaere, B. (ed.). 2002a. *Biodiversity indicators and monitoring: Moving towards implementation.* Tilburg: European Centre for Nature Conservation.

Delbaere, B. (ed.). 2002b. *An inventory of biodiversity indicators in Europe.* Tilburg: European Centre for Nature Conservation.

Faith, D. P. 2002. Quantifying biodiversity: A phylogenetic perspective. *Conservation Biology* 16:248–252.

Faith, D. P., and K. J. Williams. 2006. Phylogenetic diversity and biodiversity conservation. Pp. 233-235 in *2006 McGraw-Hill Yearbook of Science and Technology*. New York: McGraw-Hill.

FAO. 2001. *Global forest resources assessment 2000*. Main report. Rome: FAO.

FAO. 2003. *State of the world's forests 2003*. Rome: FAO.

French, D. D. 1994. Hierarchical Richness Index (HRI): A simple procedure for scoring "richness" for use with grouped data. *Biological Conservation* 69:207–212.

Govaerts, R. 2001. How many species of seed plants are there? *Taxon* 50:1085–1090.

Gregory, R. D., N. I. Wilkinson, D. G. Noble, A. F. Brown, J. A. Robinson, J. Hughes, D. A. Proctor, D. W. Gibbons, and C. A. Galbraith. 2002. The population status of birds in the United Kingdom, Channel Islands and the Isle of Man: An analysis of conservation concern 2002–2007. *British Birds* 95:410–448.

Hughes, J. R., G. C. Daily, and P. R. Erlich. 1997. Population diversity: Its extent and extinction. *Science* 278:689–692.

Hurlbert, S. H. 1971. The non-concept of species diversity: A critique and alternate parameters. *Ecology* 52:577–586.

Iszák, J., and L. Papp. 2000. A link between ecological diversity measures and measures of biodiversity. *Ecological Modelling* 130:151–156.

IUCN. 2002. *2002 IUCN Red List of threatened species*. Gland, Switzerland: IUCN.

Jenkins, M., R. E. Green, and J. Madden. 2003. The challenge of measuring global change in wild nature: Are things getting better or worse? *Conservation Biology* 17:20–23.

Karr, J. R. 2002. Understanding the consequences of human actions: Indicators from GNP to IBI. Pp. 98–110 in *Just ecological integrity: The ethics of maintaining planetary life*, edited by P. Miller and L. Westra. Lanham, MD: Rowman & Littlefield.

Kinzig, A., D. Tilman, and S. Pacala (eds.). 2001. Biodiversity and ecosystem function: Empirical and theoretical analysis of the relationship. *Monographs in Population Biology* 33. Princeton, NJ: Princeton University Press.

Kutsch, W. L., W. Steinborn, M. Herbst, R. Baumann, J. Barkmann, and L. Kappen. 2001. Environmental indication: A field test of an ecosystem approach to quantify biological self-organisation. *Ecosystems* 4:49–66.

Leakey, R., and R. Lewin. 1995. *The sixth extinction: Patterns of life and the future of humankind*. New York: Random House.

Liu, J., G. C. Daily, P. R. Ehrlich, and G. W. Luck. 2003. Effects of household dynamics on resource consumption and biodiversity. *Nature* 421:530–533.

Loh, J. (ed.). 2002. *Living planet report 2002*. Gland, Switzerland: WWF International.

Longhurst, A. 1998. *Ecological geography of the sea*. San Diego: Academic Press.

Loreau, M., S. Naeem, and P. Inchausti (eds.). 2002. *Biodiversity and ecosystem functioning: A synthesis*. New York: Oxford University Press.

MA. 2003. *Ecosystems and human well-being: A framework for assessment*. A report of the

Conceptual Framework Working Group of the Millennium Ecosystem Assessment. Washington, DC: Island Press.

MA. 2005. *Ecosystems and human well-being: Biodiversity synthesis.* A report of the Millennium Ecosystem Assessment. Washington, DC: World Resources Institute.

Magurran, A. E. 2004. *Measuring biological diversity.* Oxford: Blackwell.

Margules, C. R., and R. L. Pressey. 2000. Systematic conservation planning. Nature 405:243–253.

Moldan, B. 1997. Geobiosphere load: A tentative proposal for a comprehensive set of policy-relevant indicators. Pp. 164–169 in *Sustainability indicators,* SCOPE 58, edited by B. Moldan and S. Billharz. Chichester, UK: Wiley.

Myers, N., R. A. Mittermeier, C. G. Mittermeier, G. A. B. da Fonseca, and J. Kent. 2000. Biodiversity hotspots for conservation priorities. *Nature* 403:853–858.

Naeem, S. 2002. Disentangling the impacts of diversity on ecosystem functioning in combinatorial experiments. *Ecology* 83:2925–2935.

Nei, M. 1972. Genetic distance between populations. *American Naturalist* 106:283–292.

Noss, R. F. 1990. Indicators for monitoring biodiversity: A hierarchical approach. *Conservation Biology* 4:355–364.

Pankhurst, C. E., B. M. Doube, and V. V. S. R. Gupta (eds.). 1997. Biological indicators of soil health: Synthesis. Pp. 419–435 in *Biological indicators of soil health,* edited by C. E. Pankhurst, B. M. Doube, and V. V. S. R. Gupta. Wallingford, UK: CAB International.

Pielou, E. C. 1969. *An introduction to mathematical ecology.* New York: Wiley Interscience.

Prance, G. T., H. Beentje, J. Dransfield, and R. Johns. 2002. The tropical flora remains undercollected. *Annals of the Missouri Botanical Garden* 87:67–71.

Reid, W. V., J. A. McNeely, D. B. Tunstall, D. A. Bryant, and M. Winograd. 1993. *Biodiversity indicators for policy makers.* Washington, DC: World Resources Institute.

Rojstaczer, S., S. M. Sterling, and N. J. Moore. 2001. Human appropriation of photosynthesis products. *Science* 294:2549–2552.

Royal Society. 2003. *Measuring biodiversity for conservation.* Policy document 11/03. London: The Royal Society.

Sanderson, E. W., M. Jaiteh, M. A. Levy, K. H. Redford, A. V. Wannebo, and G. Woolmer. 2002. The human footprint and the last of the wild. *BioScience* 52:891–904.

Santini, F., and A. Angulo. 2001. Assessing conservation biology priorities through the development of biodiversity indicators. *Rivista di Biologia* 94:259–276.

Scholes, R. J., and R. Biggs (eds.). 2004. *Ecosystem services in southern Africa: A regional assessment.* Pretoria, South Africa: Council for Scientific and Industrial Research. Available at www.millenniumassessment.org.

Scholes, R. J., and R. Biggs. 2005. A biodiversity intactness index. *Nature* 434:45–49.

Scotland, R. W., and A. H. Wortley. 2003. How many species of seed plant are there? *Taxon* 52:101–104.

Sherman, K., L. M. Alexander, and B. D. Gold (eds.). 1990. *Large marine ecosystems: Patterns, processes, and yields.* Washington, DC: AAAS Publications.

Smith, T. M., and H. Schugart (eds.). 1996. *Plant functional types.* Cambridge: Cambridge University Press.

Stockwell, D. R. B. 1994. Genetic Algorithm for Rule-set Production (GARP). Available at kaos.erin.gov.au/general/biodiv_model/ERIN/GARP/home.html.

Takezaki, N., and M. Nei. 1996. Genetic distances and reconstruction of phylogenetic trees from microsatellite DNA. *Genetics* 144:389–399.

ten Brink, B. J. E. 2000. *Biodiversity indicators for the OECD environmental outlook and strategy.* RIVM Feasibility Study Report 402001014. Bilthoven, the Netherlands: RIVM.

Tienderen, P. H., A. A. van de Haan, C. G. van der Linden, and B. Vosman. 2002. Biodiversity assessment using markers for ecologically important traits. *Trends in Ecology and Evolution* 17:577–582.

UNDP. 2003. *Human development report 2003.* New York: United Nations Development Programme.

UNEP. 2002. *Global environmental outlook 3.* London: Earthscan.

Vitousek, P. M., P. R. Ehrlich, A. H. Ehrlich, and P. Matson 1986. Human appropriation of the products of photosynthesis. *BioScience* 36:368–373.

Wackernagel, M., C. Monfreda, and D. Deumling. 2002. *Ecological Footprint of the Nations: November 2002 update.* Oakland, CA: Redefining Progress.

WCMC. 2000. *Global biodiversity: Earth's living resources in the 21st century,* by B. Goombridge and M. D. Jenkins. Cambridge, UK: World Conservation Press.

Wilkinson, C. (ed.). 2000. *Status of coral reefs of the world: 2000.* Townsville: Australian Institute of Marine Science. Available at www.reefbase.org/pdf/GCRMN/GCRMN2000.pdf.

Williams, P. H., K. J. Gaston, and C. J. Humphries. 1998. Mapping biodiversity value world-wide: Combining higher-taxon richness from different groups. *Proceedings of the Royal Society, Biological Sciences* 264:141–148.

17

Human Appropriation of Net Primary Production (HANPP) as an Indicator for Pressures on Biodiversity

Helmut Haberl, Karl-Heinz Erb, Christoph Plutzar, Marina Fischer-Kowalski, and Fridolin Krausmann

Why We Need Pressure Indicators for Biodiversity Loss

The loss of biodiversity resulting from human activities is thought to be one of the most pressing problems of global environmental change. Nevertheless, our understanding of biodiversity loss is hampered by significant knowledge gaps. At present, there is not even an agreement on how many species inhabit the earth and how fast biological diversity is being depleted (Groombridge 1992).

Slowing down the rate of human-caused biodiversity loss requires indicators according to the drivers, pressures, states, impacts, and responses scheme (EEA 2003); that is, we need to know which socioeconomic processes result in which pressures on biodiversity, how biodiversity changes, what impacts these changes have on society, and which measures are taken to mitigate pressures or to cope with impacts. Each of these indicator types has specific functions: Indicators of socioeconomic drivers and pressures are needed to support policies to change socioeconomic trajectories in a more biodiversity-friendly direction, and state indicators are needed to monitor changes in biodiversity (Chapter 16, this volume). Response indicators monitor conservation measures, and impact indicators judge the socioeconomic significance of biodiversity changes. This chapter focuses on pressure indicators.

Whereas a host of maps, data, or assessments are needed to support specific conservation plans at local levels (an issue not discussed here), a limited number of valid aggregate pressure indicators is needed to support the development of large-scale (e.g., national, global) strategies to achieve social and economic progress while decreasing pressures on biodiversity at the same time. To validate such indicators, it has to be shown that they are unambiguously related to biodiversity loss, and they reliably represent defined socioeconomic activities. If we simply regard gross domestic product (GDP) or population as indirect pressures on biodiversity, we remain in a deadlock where we are forced to choose between adequately nourishing the growing world population or conserving biodiversity, or between economic development needed to combat poverty and biodiversity protection. By contrast, sustainable development should reconcile food production, economic development, and biodiversity conservation, and developing such strategies requires aggregate pressure indicators that are currently lacking (Eurostat 1999).

This chapter discusses theoretical considerations and the available empirical evidence (which unfortunately is limited to only one component of biodiversity, species richness) suggesting that human appropriation of net primary production (HANPP) may be such an indicator. Of course, HANPP will have to be complemented by other indicators. Some of them may be closely related (e.g., indicators that evaluate human interference with hydrological cycles; Postel et al. 1996); others may focus on different system qualities.

HANPP: An Introduction

Plants absorb solar radiation and, through photosynthesis, transform it into chemically stored energy. This process is called primary production. A part of the fixed energy is used for the plant's metabolism; the remainder either results in an accumulation of biomass stocks or nourishes humans, animals, fungi, or microorganisms; that is, it becomes part of heterotrophic food chains (Odum 1971). Net primary production (NPP) is the net amount of primary production after the costs of plant respiration (i.e., the energy needed for the plant's metabolism) are included; it equals the amount of biomass produced. HANPP is the fraction of NPP appropriated by humans and has been used to assess human domination of the earth's ecosystems (Vitousek et al. 1986, 1997; Whittaker and Likens 1973).

Definition of HANPP

Vitousek et al. (1986) calculated HANPP using three different definitions. The most narrow definition regarded only biomass used by society (e.g., food, timber) as appropriated, the intermediate definition additionally included the NPP of human-dominated ecosystems (e.g., cropland), and the third definition also considered an assessment of the NPP foregone because of human-induced changes in ecosystem productivity (e.g., ecosystem degradation).

Vitousek's first and second definition could lead to problematic results, however. As demonstrated for Austria, changes in agricultural technology increased aboveground productivity on agricultural land by a factor of 2.6 from 1830 to 1995 (Krausmann 2001). Consider, for example, 1 hectare of this cropland: According to Vitousek's intermediate definition (also used by Rojstaczer et al. 2001), one would find an increase in HANPP of about 2.6 because of the increase in harvest, although the NPP remaining in the ecosystem stayed near zero, because the increase in the agroecosystem's productivity was compensated for by a similar increase in harvest.

On the other hand, regarding all NPP of human-dominated ecosystems as appropriated is also problematic: In forest plantations and grasslands a large fraction of the NPP remains in the ecosystem and supports food chains not directly controlled by humans. This argument has already been used to question the HANPP concept altogether (Davidson 2000).

Wright (1990), who was interested primarily in the possible impact of HANPP on biodiversity, proposed to define HANPP as the difference in NPP available in (hypothetical) undisturbed ecosystems and the amount of NPP actually available to support heterotrophic food chains. This definition seems to overcome some of the problems associated with Vitousek's approach. However, Wright excluded activities such as logging and biomass burning in forests on the ground that harvest in forests, though removing energy, does not result in a long-term reduction of productivity of the land for wild species if forests are allowed to regrow. Although this argument may be correct as long as nutrient-rich parts (e.g., leaves) remain in the forest, it does not justify the exclusion of wood harvests from the definition of HANPP because harvest and biomass burning are very important for forest ecology (Harmon et al. 1986).

We have therefore defined HANPP (Haberl 1997) by measuring changes in the availability of NPP for ecological processes induced by alterations of the productivity of vegetation that result from land use and extraction of NPP from ecosystems through biomass harvest, including wood harvest in forests. HANPP is thus the difference between NPP_0, the NPP of potential vegetation (Tüxen 1956), and NPP_t, the part of the NPP of actual vegetation (NPP_{act}) remaining in ecosystems after harvest (NPP_h)[1]

$$HANPP = NPP_0 - NPP_t, \text{ with } NPP_t = NPP_{act} - NPP_h$$

HANPP can be expressed as material (kilograms dry matter), substance (kilograms carbon), or energy flow (joules) or as a percentage of NPP_0 (%HANPP = $HANPP/NPP_0 \times 100$).

This definition of HANPP is appropriate for interregional comparisons and time series analysis. By monitoring HANPP and its various components, such as NPP_{act}, NPP_t, and NPP_h, we can evaluate the impacts of different land use practices on ecosystem energetics and their socioeconomic performance: land use may increase or reduce productivity, it may leave more or less energy in the ecosystem, it may yield rich or poor harvests, and so on. Thus, we are also able to observe a possible decoupling of biomass harvest and HANPP (Krausmann 2001). This definition of HANPP does not exaggerate human impact by including all NPP of human-dominated ecosystems as appropriated. HANPP includes only the amount of biomass actually harvested, on top of the NPP prevented by human land use. It is possible to assess HANPP in great spatial detail by combining statistical data with land cover data derived from remote sensing (Haberl et al. 2001).

Meaning and Interpretation of HANPP

HANPP indicates how intensively a defined area of land is being used in terms of ecosystem energetics (Haberl et al. 2004b). With reference to a given territory, it reveals how much energy is diverted by humans as compared to the energy potentially available. This can be interpreted as a measure of how strongly human use of a defined land area affects its primary productivity and how much of the NPP is diverted to human uses and thus is not available for processes within the ecosystem.

HANPP has been developed in the context of the debate on global ecological changes caused by humans and their activities (Vitousek et al. 1986, 1997), and it has been linked to the issue of human influence on biodiversity (Wright 1990). It has been used in ecological economics as a biophysical indicator of strong sustainability (Martinez-Alier 1998; Sagoff 1995), although the initial idea that HANPP was a straightforward indicator for ecological limits (Costanza et al. 1998; Meadows et al. 1992) was proven wrong because biomass harvest can be increased without increasing HANPP (Davidson 2000; Krausmann 2001), and neither GDP nor population size is directly constrained by HANPP (Haberl and Krausmann 2001). Such decoupling of HANPP and biomass harvest requires fossil energy input into agroecosystems (Krausmann et al. 2003; Pimentel et al. 1990) and may be associated with environmentally detrimental impacts (e.g., pesticides, nitrogen leaching, and soil deterioration). Economic growth, particularly in the transition from agricultural to industrial society, can to a large extent be decoupled from growth in biomass consumption because industrial economies rely much more on minerals and fossil energy than agricultural societies (Krausmann and Haberl 2002). On the other hand, a larger proportion of land is devoted to settlement, industry, and transport, resulting in HANPP increases. But HANPP does not generally rise with industrialization because agricultural intensification may raise the productivity of agricultural land more than biomass harvest increases (Krausmann 2001).

A straightforward interpretation of HANPP is that it is a measure of the human domination (Vitousek et al. 1997) or colonization (Fischer-Kowalski and Haberl 1997) of terrestrial ecosystems. Current dynamics of land use–induced changes in ecosystem processes are best conceived of as a process of intensification driven by population growth, changes in technology, and increasing demand for ecosystem services. But there are also areas where human use is deintensified (marginal land, reforestation). By comparing patterns and processes to be expected in the (hypothetical) potential vegetation with those that can currently be observed, the impact of human colonization on terrestrial ecosystems can be assessed.

HANPP Components and HANPP-Related Indicators

Assessments of HANPP require the calculation of several components of primary production and the changes induced by land use. These components allow additional insights. For example, if land management in a country results in a downward trend of NPP_{act}, this indicates environmental degradation (Munasinghe and Shearer 1995). The relationship between NPP_0 and NPP_{act} shows how well agriculture uses the production potential of a region and therefore is an indicator of agricultural area efficiency. The relationship between NPP_h and HANPP reveals how much of the HANPP results from harvest and how much from changes in productivity. Land use may reduce (e.g., urban settlements, infrastructure, erosion) or increase (e.g., irrigation, fertilization) productivity.[2]

Similar indicators for human-induced changes in the production ecology of terrestrial ecosystems have been proposed. Land use influences the standing crop (the amount of live biomass) of ecosystems, which is relevant for ecological carbon balances (Erb 2004). Land use may also accelerate biomass turnover (= NPP/standing crop/yr) by up to fifty times (Erb 2004; Haberl et al. 2001), with currently largely unknown consequences.

HANPP and Biodiversity

In contrast to indicators such as the ecological footprint (Haberl et al. 2004b; Wackernagel et al. 2002), there is no clear-cut sustainability threshold referring to HANPP. One hundred percent HANPP clearly would be destructive because this would leave no resources for other species except those used directly by humans. It is a matter of debate how to set a meaningful lower threshold. It has been argued that human impact should be small compared with natural processes, resulting in a proposed threshold of 20 percent HANPP (Weterings and Opschoor 1992), but this number cannot be justified scientifically. This section takes stock of our knowledge on effects of HANPP on biodiversity, thus aiming at a more rational discussion on this issue.

What Are Pressures on Biodiversity?

Three principal levels of biodiversity are generally recognized: genes, organisms, and ecosystems (Heywood et al. 1995). According to Article 2 of the Convention on Biological Diversity (CBD), "Biological diversity means the variability among living organisms from all sources, including, inter alia, terrestrial, marine and other aquatic ecosystems and the ecological complexes of which they are part; this includes diversity within species, between species and of ecosystems."

Species diversity is only one component of biodiversity, but this notion also encompasses several parameters, the most important of which are species richness (number of species of a defined taxon occurring in a defined area) and the relative abundance of species (Magurran 1988). The diversity of small, homogenous habitats is denoted as α diversity, the diversity of a landscape comprising several different habitats is β diversity, and γ diversity is a measure for the differences in species composition of the habitats (Whittaker 1960).

What is regarded as a pressure on biodiversity very much depends on which aspects of biodiversity one would like to conserve. Possible objectives of biodiversity conservation include conserving genetic information, maximizing (endemic, originally present, or all) species numbers, preventing species from going extinct, maintaining a representative set of habitats, maintaining diverse landscapes, maintaining the capacity of ecosystems to adapt to change (resilience), maintaining or improving ecosystem functions and services, maintaining particular biological states, or preserving defined natural processes (Heywood et al. 1995; Miller et al. 1995).

Which of these goals are pursued is less a question of biology than of social, economic, and political factors (Miller et al. 1995), but it will determine which properties have to be related to HANPP in order for HANPP to be relevant for biodiversity conservation. The work we report in this chapter focuses on the relationships between HANPP and species richness and on the relationship between HANPP and land cover diversity. It would be desirable to explore links between HANPP and other aspects of biodiversity, but no such evidence is available.

HANPP and Biodiversity: Theoretical Considerations

Mechanisms of human impacts on biodiversity have been grouped into overexploitation of wild living resources; expansion of agriculture, forestry, or aquaculture; habitat loss and fragmentation; indirect negative effects of species introduced by humans; pollution; and global climate change (McNeely et al. 1995). Because HANPP is an indicator for changes in terrestrial ecosystems caused by land use, it refers mostly to expansion of agriculture, forestry, or aquaculture and habitat loss and fragmentation, which are closely related.

On an abstract level it is obvious why HANPP is relevant for biodiversity. Biomass is the mass of living or dead organisms present in a system. The very idea of the production–ecological (or trophic–dynamic) process in ecosystems (Lindemann 1942) is an abstract notion for organisms coming into being, growing, and dying. This process is fueled by various metabolic processes taking place within organisms. Energy enters organisms above all through two processes: photosynthesis and ingestion of dead or living organisms or parts thereof. Human-induced changes in this process affect patterns (including biodiversity), processes, functions, and services of ecosystems almost by definition.

Vitousek et al. (1986:368) put it as follows: "*Homo sapiens* is only one of perhaps 5–30 million animal species on Earth, . . . yet it controls a disproportionate share of the planet's resources. . . . NPP provides the basis for maintenance, growth, and reproduction of all heterotrophs (consumers and decomposers); it is the total food resource on Earth. We are interested in human use of this resource . . . for what it implies for other species, which must use the leftovers." Discussing their finding that humans appropriate about 40 percent of global terrestrial NPP, they add, "People and associated organisms use this organic material largely, but not entirely, at human direction, and the vast majority of other species must subsist on the remainder. An equivalent concentration of resources into one species and its satellites has probably not occurred since land plants first diversified. The co-option, diversion, and destruction of these terrestrial resources clearly contributes to human-caused extinctions of species and genetically distinct populations."

It has been stated that "'sustainability' might be interpreted as the maintenance of a level of biological diversity that will guarantee the resilience of ecosystems that sustain human society. The goal of a conservation strategy should be to protect not all biodiversity in some areas, but biodiversity thresholds in all areas" (Folke et al. 1996:1021). Theoretical considerations indicate that a sufficient amount of energy remaining in the ecosystem is necessary for ecosystems to be resilient (Kay et al. 1999). HANPP might impede ecosystem services and thus sustainability: "To the extent that . . . natural systems, species and populations provide goods or services that are essential to the sustainability of human systems, their shrunken base of operations must be a cause of concern" (Vitousek and Lubchenko 1995:60).

It is not easy to be more specific, though: How exactly are biodiversity, resilience, or other properties of ecosystems related to HANPP? Almost 20 years after Vitousek's famous article, disappointingly little is known, mostly because most ecological work is focused on systems with little human impact (McDonnell and Pickett 1997) and because of the lack of generally agreed-upon ecological theories in that field (Brown 1995). Nevertheless, attempts have been made, based on the so-called species–energy hypothesis, to evaluate the potential effect of HANPP on species richness (Wright 1987, 1990).

The species–energy hypothesis (Brown 1981, 1995; Gaston 2000; Hutchinson 1959; Wright 1983; Wright et al. 1993) suggests that more available energy should allow more species to coexist, resulting in a positive relationship between energy availability and species diversity. Mechanisms behind this pattern could be a finer subdivision of resources (specialization) in richer environments, density-dependent regulation of population size (costs of commonness; Brown 1981), or the fact that more resources allow more organisms to live in a defined location. This greater number of organisms is more likely to belong to a larger number of species than fewer organisms would (Hubbell

2001). Irrespective of the mechanism, the species–energy hypothesis implies that the number of heterotroph species present in an ecosystem is related to the amount of energy remaining in the system (i.e., NPP_t) because this is the amount of energy potentially available for all food chains. According to the species–energy hypothesis, HANPP contributes to species loss because it reduces NPP_t (Wright 1987, 1990).

One problem is that the specific mathematical relationship between species diversity and energy flows is uncertain. Some believe that there is a monotonous relationship such as $S = c. E^z$ between energy flow (E) and species richness (S) (Currie and Paquin 1987; Lennon et al. 2000; Weiher 1999; Wright 1983; Wright et al. 1993), whereas others favor unimodal (hump-shaped) species–energy curves (Rosenzweig and Abramsky 1993; Rapson et al. 1997). A recent review found that linear and unimodal patterns seem to be found about equally often (Waide et al. 1999). In the first case, HANPP should always result in species loss, whereas in the second case intermediate levels of HANPP could increase species richness.

Although the ability of HANPP to aggregate various processes increases its utility for many purposes, it also means that it is associated with a host of different changes in ecosystems. For example, in central Europe, where climax vegetation is mostly forest, introducing agriculture increases habitat diversity and thus should favor species richness according to the habitat diversity hypothesis. This hypothesis claims that environmental heterogeneity promotes species richness (Gaston and Blackburn 2000; Hubbell 2001; MacArthur and MacArthur 1961; Levin and Paine 1974). The interpretation of correlations between HANPP and species richness is hampered by such effects, which are difficult to control.

HANPP and Landscape Diversity

A recent empirical analysis focused on a study area of 2,864 km^2 around St. Pölten, the capital of lower Austria. This study was conducted at the scale of 1 × 1 km plots and asked to what extent a variety of landscape ecological indicators depended on HANPP (Wrbka et al. 2004).

The study showed that HANPP was clearly, monotonously, and highly significantly correlated with two indicators of landscape naturalness: hemeroby and urbanity (Figure 17.1). The urbanity index analyzes the domination of landscapes by strongly human-altered systems (O'Neill et al. 1988). It is defined as $\log_{10}$ of $(U + A)/(F + W + B)$ where U denotes urban, A agricultural, F forest, W water and wetland area, and B natural or seminatural biotopes. Hemeroby was introduced to describe gradients of human influence on landscape and flora (Jalas 1955) and is defined on an ordinal scale ranging from 1 (without actual human impact) to 7 (artificial landscape elements completely human dominated).

The study revealed that the relationship between HANPP and landscape diversity and landscape richness followed a hump-shaped curve (Figure 17.1). Landscape richness was defined as the number of different land cover classes present in each 1 × 1 km^2

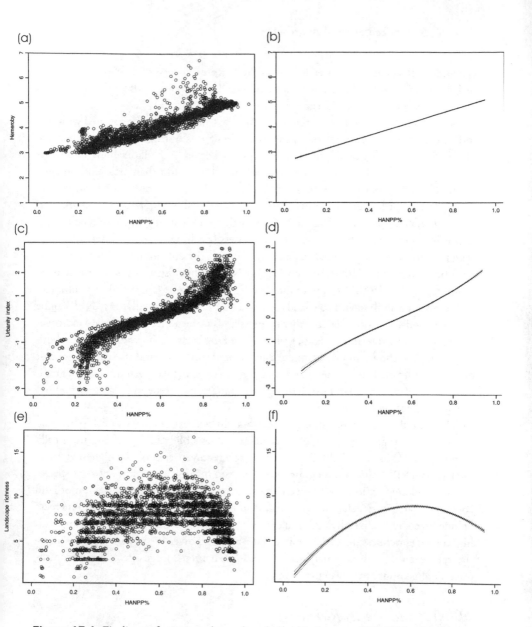

Figure 17.1. Findings of a case study on the relationship between HANPP and various landscape ecological indicators in lower Austria: *(a, b)* the scatter plot and response function of the correlation between HANPP and hemeroby (r^2 = .84), *(c, d)* the correlation between HANPP and urbanity (r^2 = .87), and *(e, f)* the correlation between HANPP and landscape richness (r^2 = .35). The pattern for landscape diversity (not shown) was almost identical. In the regression, polynomials (restricted cubic splines) were fitted to the data using ordinary least-square techniques. All correlations were significant at p = .001. r^2 are "corrected r^2" values obtained in a bootstrap model validation with ≤ 100 runs. For details see Wrbka et al. (2004). Reprinted from *Land Use Policy*, Vol. 21, "Linking pattern and process in cultural landscapes" pp. 289–306, with permission from Elsevier.

cell and landscape diversity was defined as the Simpson diversity index of the number of land cover classes. The relationship between HANPP and landscape diversity was almost the same as that of landscape richness.

We interpret these findings as follows: HANPP is almost linearly correlated with two indicators of the naturalness of landscapes. Conceptually this is nontrivial because HANPP is an indicator for human activities, whereas both naturalness indicators evaluate the state of landscapes. HANPP is much higher in urban areas and on agricultural land than in forests or natural areas, so for urbanity the result is not surprising. This is less so for hemeroby because hemeroby is defined with reference to changes in plant species composition. Because there is evidence of a strong negative relationship between hemeroby and bryophyte species richness (Zechmeister and Moser 2001) this is an indication for the potential value of HANPP as a pressure indicator for species loss.

Intuitively, the relationship between HANPP and landscape diversity or richness is also plausible: HANPP is low in forests (which tend to be large) and high in large-scale croplands, both having low landscape heterogeneity. Intermediate HANPP can be found in landscapes dominated either by grasslands or by a mix of different land cover classes, including forests, cropland, grassland, and urban areas. It is quite obvious, both empirically (as shown in our example) and theoretically, that land use can increase or decrease landscape heterogeneity and that there may be regular patterns along a gradient of intensification of land use, although the pattern may depend on geomorphological and socioeconomic conditions.

This may also have implications for species richness: A positive relationship between landscape heterogeneity and species richness is plausible and has been empirically demonstrated (Moser et al. 2002). Therefore, species diversity could be highest at intermediate HANPP values on a landscape scale, even if the relationship between species richness and energy flow (NPP_t) were linear at small scales: The introduction of land cover types such as grassland or even cropland in large-scale forests is likely to increase habitat diversity (α diversity). This may result in a more heterogeneous landscape with higher species richness than in the initial state because β diversity may increase if γ diversity rises, even if α diversity may decrease in some habitats. Edge effects and the introduction of ecotones may also play a role.

HANPP and Species Richness

Until very recently, only two studies existed that used the species–energy hypothesis to evaluate the possible significance of HANPP for species endangerment. The first studied estimates of species numbers on a continental scale (Wright 1987); the second studied extinctions on a global scale since the year 1600 (Wright 1990). Although the patterns found in these two studies were consistent with the species–energy hypothesis, their usefulness was limited by the extremely coarse spatial resolution.

In an attempt to present more convincing evidence, we summarize results of two recent correlation analyses aiming to test the utility of HANPP as a pressure indicator for species loss. A direct assessment proved to be impossible: According to the species–energy hypothesis, a reduction in NPP_t should result in a decline in species numbers. To test whether HANPP is a valid pressure indicator for biodiversity loss, we should have tested the ability of HANPP to predict species loss (ΔS), not its relationship to actual species richness (S_{act}). Because there is no information on potential species richness (S_0), there are no data on the change in species richness (ΔS) as compared with the potential state. Moreover, there is no linear relationship between HANPP and NPP_t, the factor that should influence the pattern of S_{act}. NPP_t can be low because of high HANPP but also because of low NPP_0. Without data on ΔS it is not possible to test HANPP directly. Indirect tests of HANPP assume that if S_{act} is correlated to NPP_t, this would be evidence that a reduction in NPP_t should also lower species richness. This is exactly what we found.

The first study (Haberl et al. 2004a) was based on a transect of 38 squares sized 600 × 600 m in east Austria. Species numbers of seven taxonomic groups (vascular plants, bryophytes, orthopterans, gastropods, spiders, ants, and ground beetles) were determined (Sauberer et al. 2004) and correlated with HANPP and its components. Both a linear and a quadratic polynomial function were fitted to the data; the choice of model was based on the Akaike Information Criterion (AIC). The study found a highly significant correlation between NPP_t and species richness ($.13 < r^2 < .76$, depending on taxon). The AIC confirmed that the relationship between NPP_t and species richness was

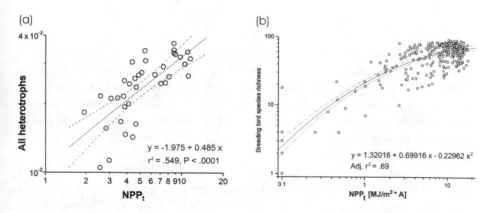

Figure 17.2. Correlation analyses between $\log(NPP_t)$ and the logarithm of species numbers of various groups. (a) NPP_t and all heterotrophs (5 taxonomic groups) on 38 east Austrian plots sized 600 × 600 m (Haberl et al. 2004a and additional unpublished data). (b) NPP_t (NPP remaining in ecosystems) and bird species number on 328 plots sized 1 × 1 km randomly selected from Austria's total area of about 83,000 km² (Haberl et al. 2005).

linear. Figure 17.2a displays the regression between NPP$_t$ and an index of the species numbers of all five heterotroph groups analyzed in this study. The scatter diagrams looked similar for all seven groups.

In Figure 17.2b we present findings of a recent study (Haberl et al. 2005) on the relationships between HANPP and bird species richness in Austria. Bird species numbers for Austria's total area were extrapolated from Austria's bird inventory (Dvorak et al. 1993) on a 250 × 250 m grid (N = 1.3 million. grid cells) using an expert system (C. Plutzar and M. Pollheimer, personal communication); HANPP data were recalculated from Haberl et al. (2001). Some simple measures of land cover heterogeneity and landscape heterogeneity were also assessed based on a land cover classification and a landscape type classification. Four different plot sizes were considered: 0.25 × 0.25 km, 1 × 1 km, 4 × 4 km, and 16 × 16 km. A nested representative sample of N = 328 squares of each size was randomly chosen. As in the previous study, both linear and quadratic polynomials were tested, and the AIC was used to decide between the two models.

The results suggest that NPP variables generally do a much better job of explaining bird species richness than all available landscape heterogeneity indicators. Consistent with the species–energy hypothesis, we found highly significant and almost always monotonous (but not linear) positive correlations between NPP$_t$ and bird species numbers (e.g., see the correlation found at the 1 × 1 km scale, Figure 17.2b).

Although direct tests of the ability of HANPP to predict species loss (ΔS) would be desirable, this indirect evidence supports the line of reasoning outlined in this chapter. We cannot exclude additional effects of other factors influencing biodiversity, such as possible effects of disturbance frequency or intensity (Wrbka et al. 2004).

Conclusions

HANPP is an aggregate indicator of human-induced changes in ecosystem processes resulting from human activities that aim at shaping terrestrial ecosystems according to human needs and wants. HANPP is an interesting indicator for the intensity of human colonization of terrestrial ecosystems and the extent of human domination of ecosystems. Moreover, because changes in HANPP are also relevant for carbon stocks and flows in ecosystems (Erb 2004), it is also relevant for the discussion of human-induced changes in the carbon cycle (Haberl 2001).

HANPP can be related to specific socioeconomic activities. Using input–output analysis (Duchin 1998) it would be possible to calculate the amount of HANPP resulting from the economic activity in every economic sector of a country under consideration. It would also be possible to calculate HANPP associated to many relevant products. It would thus be possible to relate HANPP to final consumption or to GDP. HANPP can be aggregated, and it can be calculated in great spatial detail. These features make HANPP an attractive pressure indicator because it allows a host of relevant analyses (e.g., assessments of the effect of changes in consumer behavior on HANPP, evaluations of the impacts of policy changes on HANPP). This contrasts with landscape ecological indicators that can be qualitatively explained only by regional characteristics in land use and must take into account geomorphological, cultural, and other peculiarities of a study region (Wrbka et al. 2004) and can thus not really be linked to aggregate socioeconomic trends.

Based on statistical data, remote sensing data, and appropriate models, HANPP can already be assessed easily. Appropriate extensions of current vegetation models such as the Lund–Potsdam–Jena (LPJ) model (Sitch et al. 2003) could facilitate such assessments. These models would also allow us to relate changes in production and consumption patterns and climate change in scenario-based approaches and therefore could be very useful for the projection of possible future pressures on biodiversity.

Both the theoretical considerations and the empirical analyses reported show that HANPP is a good candidate for an aggregate pressure indicator for biodiversity loss. Significant empirical evidence substantiates the previously rather general considerations on its potential usefulness (Wright 1987, 1990), although more evidence, particularly from outside Europe, is desirable. Even without such evidence we believe the HANPP concept and method are a well-founded basis for a pressure indicator for biodiversity loss.

Acknowledgments

The research was done within several projects of the Kulturlandschaftsforschung program of the Austrian Federal Ministry of Education, Science and Culture and also funded by the Austrian Science Fund (FWF), project P-P16692. It is part of the LUCC-endorsed project "Land Use Change and Socio-Economic Metabolism" and serves as an input to ALTER-Net, a network of excellence funded by FP6 of the EU. We thank V. Gaube, C. Hahn, T. Hak, W. Loibl, B. Moldan, D. Moser, J. Peterseil, M. Pollheimer, N. Sauberer, B. Scholes, N.-B. Schulz, H. Weisz, T. Wrbka, H. Zechmeister, P. Zulka, and an anonymous reviewer.

Notes

1. HANPP is related to biomass extraction but not identical. In perennial plant associations such as forests, the ratio of extraction can exceed NPP (e.g., deforestation) and eventually result in a permanent reduction of the biomass stock of vegetation. This can be taken into account separately. Plant residues plowed into the soil were regarded as appropriated in the calculations presented in this study because the data we present here were limited to the aboveground compartment for reasons of data availability. Biomass returned on-site (e.g., feces dropped by grazing animals) was also accounted for as appropriated. We are aware that some applications of the HANPP concept might treat these flows differently.

2. Irrigation of arid land or intensive agricultural use (application of large amounts

of fertilizers) may raise NPP_{act} to levels above those of NPP_0. On intensively used agricultural land in humid climates, HANPP is mostly still positive because of the high amounts of biomass harvested. In the case of arid land, however, natural productivity may be extremely low. If such land is irrigated, this may raise NPP sufficiently to result in negative HANPP values, even in the case of high harvest levels.

Literature Cited

Brown, J. H. 1981. Two decades of hommage to Santa Rosalia: Toward a general theory of diversity. *American Zoologist* 21:877–888.

Brown, J. H. 1995. *Macroecology*. Chicago: Chicago University Press.

Costanza, R., J. Cumberland, H. E. Daly, R. Goodland, and R. Norgaard. 1998. *An introduction to ecological economics*. Boca Raton, FL: CRC Press.

Currie, D. J., and V. Paquin. 1987. Large-scale biogeographical patterns of species richness in trees. *Nature* 329:326–327.

Davidson, C. 2000. Economic growth and the environment: Alternatives to the limits paradigm. *BioScience* 50:433–440.

Duchin, F. 1998. *Structural economics. Measuring change in technology, lifestyles, and the environment*. The United Nations University. Washington, DC: Island Press.

Dvorak, M., A. Ranner, and H.-M. Ber. 1993. *Atlas der Brutvögel Österreichs. Ergebnisse der Brutvogelkartierung 1981–1985 der Österreichischen Gesellschaft für Vogelkunde*. Vienna: Federal Environment Agency.

EEA. 2003. *Europe's environment: The third assessment*. Environmental assessment report no. 10. Copenhagen: European Environment Agency.

Erb, K.-H. 2004. Land use–related changes in aboveground carbon stocks of Austria's terrestrial ecosystems. *Ecosystems* 7:563–572.

Eurostat. 1999. *Towards environmental pressure indicators for the EU*. Luxembourg: Office for Official Publications of the European Communities.

Fischer-Kowalski, M., and H. Haberl. 1997. Tons, joules, and money: Modes of production and their sustainability problems. *Society and Natural Resources* 10:61–85.

Folke, C., C. S. Holling, and C. Perrings. 1996. Biological diversity, ecosystems, and the human scale. *Ecological Applications* 6:1018–1024.

Gaston, K. L. 2000. Global patterns in biodiversity. *Nature* 405:220–227.

Gaston, K. L., and T. M. Blackburn. 2000. *Pattern and process in macroecology*. Oxford: Blackwell Science.

Groombridge, B. 1992. *Global biodiversity, status of the earth's living resources*. London: World Conservation Monitoring Centre and Chapman & Hall.

Haberl, H. 1997. Human appropriation of net primary production as an environmental indicator: Implications for sustainable development. *Ambio* 26:143–146.

Haberl, H. 2001. The energetic metabolism of societies, Part II: Empirical examples. *Journal of Industrial Ecology* 5(2):71–88.

Haberl, H., and F. Krausmann. 2001. Changes in population, affluence and environmental pressures during industrialization. The case of Austria 1830–1995. *Population and Environment* 23:49–69.

Haberl, H., K.-H. Erb, F. Krausmann, W. Loibl, N. Schulz, and H. Weisz. 2001. Changes in ecosystem processes induced by land use: Human appropriation of net primary production and its influence on standing crop in Austria. *Global Biogeochemical Cycles* 15:929–942.

Haberl, H., N. Schulz, C. Plutzar, K.-H. Erb, F. Krausmann, W. Loibl, D. Moser, N. Sauberer, H. Weisz, H. G. Zechmeister, and P. Zulka. 2004a. Human appropriation of net primary production and species diversity in agricultural landscapes. *Agriculture, Ecosystems & Environment* 102:213–218.

Haberl, H., M. Wackernagel, F. Krausmann, K.-H. Erb, and C. Monfreda. 2004b. Ecological footprints and human appropriation of net primary production: A comparison. *Land Use Policy* 21:279–288.

Haberl, H., C. Plutzar, K.-H. Erb, V. Gaube, M. Pollheimer, and N. B. Schulz. 2005. Human appropriation of net primary production and avifauna diversity in Austria. *Agriculture, Ecosystems & Environment* 110(3–4):119–131.

Harmon, M. E., J. F. Franklin, F. J. Swanson, P. Sollins, S. V. Gregory, J. D. Lattin, N. H. Anderson, S. P. Cline, N. G. Aumen, J. R. Sedell, G. W. Lienkaemper, K. Cromack, and K. W. Cummins. 1986. Ecology of coarse woody debris in temperate ecosystems. *Advances in Ecological Research* 15:133–302.

Heywood, V. H., I. Baste, K. A. Gardner, K. Hindar, B. Jonsson, and P. Schei. 1995. Introduction. Pp. 5–19 in *Global biodiversity assessment*, edited by V. H. Heywood and R. T. Watson. Cambridge: Cambridge University Press and UNEP.

Hubbell, S. P. 2001. *The unified neutral theory of biodiversity and biogeography*. Princeton, NJ: Princeton University Press.

Hutchinson, G. E. 1959. Hommage to Santa Rosalia, or why are there so many kinds of animals? *American Naturalist* 93:145–159.

Jalas, J. 1955. Hemerobe und hemerochore Pflanzenarten. Ein terminologischer Reformversuch. *Acta Fauna Flora Fennica* 72(11):1–15.

Kay, J. J., H. A. Regier, M. Boyle, and G. Francis. 1999. An ecosystem approach for sustainability: Addressing the challenge of complexity. *Futures* 31:721–742.

Krausmann, F. 2001. Land use and industrial modernization, an empirical analysis of human influence on the functioning of ecosystems in Austria 1830–1995. *Land Use Policy* 18:17–26.

Krausmann, F., and H. Haberl. 2002. The process of industrialization from the perspective of energetic metabolism: Socioeconomic energy flows in Austria 1830–1995. *Ecological Economics* 41:177–201.

Krausmann, F., H. Haberl, N. Schulz, K.-H. Erb, E. Darge, and V. Gaube. 2003. Land-use change and socioeconomic metabolism in Austria, Part I: Driving forces of land-use change 1950–1995. *Land Use Policy* 20:1–20.

Lennon, J. J., J. J. D. Greenwood, and J. R. G. Turner. 2000. Bird diversity and environmental gradients in Britain: A test of the species–energy hypothesis. *Journal of Animal Ecology* 69:581–598.

Levin, S. A., and R. T. Paine. 1974. Disturbance, patch formation, and community structure. *Proceedings of the National Academy of Sciences* 71:2744–2747.

Lindemann, R. L. 1942. The trophic-dynamic aspect of ecology. *Ecology* 23:399–418.

MacArthur, R. H., and J. MacArthur. 1961. On bird species diversity. *Ecology* 42:594–598.

Magurran, A. 1988. *Ecological diversity and its measurement.* London: Chapman & Hall.

Martinez-Alier, J. 1998. *Ecological economics as human ecology.* Madrid: Fundación César Manrique.

McDonnell, M. J., and S. T. A. Pickett. 1997. *Humans as components of ecosystems: The ecology of subtle human effects and populated areas.* New York: Springer.

McNeely, J. A., M. Gadgil, C. Leveque, C. Padoch, and K. Redford. 1995. Human influences on biodiversity. Pp. 711–821 in *Global biodiversity assessment,* edited by V. H. Heywood and R. T. Watson. Cambridge: Cambridge University Press and UNEP.

Meadows, D., D. Meadows, and J. Randers. 1992. *Beyond the limits: Global collapse or a sustainable future.* London: Earthscan.

Miller, K., M. H. Allegretti, N. Johnson, B. Jonsson, R. Hobbs, E. Lleras, M. Wells, and C. de Klemm. 1995. Measures for conservation of biodiversity and sustainable use of its components. Pp. 915–1061 in *Global biodiversity assessment,* edited by V. H. Heywood and R. T. Watson. Cambridge: Cambridge University Press and UNEP.

Moser, D., H. G. Zechmeister, C. Plutzar, N. Sauberer, T. Wrbka, and G. Grabherr. 2002. Landscape shape complexity as an effective measure for plant species richness in rural landscapes. *Landscape Ecology* 17:657–669.

Munasinghe, M., and W. Shearer. 1995. An introduction to the definition and measurement of biogeophysical sustainability. Pp. xvii–xxxiii in *Defining and measuring sustainability: The biogeophysical foundations,* edited by M. Munashinghe and W. Shearer. Washington, DC: United Nations University and The World Bank.

Odum, E. P. 1971. *Fundamentals of ecology.* Philadelphia: Saunders.

O'Neill, R. V., J. R. Krummel, R. H. Gardner, G. Sugihara, B. Jackson, D. L. DeAngelis, B. T. Milne, M. G. Turner, B. Zygmunt, S. W. Christensen, V. H. Dale, and R. L. Graham. 1988. Indices of landscape patterns. *Landscape Ecology* 1:153–162.

Pimentel, D., W. Dazhong, and M. Giampietro. 1990. Technological changes in energy use in U.S. agricultural production. Pp. 305–321 in *Agroecology: Researching the ecological basis for sustainable agriculture,* edited by S. R. Gliessmann. New York: Springer.

Postel, S. L., G. C. Daily, and P. R. Ehrlich. 1996. Human appropriation of renewable fresh water. *Science* 271:785–788.

Rapson, G. L., K. Thompson, and J. G. Hodgson. 1997. The humped relationship between species richness and biomass: Testing its sensitivity to sample quadrat size. *Journal of Ecology* 85:99–100.

Rojstaczer, S., S. M. Sterling, and N. Moore. 2001. Human appropriation of photosynthesis products. *Science* 294:2549–2552.

Rosenzweig, M. L., and Z. Abramsky. 1993. How are diversity and productivity related? Pp. 52–65 in *Species diversity on ecological communities*, edited by R. E. Ricklefs and D. Schluter. Chicago: University of Chicago Press.

Sagoff, M. 1995. Carrying capacity and ecological economics. *BioScience* 45:610–620.

Sauberer, N., K. P. Zulka, M. Abensberg-Traun, H.-M. Berg, G. Bieringer, N. Milasowsky, D. Moser, C. Plutzar, M. Pollheimer, C. Storch, R. Tröstl, H. G. Zechmeister, and G. Grabherr. 2004. Surrogate taxa for biodiversity in agricultural landscapes of eastern Austria. *Biological Conservation* 117:181–190.

Sitch, S., B. Smith, I. C. Prentice, A. Arneth, A. Bondeau, W. Cramer, J. O. Kaplan, S. Levis, W. Lucht, M. T. Sykes, K. Thonicke, and S. Venevsky. 2003. Evaluation of ecosystem dynamics, plant geography and terrestrial carbon cycling in the LPJ dynamic global vegetation model. *Global Change Biology* 9:161–185.

Tüxen, R. 1956. Die heutige potentielle natürliche Vegetation als Gegenstand der Vegetationskartierung. *Angewandte Pflanzensoziologie* 13:5–42.

Vitousek, P. M., and J. Lubchenko. 1995. Limits to sustainable use of resources: From local effects to global change. Pp. 57–64 in *Defining and measuring sustainability: The biogeophysical foundations*, edited by M. Munashinghe and W. Shearer. Washington, DC: United Nations University and The World Bank.

Vitousek, P. M., P. R. Ehrlich, A. H. Ehrlich, and P. A. Matson. 1986. Human appropriation of the products of photosynthesis. *BioScience* 36:368–373.

Vitousek, P. M., H. A. Mooney, J. Lubchenko, and J. M. Melillo. 1997. Human domination of Earth's ecosystems. *Science* 277:494–499.

Wackernagel, M., N. B. Schulz, D. Deumling, A. C. Linares, M. Jenkins, V. Kapos, C. Monfreda, J. Loh, N. Myers, R. Norgaard, and J. Randers. 2002. Tracking the ecological overshoot of the human economy. *Proceedings of the National Academy of Sciences* 99:9266–9271.

Waide, R. B., M. R. Willig, C. F. Steiner, G. Mittelbach, L. Gough, S. I. Dodson, G. P. Juday, and R. Parmenter. 1999. The relationship between productivity and species richness. *Annual Review of Ecology and Systematics* 30:257–300.

Weiher, E. 1999. The combined effects of scale and productivity on species richness. *Journal of Ecology* 87:1005–1011.

Weterings, R. A. P. M., and J. B. Opschoor. 1992. *The ecocapacity as a challenge to technological development*. Rijswijk, the Netherlands: Advisory Council for Research on Nature and Environment.

Whittaker, R. H. 1960. Vegetation of the Siskiyou mountains, Oregon and California. *Ecological Monographs* 30:277–332.

Whittaker, R. H., and G. E. Likens. 1973. Primary production: The biosphere and man. *Human Ecology* 1:357–369.

Wrbka, T., K.-H. Erb, N. B. Schulz, J. Peterseil, C. Hahn, and H. Haberl. 2004. Linking pattern and processes in cultural landscapes. An empirical study based on spatially explicit indicators. *Land Use Policy* 21:289–306.

Wright, D. H. 1983. Species–energy theory, an extension of species–area theory. *Oikos* 41:495–506.

Wright, D. H. 1987. Estimating human effects on global extinction. *International Journal of Biometeorology* 31:293–299.

Wright, D. H. 1990. Human impacts on the energy flow through natural ecosystems, and implications for species endangerment. *Ambio* 19:189–194.

Wright, D. H., D. J. Currie, and B. A. Maurer. 1993. Energy supply and patterns of species richness on local and regional scales. Pp. 66–76 in *Species diversity in ecological communities*, edited by R. E. Ricklefs and D. Schluter. Chicago: University of Chicago Press.

Zechmeister, H. G., and D. Moser. 2001. The influence of agricultural land-use intensity on bryophyte species richness. *Biodiversity and Conservation* 10:1609–1625.

Part V
Case Studies
Arthur Lyon Dahl

The final set of chapters in this volume presents case studies that illustrate many of the themes raised in the earlier chapters and demonstrate the present state of the art and challenges of assessing sustainability. The first three chapters provide case studies for specific geographic entities (countries or provinces), and the last three discuss specific indicator initiatives.

In Chapter 18 Stephen Hall describes the pioneering work in the United Kingdom to monitor progress in their sustainable development strategy with quality-of-life indicators. Annual reports on progress are complemented by 5-year reviews of policy and indicators. Starting with 120 mostly environmental indicators in 1996, they considered more than 400 indicators by 1999, settling on a core set of 150 indicators with 15 headline indicators in a quality-of-life barometer. The latter, presented graphically in a leaflet with a traffic light system, has been most useful in building public perceptions of sustainability issues. The indicators worked best where policies were already well established and were less successful in driving and influencing policy. They also tended to focus attention on problem areas rather than giving an integrated overall view. Indicator summaries proved more useful than massive compendiums. One issue was to determine how much change in an indicator was significant. For the 2004 review aiming for a new indicator set, some 5,500 indicators were proposed for consideration, showing how rapidly the field is evolving.

In Chapter 19, Knippenberg et al. describe the work of the Telos Brabant Centre for Sustainability Issues in developing tools for the assessment of sustainable development at the provincial level in the Netherlands, starting with the province of Brabant. These were structured by the three pillars or types of capital (ecological, sociocultural, and economic) and assessed by three criteria of an integrated approach that does not favor one pillar at the expense of the others, is sustainable over time and across generations, and is sustainable at the global level and not at the expense of other places elsewhere. The aim is to achieve social solidarity, economic efficiency, and ecological resilience. The types of capital were made up of a number of stocks for which several sustainability requirements were defined and measured with indicators. The methods included scientific criteria, expert judgment, and wide stakeholder consultations. Although the aim was to develop a standard provincial approach, they found major divergences between provinces, with less than a third of requirements common to all provinces and another third completely province specific. Implementing the indicators was difficult and labor intensive because of the lack or inaccessibility of appropriate data at the provincial level. The method showed the interconnections fundamental to sustainable development, although it was difficult to relate the ecological, social, and economic pillars. Also, the subjectivity of expert judgments and the variability produced by the weight given to often changing stakeholder views prevented comparisons in time and space. Nevertheless, the learning process itself, with the wide involvement of stakeholders, had an important impact.

In the third geographic case study, in Chapter 20, Wang and Paulussen review the development of sustainability assessment indicators in China. Starting with a framework that bridges and incorporates traditional Chinese concepts and modern natural, economic, and social subsystems, China has addressed the challenge to frame the complicated interactions and integrate the diverse relationships of these subsystems to produce practical instruments for promoting sustainable development. The models used include the mechanisms of competition, symbiosis, circulation, and self-reliance; temporal and spatial processes and patterns; the balance between the driving forces of energy, money, power, and spirit and the human interferences of technology, institutions, and culture; and various planning and management models at multiple scales. The aim is the overall harmony of productive wealth, functional health, and people's faith. The problems faced are common to many indicator initiatives, including definitions, the imbalance of economic and environmental evaluations, the challenge of linking ecology and gross domestic product, the tendency for a local and short-term orientation, and issues of quantitative and qualitative indicators, aggregation, weighting, substitution, and data availability. Some examples are given. There is a need to reduce and simplify indicators, address the shortcomings of the three-pillar approach, and resolve the problems of data availability and the imbalance of human and environmental data. Other challenges identified are to overcome the focus on hard numeric data and the low transparency that comes with high aggregation.

In Chapter 21, van Woerden et al. describe the use of indicators in the UNEP Global Environment Outlook (GEO) process. Building on the GEO Data Portal, with 400 variables, a GEO Core Data/Indicator matrix has been developed with seventy indicators aiming to give a concise picture of the global environment. However, the indicators are used only as illustrations and are not comprehensive or integrated. With the problems of data adequacy at the global level, it is impossible to illustrate all environmental issues adequately, and some important issues cannot be shown with indicators. The present results are sketchy and unbalanced, with much reliance on proxy indicators, but the GEO process is working on continuing improvements.

In Chapter 22 Gutiérrez-Espeleta provides a useful developing country perspective on the further work that is needed. He starts from the challenge to development concepts from the holistic perspective of sustainable development, which shows that sectoral policies are not all useful and that there are not just economic or environmental solutions, necessitating a new approach. He then classifies the various generations of sustainable development indicators from single measures to composite cross-sectoral indices and shows that most present indicator sets are first- or second-generation measures that fail to show the relationships between environment and society. Examples given are a Latin America and Caribbean indicator set and the indicators from the UN Commission on Sustainable Development. Although the present state of social indicators is bad, the situation for the environment is even worse, and existing indicators present a fragmented view of development. The task is to develop high-level integrated indicators. The chapter reviews two interesting initiatives in this direction: the Environmental Vulnerability Index (EVI) and the Environmental Sustainability Index (ESI), which are politically relevant, innovative, and integrative and give a better understanding of national development dynamics. These are important steps toward a more holistic approach.

Finally, in Chapter 23, Hák reviews the 2005 version of the ESI, which is designed for use as a policy tool to identify issues deserving greater attention within national environmental programs and across societies. Given that sustainability deals with dynamic systems, the aim is long-term maintenance of environmental resources. The index does not define sustainability but measures present environmental quality and the capacity to maintain it in the short and medium term. It uses a driving force, pressure, state, impact, response framework with variables that range from local to global in scope and are relevant to all countries despite differing national priorities. There is inevitably a hidden weighting from the selection of the numbers of variables, which is imperfect but provides a clear starting point for analysis and can be modified in a planned interactive version. The ESI country rankings attract attention but are hard to interpret. The index is most useful for identifying best practices at the indicator or variable level. The results show the distinct environmental challenges at different levels of development and reveal that economic success can contribute to environmental success but does not guarantee it. Although the ESI, even in its improved form, suffers from data gaps and is still open to criticism, the latest report identifies the additional work needed to improve it.

18

The Development of UK Sustainable Development Indicators: Making Indicators Work

Stephen Hall

After more than 10 years of experience in developing and using sustainable development indicators, in 2005 the United Kingdom established its third generation of indicators to support a new sustainable development strategy.

When sustainable development indicators were first established, the proactive use of indicators and targets in government was in its infancy. The first set of indicators therefore was breaking new ground. By the second generation of indicators, there had been a proliferation of indicators and performance targets across the machinery of government. It was therefore possible to strengthen sustainable development indicators with a commitment to make progress, and a set of headline indicators was established to be drivers for action and to highlight where policies needed to be adjusted.

However, with performance measures across every area of government and every new policy initiative generating more targets, the approach to developing a set of sustainable development indicators is more challenging, and their perceived role has changed. In one respect, with a multitude of indicators and targets already in place, establishing a set of sustainable development indicators ought to be easier because it should be possible to cherry-pick the best indicators from a wider variety of existing ones. However, the challenge now is to identify and develop indicators that are adding value and bringing a sustainable development perspective rather than simply reusing existing performance measures and giving them a sustainable development badge.

There is now greater sensitivity about what messages a set of sustainable development indicators might convey and, in many cases, a desire by policymakers and politicians that these should be consistent with the indicators and targets already adopted in specific policy areas. Consequently, there is a danger that a set of sustainable development indicators may be only a repackaging exercise. However, where a set of sustainable development indicators is closely allied with a sustainable development strategy, as is the case for the UK, perhaps greater pressure can be applied through the commitments in the strategy to extract agreement for more challenging sustainable development indicators, perhaps reducing some of the repackaging.

It has long been desired that sustainable development indicators be fully integrated into policymaking and directly influencing policy decisions. However, there are very few examples in which this has happened. The problem is that the principal role of indicators is in communication, particularly to the public and to ministers, who do not necessarily need lots of detail.

Most indicators therefore provide only a broad overview of an issue and are of little use for detailed policy considerations. They are often too broad for policymakers to identify other policy areas where their decisions may have effects.

Some stakeholders call for a set of indicators that are better integrated internally (i.e., with all the linkages identified and quantified), but we are a long way from being able to construct models that allow us to know what impact a change in one indicator will have on another. Other stakeholders believe that holistic sustainable development measures are needed, and a growing number of aggregate indices have been promoted internationally that profess to be measures of sustainability. However, there is also a high degree of skepticism about their methods and meaningfulness. Although aggregate indices may have their place in a package of communication tools, there is a concern that they are more likely to mislead than to lead to discernible progress. However, the idea of condensing the messages is valid, and indicators sets might be reduced to be more manageable for those trying to understand the messages and those trying to maintain them.

The immediate focus should be on raising the profile of indicators and making them more effective as communication tools in order to raise awareness and understanding of sustainable development.

First-Generation Indicators

In 1994 the UK became one of the first countries to produce a sustainable development strategy (HM Government 1994) in response to the 1992 Earth Summit in Rio de Janeiro. The strategy led the government to pursue, via an interdepartmental working group, a set of indicators with which to monitor progress. In 1996 a preliminary set of 120 indicators, *Indicators of Sustainable Development for the United Kingdom* (Department of the Environment 1996), was published for discussion and consultation.

In reflecting the structure and hence also the inadequacies of the strategy, the indicators unfortunately focused too heavily on economic and environmental issues and also preempted, by a few months, the UN Commission for Sustainable Development draft menu of indicators. However, the UK was subsequently one of twenty-two countries to volunteer to pilot test the applicability of the commission's indicators.

Second-Generation Indicators

Following a change of government in 1997, a new strategy, "A better quality of life" (DETR 1999a), was published in 1999. The establishment of indicators was an integral part of the development of the new strategy, with work on indicators going alongside and sometimes ahead of discussions on the content of the strategy.

One of the strengths of this approach was that the indicators helped to focus people's minds on the issues that should be covered by the strategy. In some cases indicators led to the inclusion of issues in the strategy that might not otherwise have been included, or at least not in the same way, such as indicators on wild bird populations and air quality. However, some of the indicator work (e.g., on social indicators) was not used in the final set, or the experts engaged in the exercise felt unable to contribute constructively without knowing the direction of the strategy.

Working to some extent blind, without a strong policy lead, may have resulted in a much larger volume of candidate indicators than would have been the case if indicator development had awaited finalization of the policy framework.

Furthermore, and perhaps inevitably, when the debate on indicators was opened to stakeholders, they tended to be strongly motivated to see their own areas of concern covered by an indicator. This was often on the erroneous assumption that if it was not an "indicator of sustainable development" then it was not monitored at all. Another motivation may have been that in their view a particular issue had to be seen as contributing to sustainable development through the indicators, possibly in anticipation of potential funding or for political or presentational elevation.

Though undoubtedly eliciting wider support and ensuring a more robust set of indicators, stakeholder involvement, with a still-evolving policy framework, had the potential to hamper the establishment of a coherent set. For example, in one particular workshop event, the aim was to reduce an already large list of indicators, some 200 or so, down to perhaps as few as 50. By the end of the day's deliberations, rather than reducing the list, stakeholders had argued the need for more candidate indicators, and the list had grown to more than 400.

Second-Generation Indicators: Headline Indicators and *Quality of Life Counts*

With a potentially large set of indicators, it was clear that it would be very difficult to answer the question, "Are we becoming more or less sustainable?" Each indicator would give a different answer for a specific area. Ministers therefore asked that some headline indicators be established that might provide a broad overview of progress. Responses to a public consultation paper, *Sustainability Counts* (DETR 1998), showed wide support for a set of headline indicators.

Some 6 months after the publication of the strategy document, *Quality of Life Counts* (DETR 1999b) was published. This provided a baseline assessment of 15 headline indicators and 132 core sustainable development indicators. The headline indicators were described as a quality-of-life barometer "to provide a high level overview of progress, and be a powerful tool for simplifying and communicating the main messages for the public."

The headline indicators were to play a key role in the promotion of sustainable development and were at the center of four successive UK government annual reports on progress, *Achieving a Better Quality of Life* (DEFRA 2004a).

The wider *Quality of Life Counts* proved to be very influential in other indicator initiatives throughout the UK and internationally. However, with hindsight it is questionable whether such a large set of indicators, 147 including the headline indicators, was practical to maintain and effective in communicating or in influencing policy.

Third-Generation Indicators: Public Consultation

The 1999 strategy document included a commitment to review the strategy and its supporting indicators after 5 years. In 2004, the UK government launched a public consultation document, *Taking It On* (DEFRA 2004c), which sought views on the direction of sustainable development strategy and future monitoring of progress through indicators. The questions on how progress should be reviewed and communicated were as follow:

• What are the strengths and weaknesses of the current sustainable development indicators, and how they are used in general? And, more specifically, what about indicators used in the UK government's headline set; in the wider UK core set in *Quality of Life Counts*; in Scotland, Wales, and Northern Ireland; in the English regions; in local authorities; and elsewhere (e.g., sectoral indicators)?
• What needs to be monitored and measured across the UK?
• Who are the audiences for indicators, and how can we better meet their needs?
• Should any set of indicators supporting the new strategy concentrate on just the main priorities in the strategic framework or be wider and more comprehensive?
• Should important high-level sustainable development indicators focus on monitoring general progress toward final outcomes, specific delivery actions and targets, or both?

Despite best efforts to have indicator questions positioned early in the consultation document, they were relegated to the end. Many of the preceding questions required respondents to provide detailed answers, so there was an inevitable decline in the responses for later questions. However, in practice monitoring and indicators were important threads running through responses to many of the questions in the consultation document. More than 700 individuals and organizations responded, and their responses included 1,500 references to monitoring or indicators.

Ninety-five percent of respondents supported a set of headline indicators, but only 11 percent specifically favored the existing headline set with no change, and 25 percent supported the existing set with some modification.

Eleven percent of all indicator responses were specifically about gross domestic product (GDP) as a measure of sustainable development, with the majority of these advocating its exclusion from the set or changing it radically.

A wide variety of candidate indicators were proposed for a headline set, including a number of aggregate indices, with some people suggesting that there should be no more than five headline indicators and that these should be aggregate measures. Eight percent of all indicator responses strongly supported the inclusion of an ecological footprint. There was also strong support for other measures that encapsulate a concept of well-being.

In addition to the consultation, a review was undertaken of indicators used directly to monitor sustainable development and indicators in closely related national strategies. The exercise was then extended to a wider array of indicator sets used nationally and internationally. In total, more than 5,000 indicators were identified. These were then grouped into broad themes and into economic, social, and environmental impacts and drivers. This exercise was not as useful as was hoped but provided a useful insight into indicators used elsewhere and a reassurance that the UK set was not missing important measures used by others.

Similar to the situation for the earlier *Quality of Life Counts* report, there was the challenge of trying to establish a policy-relevant set of indicators in time for inclusion in the new strategy while the policy thinking for the new strategy was still being developed. A degree of pragmatism was needed, along with constructive dialogue with policy colleagues to negotiate an acceptable set of indicators.

Third-Generation Indicators: The Final Set

In 2005 the new UK government sustainable development strategy, *Securing the Future* (DEFRA 2005a), was published. Twenty "UK Framework Indicators" were outlined that reflected the broad priorities set out in a framework for sustainable development agreed between the UK government and the devolved administrations in Scotland, Wales, and Northern Ireland. These broadly take on the role of headline indicators, for which the devolved administrations and the UK government have shared responsibility.

In addition to the "UK Framework Indicators," the UK government sustainable development strategy outlined another forty-eight indicators related to the priority policy areas covered by the strategy.

The new indicator set included eight that needed development, and the most challenging of these is well-being, which had been so strongly suggested in the consultation responses. A number of surveys have asked people to rate their life satisfaction, but the degree of satisfaction is surprisingly high and has changed little over many years. Research has been commissioned to investigate the concept of well-being, its relevance for policy and sustainable development, and how it might be measured in a meaningful way.

The sixty-eight indicators include the previous fifteen headline indicators, although not all of them are included in the twenty "UK Framework Indicators."

This means that GDP has been retained. Arguments for its retention included recognition that GDP provides essential context for a number of the other indicators, it is a driver for many environmental pressures, and economic growth is an essential aspect of sustainable development in terms of supporting environmental and social development.

A number of the indicators in the new set were decoupling indicators, which attempt to show whether impacts (predominantly environmental) are being decoupled from their potential drivers (predominantly economic growth and demographic changes).

The International Dimension

There was much debate about the new indicator set featuring a number of international indicators. Indeed, policy colleagues and some of the *Taking It On* consultation responses suggested the inclusion of indicators that capture the UK's global footprint or other international indicators.

This was fine in theory, but it was not clear what people meant by "international indicators" because they could include indicators of the UK's performance compared with other countries, indicators highlighting global trends, and indicators trying to capture the UK's international impacts. Given the aim of reducing the size of the national indicator set to improve its manageability and communication, there was a danger that the indicator set could be swamped with international indicators, which would be difficult to maintain and would duplicate reporting being done by many reputable international organizations.

A more practical approach therefore was needed, and it was agreed that the new set of indicators would not formally include international indicators. Instead, commitments were made to make international comparative information available via links to international Web sites and in due course to explore how the UK's international impacts might be measured for particular sectors.

Indicator Frameworks and Selection of Indicators

Much work has been undertaken nationally and internationally to determine frameworks for sustainable development indicators. Sometimes perhaps too much effort is expended in theorizing about frameworks. They may help to ensure that cause and effect can be monitored, and they may help to ensure that significant gaps in monitoring are filled. So it is clear that some structure is needed.

However, as seen in the experience of *Quality of Life Counts*, the strength of the indicator structure was that it was precisely the same as the policy framework, with direct links to both broad and specific policy objectives in the strategy. It meant that the indicators were seen not as an academic or statistical exercise but as core components of the overall policy approach. Ensuring their policy relevance in structure and coverage also meant that strong government commitments were associated with the indicators.

In the third generation of indicators, the approach was not as meticulous, and indicators were selected that related to the four broad priority areas identified in the strategy. The links to policy were not necessarily specific but came through the preexistence of policy targets that if achieved would directly or indirectly contribute to progress in the broad policy area. This approach reflected in part a stronger focus in the new strategy on tangible delivery of sustainable development through outcomes rather than laudable but vaguely defined objectives.

Compared with a detailed list of criteria used to select indicators for *Quality of Life Counts*, which often had to be compromised, criteria for the new set of indicators were less ambitious, and wherever possible indicators were linked to the purpose and priorities in the UK strategy, were held as high priorities by the UK government, had UK coverage, had trends available, highlighted challenges, and were statistically robust and meaningful.

There was also an overall aim of having about fifty indicators in the final set. Although they did not quite achieve this goal, the sixty-eight indicators in the new set are less than half the number in the *Quality of Life Counts* set.

Indicators Influencing Policy

It is unlikely that many of the indicators have influenced policy because they are part of a sustainable development set. In most cases the indicators selected were already well-established measures. One of the exceptions to this was the indicator on populations of wild birds. The media initially made much of the novelty of the government measuring people's quality of life by counting birds, but the messages conveyed by the indicator demanded action. Although overall the population of birds had not changed significantly, the populations of farmland species had fallen dramatically since the 1970s. As a direct result of the indicator, a policy response was put in place to halt the decline and stabilize populations.

Assessing Progress

The UK has a decentralized statistical system, with statistics collected and published by all principal ministries and their agencies. The Department for Environment, Food and Rural Affairs (DEFRA) is responsible for coordinating efforts across the government for sustainable development. Statisticians in DEFRA therefore have the task of establishing and maintaining the UK sustainable development indicators.

DEFRA statisticians have been at the forefront of negotiations with other ministries to establish the indicators, agree on presentations, and in some cases persuade them to initiate new data collection. DEFRA statisticians have collated all the indicator data and have had responsibility for assessing and publishing the indicator set. Although coordinating the indicators is a logistical challenge, the work is done under the auspices of National Statistics, which is the independent framework under which statistics are produced in the UK, thus enabling the indicators to be compiled and reported without policy or ministerial interference.

Only a handful of countries and institutions have actively made summary assessments of indictors; in most cases the indicators are presented only as charts and commentary. For the UK's *Quality of Life Counts*, early attempts were made to have targets associated with the indicators, but it was concluded that in most cases there was no easily identified point at which a trend was sustainable. Therefore the approach of assessing progress over baselines was established and reported using "traffic lights."

With hindsight, there are some arguments for why it might have been better to avoid making summary assessments. Policymakers and ministers undoubtedly are sensitive about what color traffic light is reported for their particular policy areas, and the media can become very focused on the traffic lights and not on the wider issues behind the indicators. However, on balance, symbol assessments probably are useful to help people understand what the charts are saying and to learn at a glance whether things are improving or getting worse. Now that traffic light assessments have been in use for 5 years, it is doubtful that stakeholders and the media would accept UK indicators without assessments.

Problems surrounding this means of assessment include the fact that baselines are arbitrary, with the danger that a different baseline could result in a very different assessment of progress and the difficulty of determining whether change in an indicator should be regarded as significant. Pressure has been applied by the National Audit Office and others for the basis of the assessments to be made much more transparent, with clear justifications for the traffic lights.

This has remained difficult, not least because for many of the data sets limited statistical information on significance was available. Assessments hitherto had been made based on the experience and knowledge (and sometimes gut feeling) of the statisticians involved, but it was very difficult to robustly justify the assessments beyond saying what the latest data were and what the baseline figures were.

To try to make the assessments more rigorous, a threshold percentage change in the indicators was determined, above which a change was considered significant. This work was undertaken as part of an update of *Quality of Life Counts* in 2004. The determination of the threshold was still arbitrary to some extent but was based on what percentage change would support the assessments previously made for most if not all indicators. So it was an a priori judgment rather than one based on statistical rigor. The main benefit was that although debates could be had about the threshold, there was at least greater transparency in and defense of the traffic light assessments. For most indicators a 3 percent change was regarded as sufficient for a green or red traffic light. Where the value of an indicator was already very high and could not be expected to change greatly, a smaller amount of change might be regarded as significant, so there remained some latitude for common sense to prevail.

In the new set of indicators, attempts have been made to reduce the effect of the baseline year by making the baseline figure, against which the latest data are assessed, a 3-year average around the baseline year.

The more transparent method of assessing and reporting the new indicators has recently been endorsed by the National Audit Office and the independent UK Statistics Commission.

Communication Products: "Quality of Life Barometer" Leaflet

In the initial years of *Quality of Life Counts* there was frustration among ministers that the headline indicators were not making headlines in the media, and awareness of sustainable development was low. The main approach to highlighting the indicators was through the government's sustainable development Web site and through annual reports, but these were eliciting little interest from the media. It was clear that a more succinct way of getting the indicators across to audiences beyond the cognoscenti was needed.

A leaflet, the "Quality of Life Barometer," attempted to present the indicators in simplified form, stripping out unnecessary detail and providing very short commentary and traffic light assessments. Information on all fifteen headline indicators was condensed onto two sides of A4 paper. (See Annex 18.1 for an example of the leaflet.)

The leaflet proved to be extremely effective in promoting the headline indicators to wider audiences, not least because it could be updated regularly, produced in bulk, and easily distributed. It was applauded by the UK's independent Sustainable Development Commission and EU indicator experts and was described as "the single most important development in communicating sustainable development" (Professor Anne Power, UK Sustainable Development Commissioner, 2001).

At media briefings, it was often the "Quality of Life Barometer" leaflet that the journalists turned to rather than the weighty tome that was the main focus of the event. Many of their questions directed at ministers were then based on the headline indicators and traffic light assessments shown in the leaflet.

The leaflet was particularly successful at one media briefing. It resulted in a healthy debate in newspapers and television news programs on what quality of life means, how it should be measured, and whether the government's assessments of progress were the right ones. Examples of the newspaper headlines were as follow:

Evening Standard: "Crime up, roads worse but life is better says Labour"
The Times: "Life is better despite crime, illness and cars, says Labour"
The Express: "Quality of life is better? But what about all the thuggery and the jams"
The Guardian: "Quality of life 'getting better'"

The leaflet has inspired similar documents to be produced by, for example, the European Commission, the Environment Agency (England and Wales), and the Finnish Environment Institute and has been emulated more widely since.

Communication Products: Pocket-Sized Booklets

The *Quality of Life Counts* set was not intended to be updated as frequently as the fifteen headline indicators; to have done so would be impractical, and most trends would not be expected to change dramatically annually. An updated compendium of the indicators, *Quality of Life Counts: Update 2004*, was published on the sustainable development Web site but received little stakeholder and media recognition.

A month later a new publication, *Sustainable Development Indicators in Your Pocket 2004* (DEFRA 2004b), was published and was a great success. This pocket-sized booklet (A6 in size) contained a selection of fifty indicators to help illustrate the breadth of issues covered by the sustainable development agenda but without overloading the reader with too many indicators. Orders for the booklet surpassed expectations, and a reprint had to be done to meet demand from, in particular, schools and other educational institutions. This success thus reinforced the assumption that pocket summaries of indicators would be more useful and attract wider audiences than large statistical volumes.

This in part influenced the aim for the third generation of indicators to try to reduce the number of indicators in the set and thereby make them more manageable in communication terms. A new booklet, *Sustainable Development Indicators in Your Pocket 2005* (DEFRA 2005b), provides baseline assessments for the new indicator set and contains all sixty-eight indicators in one small volume. It has proved very popular and has been applauded by a wide variety of stakeholders.

Regional and Local Indicators

Once *Quality of Life Counts* was released, there were demands for indicators that were more relevant to local experiences. *Regional Quality of Life Counts* therefore was produced and updated annually, providing regional versions of the headline indicators, where data were available, for the English Regions. These were intended to help raise awareness of sustainable development, provide a useful input into regional sustainable development frameworks, and help direct policies where there are regional disparities.

Inevitably, producing regional indicators led to comparisons between regions, and in England the media often assume that things are better in the south of the country than in the north. The *Regional Quality of Life Counts* (DEFRA 2002) publication generated some interesting newspaper headlines:

The Daily Telegraph: "It's grim up North, say life quality statistics"
Daily Express: "Great divide" Head south if you want a longer life northerners told"
The Guardian: "Poverty and crime make it tough up north—but more birds are singing"
The Times: "Life sounds sweet in poorer North"

In December 2005, new regional versions of forty-four of the sixty-eight national indicators were published. In terms of interest, they generated possibly the best media coverage ever in the UK of sustainable development and indicators. Articles featured in both the national and the regional press, particularly regional newspapers, produced analyses of the indicators for their regions and highlighted the successes and the challenges.

Work has been done at the local level, too. In 2000 a menu of twenty-nine indicators was developed, which local authorities were encouraged to consider using for their strategies and other local monitoring. The menu *Local Quality of Life Counts* (DETR 2000) was developed jointly by Central Government, local government bodies, the Audit Commission, and Local Agenda 21 groups and tested by thirty local authorities. The development of local indicators was then taken forward by the Audit Commission, and in collaboration with DEFRA and other ministries a new set of local indicators have been produced, which are related where possible to the national indicators.

Annex 18.1.

Quality of Life Barometer

For Annex 18.1, see following pages.

Quality of Life Barometer

Updated October 2004

Sustainable development is about ensuring a better quality of life for everyone, now and for generations to come.

The 15 Headline indicators of sustainable development – a quality of life barometer – provide an overview of progress in meeting the objectives of the UK Sustainable Development Strategy - *A better quality of life* (May 1999).

Headline indicators – assessment of progress

	since 1990	since Strategy
Economic output	☺	☺
Investment	☺	☺
Employment	☺	☺
Poverty & social exclusion	☺	☺
Education	☺	☺
Health	☺	☺
Housing - conditions	☺	☺
Crime - robbery	☹	☹
- vehicle & burglary	☺	☺
Climate change	☺	☺
Air quality	☺	☹
Road traffic - total traffic volumes	☹	☹
- traffic per GDP	☺	☺
River water quality	☺	☺
Wildlife - farmland birds	☹	◯
- woodland birds	☹	◯
Land use	☺	☺
Waste - household waste	☹	☹
- all arisings & management	⦿	◯

Key:

Significant change, in direction of meeting objective	☺
No significant change	◯
Significant change, in direction away from meeting objective	☹
Insufficient or no comparable data	⦿

Where a trend is unacceptable, the government will adjust its policies, and look to others to join it in taking action. A full assessment of progress can be found in the fourth Government Annual Report on Sustainable Development 2003: *Achieving a better quality of life*. Data and further details on the Headline and a wider core set of indicators are available on the website below.

www.sustainable-development.gov.uk

For additional copies of this leaflet, please call 020 7082 8621

H1 ECONOMIC OUTPUT
GDP per head (UK)

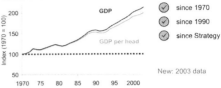

- 30% increase in real GDP per head between 1990 and 2003 2.0% per year on average.
- Real GDP per head increased by 1.8% in 2003, and has increased by 11% since 1998.

H2 INVESTMENT
Total & Social Investment (UK)

- Total real investment relative to GDP grew by 5% between 1990 and 1998 since when it has remained relatively stable.
- Social investment (railways, hospitals, schools etc.) was around 2% of GDP in 1990 and 1.9% in 2003 (only available on a current price basis).

H3 EMPLOYMENT
Percentage of people of working age in work (UK)

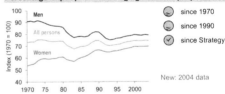

- The percentage of working age people in work was 74.6% in 2004 – about the same as in 1990.
- The percentage for 2004 was 0.1 percentage points down on 2003 but was an increase on the 1999 figure of 73.9%.

lected indicators of poverty & social exclusion

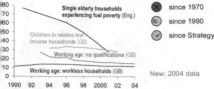

- Single elderly households experiencing fuel poverty (Eng.) ⊗ since 1970
- Children in relative low income households (GB) ⊜ since 1990
- Working age: no qualifications (GB) ⊘ since Strategy
- Working age: workless households (GB) New: 2004 data

11.6% of working age people were in workless households in 2004, reduced from 13% in 1998; 14.7% were without qualifications, down from 16.9% in 1999.

28% of children were in relatively low-income households (after housing costs; 21% before) in 2002-3, reduced from 34% (25% before) in 1996-7.

28% of single elderly households experienced fuel poverty in 2001, reduced from 77% in 1991 and 61% in 1996.

5 EDUCATION

vel 2 qualifications at age 19 (UK)

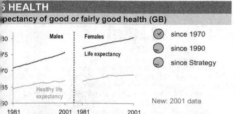

NVQ level 2 or equivalent (5 GCSEs grade C or above)

- ⦿ since 1970
- ⊘ since 1990
- ⊘ since Strategy
- New: 2004 data

In 2004, 75.4% of 19 year-olds achieved NVQ level 2 or equivalent (5 GCSEs grade C), up from 52% in 1990, and 74.4% in 1999. The 2004 figure was slightly below the 2003 figure of 76.1%.

5 HEALTH

pectancy of good or fairly good health (GB)

Males Females
Life expectancy
Healthy life expectancy

- ⊘ since 1970
- ⊖ since 1990
- ⊖ since Strategy
- New: 2001 data

Between 1990 and 2001 healthy life expectancy increased only slightly, from 66.1 to 67.0 years for men and from 68.3 to 68.8 years for women.

Overall life expectancy (75.7 years for men, 80.4 years for women) has increased more than healthy life expectancy, so an increasing proportion of those extra years are in poor health.

H7 HOUSING CONDITIONS

Households in non-decent housing (England)

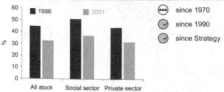

- ■ 1996   ▦ 2001
- ⦿ since 1970
- ⊘ since 1990
- ⊘ since Strategy

All stock   Social sector   Private sector

- Between 1996 and 2001, households in non-decent housing fell from 51% to 37% and from 44% to 31% in the social and private sectors respectively.
- Between 1991 and 1996 there was no significant change across a broad range of condition measures. As housing conditions have changed for the better since 1996, the overall assessment is that there has been an improvement since 1990.

H8 CRIME

Recorded crime (England & Wales)

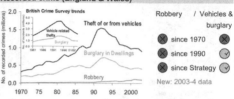

British Crime Survey trends
Theft of or from vehicles
Vehicle related thefts
Burglary
Burglary in Dwellings
Robbery

- Robbery / Vehicles & burglary
- ⊗ since 1970 ⊗
- ⊗ since 1990 ⊘
- ⊗ since Strategy ⊘
- New: 2003-4 data

- Both the British Crime Survey and recorded crime show that burglary and vehicle crimes fell substantially from the early 1990s: from 1990 such recorded crimes fell by 24% and 30% respectively (BCS indicates falls from 1991 of 32% and 45%).
- By 2003-4, recorded robbery had risen to 101,000 from 67,000 in 1998-9 but was 6% lower than the previous year.

H9 CLIMATE CHANGE

Emissions of greenhouse gases (UK)

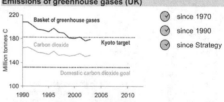

Basket of greenhouse gases
Carbon dioxide   Kyoto target
Domestic carbon dioxide goal

- ⊘ since 1970
- ⊘ since 1990
- ⊘ since Strategy

- Emissions of the 'basket' of six greenhouse gases (on which progress is assessed) fell by about 14% between 1990 and 2003 (provisional data).
- CO_2 emissions for 2003 were provisionally 7% lower than in 1990 but rose by about 1.5% between 2002 and 2003.

Days when pollution is moderate or higher (UK)

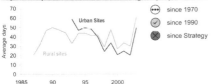

- Owing to an unusually hot summer 50 days in 2003 had moderate or higher air pollution on average at urban sites – down from 59 days in 1993 but up from 20 days in 2002.
- Rural air quality was relatively poor for 61 days in 2003 compared with 50 in 1990, but is highly dependent on the weather and there is no clear overall trend.

H11 ROAD TRAFFIC
Road traffic (GB)

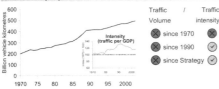

- Between 1990 and 2003, road traffic volume increased by 19% from 411 to 490 billion vehicle kilometres.
- Road traffic intensity (vehicle kilometres per GDP) fell by 12% between 1990 and 2003. This shows that, whilst traffic volumes have continued to rise, the historical link between road traffic and economic growth is weakening.

H12 RIVER WATER QUALITY
Rivers of good or fair chemical quality (UK)

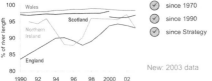

New: 2003 data

- Since 1990, English rivers have seen the biggest improvements in chemical and biological (not shown) river quality bringing more of them up to the quality seen in the rest of the UK.
- In 2003, 93% of English rivers were of good or fair chemical quality (89% in 1998) the same proportion as in Northern Ireland (96% in 1998). Over this period the chemical quality of Welsh and Scottish rivers has remained at a high level.

H13 WILDLIFE
Populations of wild birds (UK)

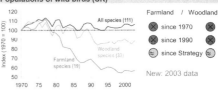

- The index of farmland bird populations has nearly halved since its 1977 peak and has fallen by 22% since 1990, but has remained at about the same level over the last five years.
- The woodland bird index fell by 11% between 1970 and 1998 since when it has remained roughly constant.

H14 LAND USE
Homes built on previously developed land (England)

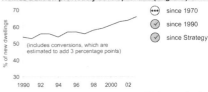

- In 2003, 66% of new dwellings were provided on previously developed land and through conversions, up from around 54% in the early 1990s.

H15 WASTE
Household waste (England & Wales)

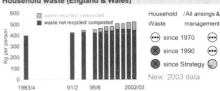

New: 2003 data

- Household waste is about a sixth of all controlled waste. Between 1991-2 and 2002-3, the amount not recycled or composted increased by 7% from 417 to 446 kg per person. However, the percentage recycled or composted increased from 3% to 14% in the same period and in 2002-3 the amount not recycled fell for the first time in recent years.
- In 1998-9 UK households, commerce and industry produced about 195 million tonnes of waste (not shown). About 50% of this went to landfill. Estimated figures for 2000-1 suggest the total amount of waste was 220 million tonnes, with 45% going to landfill. (These changes are not statistically significant.)

Literature Cited

DEFRA. 2002. *Survey of public attitudes to quality of life and to the environment: 2001.* London: DEFRA. Available at www.defra.gov.uk/environment/statistics/index.htm.

DEFRA. 2004a. *Achieving a better quality of life: Review of progress towards sustainable development. Government annual report 2003.* London: DEFRA.

DEFRA. 2004b. *Sustainable development indicators in your pocket 2004: A selection of the UK government's indicators of sustainable development.* London: DEFRA.

DEFRA. 2004c. *Taking it on: The consultation for developing new UK sustainable development strategy.* London: DEFRA.

DEFRA. 2005a. *Securing the future.* London: DEFRA.

DEFRA. 2005b. *Sustainable development indicators in your pocket 2005.* London: DEFRA.

Department of the Environment. 1996. *Indicators of sustainable development for the United Kingdom.* London: HMSO.

DETR (Department of Environment, Transport and the Regions). 1998. *Sustainability counts: Consultation paper on a set of "headline" indicators of sustainable development.* London: DETR.

DETR (Department of Environment, Transport and the Regions). 1999a. *A better quality of life: A strategy for sustainable development for the UK.* London: DETR.

DETR (Department of Environment, Transport and the Regions). 1999b. *Quality of life counts: Indicators for a strategy for sustainable development for the United Kingdom— A baseline assessment.* London: DETR.

DETR (Department of Environment, Transport and the Regions). 2000. *Local quality of life counts: A handbook for a menu of local indicators of sustainable development.* London: DETR.

HM Government. 1994. *Sustainable development: The UK strategy.* London: HMSO.

19

Developing Tools for the Assessment of Sustainable Development in the Province of Brabant, the Netherlands

Luuk Knippenberg, Theo Beckers, Wim Haarmann, Frans Hermans, John Dagevos, and Imre Overeem

Telos, the Brabant Centre for Sustainability Issues, was established in 1999. Its task is to develop and spread knowledge about sustainable development in the province of Brabant in the Netherlands. During the first years of its existence Telos devoted most of its time to developing a method to assess the degree of sustainable development in Brabant. For this, Telos adopted and adapted the three-capital model. Already in 1999, but certainly after the 2002 Johannesburg Declaration on Sustainable Development, modeling sustainable development in this way, by splitting it into three separate domains, was accepted, although not undisputed (see Prescott-Allen 2001). This will not be discussed here, although it may be useful to clarify one point. Telos uses the term *capital* instead of domain, pillar, or dimension. When we look more closely at the precise meaning of the term in the Telos approach, it becomes clear that the word *pillar* could also be used. Doing so would be consistent with the international accepted terminology; see the Plan of Implementation accepted at the World Summit on Sustainable Development at Johannesburg in 2002 (UN 2002), for example. However, the term *capital* is used in this chapter, if only because it matches the underlying idea of our model (i.e., that the capital is the score for the aggregate of stocks in each capital or pillar, as indicated in our sustainability triangle).

In this chapter we will first explain the method developed by Telos to assess sustainable development at a subnational and regional level in the Netherlands. We will then discuss some of the problems we encountered when applying the model. Since 2000, the method has been applied twice in the province of Brabant and also tested in three other provinces: Zeeland, Limburg, and Flevoland. The results of the application of the method are not presented here (see www.telos.nl).

The Telos Method

According to Telos, sustainable development can be defined as a balanced increase in quantity and quality of three forms of capital:

- Ecological capital: nature
- Sociocultural capital: the physical and mental well-being of people
- Economic capital: healthy economic improvement

According to Telos, three criteria must be met before one can speak of sustainable development:

- The approach should be *integral.* Improvement of one capital cannot take place at the expense of one or both of the other two.[1]
- The development should be *sustainable over time and throughout generations.* Our children's and grandchildren's possibilities for development should not be eroded as a consequence of our own development.
- The development must be *sustainable at the global level.* Development here cannot take place at the expense of development elsewhere.

Sustainable development has a strategic (the long-term) and a normative (responsibility for other scale levels and future generations) dimension.

Guiding Principles

It would be useful to have overall guiding principles for each type of capital in order to decide what the components should be and how they should guarantee balanced development. Telos has adopted the three principles defined by the Swiss Federal Statistical Office (SFSO 2001): social solidarity, economic efficiency, and ecological responsibility. However, there is one difference: Telos prefers the term *ecological resilience* rather than *ecological responsibility.* This idea is taken from the International Center for Integrated Assessment and Sustainable Development (ICIS) (Rotmans et al. 2001).

SOCIAL SOLIDARITY

Social solidarity consists of social equity and the quality of life. According to the SFSO (2001), each member of society is entitled to a dignified life and the free development of his or her personality, on the condition that the human dignity of other individuals and the living conditions of future generations are not compromised. Democracy, legal stability, and cultural diversity must be guaranteed. This implies a fair division of costs and benefits. Fairness is a difficult notion. The criteria to define it should be as neutral and consensual as possible. Telos has identified the following prerequisites: the existence of equal opportunities, the fulfillment of basic needs, and legal equality (Telos 2003).

Quality of life consists of the personal, physical, and social conditions that determine mental and physical well-being, including conditions with respect to housing, education, and health care.

ECONOMIC EFFICIENCY

The second principle, economic efficiency, is about the ways individual or social needs are met. According to Telos (2002b), "the level of income should be sufficient enough to provide necessary needs. The available production means should be used as efficiently as possible, and without compromising future use." This definition fits with SFSO's (2001) description: "Economic activity must effectively and efficiently meet the needs of the individual and society. The economic framework must enable and stimulate personal initiatives, self interest must be put to service of the common interest and the welfare of the present and future generations must be ensured."

ECOLOGICAL RESILIENCE

According to the SFSO, ecological responsibility implies "preservation of the natural base of human life, repair of existing damage, and protection of the dynamic diversity of nature" (SFSO 2001). In other words, it is about reaching a balance between human use and ecological regenerative capacity. ICIS formulates it as follows: "In sustainable ecological development, the development of the natural ecosystem comes first and preserving our natural resources plays a prominent part." Telos assumes a similar approach. The main concern is the preservation of the regulatory and habitat functions of natural ecosystems (Rotmans et al. 2001).

Monitoring Sustainable Development

To determine whether society is developing in a sustainable way, monitoring is needed. However, sustainable development is a multifaceted and multilevel notion. It is not only about the development of the three kinds of capital (ecological, sociocultural, and economic) but also refers to dimensions of time (now and later) and space (here and there). It is no coincidence that the first rule of the Bellagio guidelines for the assessment of sustainable development states, "Assessment of progress towards sustainable development should be guided by a clear vision of sustainable development and goals that define that vision" (Hardi and Zdan 1997).

In the case of the environment, the objectives for Brabant are clear and hardly controversial.[2] The same is true for economic issues. Here the main problems are the assessment of future needs and wants and the questions whether and how nonhuman needs can be priced. The main obstacles in Brabant are in social capital (e.g., the problem of cultural diversity versus cultural homogeneity).

Telos has chosen a multilayered method to monitor sustainable development in the Dutch province of Brabant. This approach is related to the Bossel (1999) and ICIS methods:

- *Integrated:* Economic, social, and ecological interests and considerations are all taken into account at the same time.
- *Interdisciplinary:* Expertise from different scientific disciplines is integrated into one approach.

- *Interactive:* All stakeholders are involved, not just policymakers and scientific experts but social actors and professionals as well.
- *Strategic:* The focus is not on short-term problems but on long-term sustainability requirements (i.e., goals for sustainable development).
- *Normative:* Unwanted long-term effects are made explicit, with the help of experts and stakeholders.
- *Indicative:* Undesirable consequences or outcomes are made explicit.

Relevant Terms

In the Telos method, a distinction is made between capitals, stocks, requirements, indicators, standards, and context variables. Box 19.1 gives brief definitions of these terms.

Capitals. Telos distinguishes three forms of capital: ecological capital, sociocultural capital, and economic capital. The range and development of the capitals are visualized by means of a sustainability triangle (Figure 19.1). We can discern two triangles: the main triangle and a smaller triangle.

The inside triangles indicate the actual condition of each capital. The larger ones represent the ideal situation for each capital. The ideal situation is based on science and the judgments of experts and stakeholders.

Stocks. In order to assess the condition of each capital, we have to find the components or essential elements that determine each capital. To develop an integrated perspective

---

Box 19.1. Definitions of relevant terms.

Term	Definition
Capital	The three essential aggregated subsystems of the total societal system: ecological, sociocultural, and economic.
Stocks	The essential elements that together determine the quality and quantity of the capital.
Requirements	The requirements formulated with respect to the development of a stock.
Indicators	The degree to which the requirements are met is measured by means of indicators.
Standards	Normative criteria, developed by the stakeholders, to assess the score of indicators.
Context variables	Variables that influence sustainability but cannot or will not be influenced.

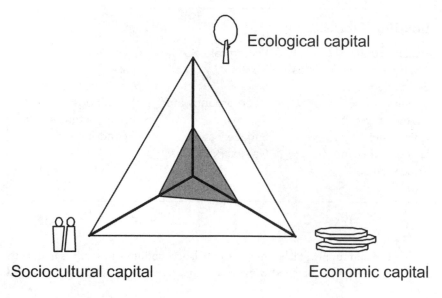

Figure 19.1. Telos sustainability triangle.

on sustainability, it is necessary to use identical notions and have a uniform analytical framework. This was achieved by introducing the concept of stocks, which is used in system dynamics. Each capital consists of a number of stocks. Together they determine the quality and quantity of the capital as a whole.

The stocks were determined with the help of desk research and in workshops with experts and stakeholders (Box 19.2).

Requirements. The next step was to make the stocks operational by introducing the notion of requirements, which are the main long-term goals. For each stock a small set of requirements was formulated. The content of a requirement was determined by asking the following question: "What should the contribution of each stock be to advance sustainable development (social solidarity, economic efficiency, and ecological resilience, here and elsewhere, now and in the future)?" This was done through expert judgment and stakeholder preferences. The first step was to consult experts; the next step was to reconcile their opinions with those of stakeholders. That was done in the form of workshops. This approach gives the method a strong normative undertone. According to Telos, this is consistent with the basic idea of sustainable development (WCED 1987; Dobson 1996, 1999; Holland 1999, 2000; Benton 1999).

For example, here are the requirements used for the labor stock in the province of Brabant:

Box 19.2. Stocks.

Sociocultural	Ecological	Economic
Health and care facilities	Nature	Labor
Solidarity	Soil	Capital goods
Safety	Deep groundwater	Knowledge
Cultural diversity	Air	Infrastructure
Citizenship	Surface water	Economic structure
Living environment	Natural resources	
Education and		
training		

- The demand and supply of labor must be in balance, in number as well as in quality.
- Labor conditions must be good with respect to safety, health, working hours, and training facilities.

Requirements (but also stocks), by definition, can and will change. The nature of the requirements depends on the place, context, stakeholders, and experts involved and the time and scale.

Indicators. The degree to which the requirements or long-term goals are met is measured by means of indicators. For each requirement, one or more indicators must be found, such as the indicators used for the requirements with respect to the labor stock in the province of Brabant:

- Ratio between employment and resident labor force
- Use of the potential labor stock
- Labor market tension
- Educational match
- Initiation and aging
- Number of jobs in information and communication technology (ICT) compared with the total number of jobs
- Number of highly educated people

To prevent an excess of indicators, a maximum of eight are established for each stock.

Figure 19.2 represents the relationship between the different elements of the method. To sum up:

- Each capital consists of stocks.
- For each stock requirements are defined.
- Indicators show the degree to which these requirements are met.

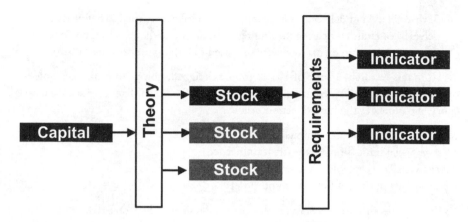

Figure 19.2. Relationship among capital, stocks, requirements, and indicators.

Determining the Value of Stocks and Capital

To ensure that forms of capital and stocks can be compared, they should be indexed, weighted, and standardized.

Indexing. Indexing involves more than calculating and trying to determine the precise value of a capital or stock by means of counting and discounting. The value of all types of capital is equally important, and their sum is the sum of everything we value or should value. Furthermore, not all stocks contribute equally in all circumstances (qualitatively and quantitatively) to the value of a capital, and not all indicators are equally important in determining what the condition of a stock is. Therefore, a weighting procedure that pays attention to quantitative and qualitative aspects was built in. The first step in developing this instrument was to set the desired score for each indictor.

The Score of an Indicator. The first step is to specify the nature of an indicator and to decide what type of data is needed. Those data can be quantitative or qualitative. The qualitative indicators are also given a numeric score. The next step is to bring the measurement in line with the agreed-upon targets or end values. This is a specific normative procedure. However, it is important to work with target values and limit values that are as objective as possible. Therefore, we followed this procedure:

1. We looked for relevant scientific documentation.
2. If that was not available or did not provide adequate answers, we looked for standards laid down in government documents or governance-oriented studies.
3. If that did not help either, we looked to see whether there was social consensus among (possible) stakeholders on target values.

4. If that still did not deliver clearly delimited target values, we tried to define them ourselves by carefully comparing the findings with the topics they were supposed to measure and the requirements they were supposed to fulfill (see also Telos 2002a).

From this research, a scale of 1–100 can be designed, going from unacceptable to optimal. This scale is divided in four categories, symbolized by different colors. Each category also indicates the kind of action needed:

Red: Socially unacceptable (immediate action needed)
Orange: Social limit (immediate attention needed)
Green: Socially acceptable (short-term purpose)
Gold: Socially optimum (long-term purpose)

For example, the ratio between employment and resident labor force is one of the indicators for the labor stock. This ratio indicates whether a region imports or exports labor. When stock and demand for labor are equal, the labor market is in balance. This optimum corresponds to a value 100. Some deficiency or surplus is acceptable, but as soon as the deficiency or surplus exceeds a certain degree, the labor market is in disorder.

Social optimum is 100.

Socially acceptable is $100 < x \leq 105$ or $95 \leq x < 100$.

Social limit is $105 < x \leq 110$ and/or $90 \ x < 95$.

Socially unacceptable is $x > 110$ or $x < 90$.

Weighting and Determination of the Extent of the Stock. Each stock receives a score on a scale from 1 to 100 for sustainable development. This is the weighted average of the indicators. The indicators have to be weighted because not every indicator is equally important when it comes to describing the stock.

For the labor stock, the indicators are weighted as shown in Box 19.3, and the outcome is illustrated in Figure 19.3.

The score of an indicator is shown by the degree to which a pie wedge is filled. The shades (or colors) represent the categories of the scores. The angle represents the weight of the indicator. The arrow represents the trend of development, if that can be determined. In this way the actual situation can be assessed immediately and compared with the desirable one for each indicator.

If the pie wedge is completely filled to the outer ring, the optimum situation is reached. This way, areas that need attention can be identified. In other words, Figure 19.3 indicates the current situation of the stock and shows where action is needed. The total score of a stock (the weighted average of the indicator scores) is represented as a bar chart.

Box 19.3. Labor stock.

Indicator	Weight
Ratio of employment to labor force	17%
Use of potential labor stock	20%
Labor market tension	9%
Educational match	17%
Initiation and aging	13%
Number of information and communication technology jobs	8%
Number of highly educated people	16%

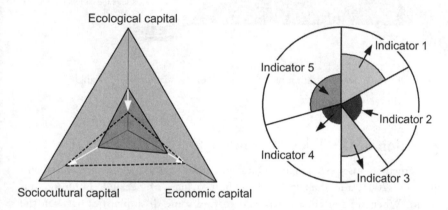

Figure 19.3. Visual representation of the actual situation and the direction of development, by indicator.

Weighting and Determining the Value of the Capitals. The score of one capital is the weighted average of the stock scores. Not all stocks are equally important, so the stocks were weighted, like the indicators. The process of determining the sustainability triangle is summarized in Figure 19.4.

By mapping out the scores per stock within a capital and by comparing them in time, we can ascertain whether there is progress in the desired direction. In each capital, progress is represented by the shaded part of the line expanding outward toward the corner of the triangle. Progress in capital as a whole can coincide with negative changes in one of its stocks if this is compensated by progress in one of the other stocks of the same capital.

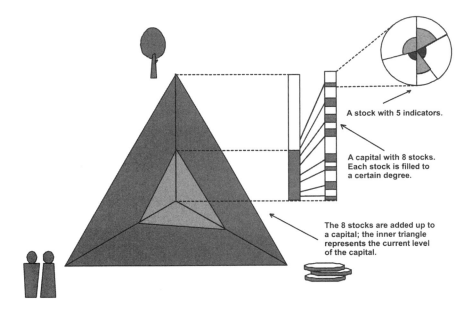

Figure 19.4. Visual representation of the structure of the sustainability triangle.

Application of the Telos Method

The Telos method was applied in four Dutch provinces: first in the province of Brabant, in 2000 and 2001, and then in 2003 in the provinces of Zeeland, Limburg, and Flevoland. When we started, we expected to find some dissimilarities between the provinces but not too many. However, the reality was different, and we found major differences. We discuss and analyze them in this section, first by looking at the process. For additional information, please see our Web site (www.telos.nl).

The Process

The Telos method had already been applied in Brabant before it was used in the other three provinces. The idea was to use the method developed for Brabant as much as possible in the other regions. The stocks, requirements, and indicators defined in Brabant were used as inputs, and the intention was to use the same procedure for the other regions (i.e., to adapt, reject, or adjust them according to expert judgment and stakeholder views).

In practice, the modus operandi in the other provinces was different. It was shaped far more by the provincial government and stakeholders' points of view than in Brabant (Figure 19.5).

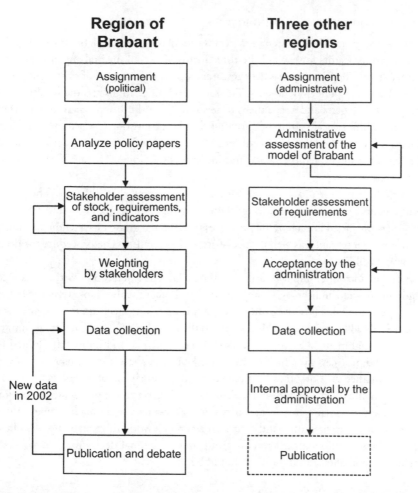

Figure 19.5. Differences and similarities between Brabant and the three other regions.

In Brabant, from the outset there was political support for the idea of assessing regional sustainable development. In the other provinces, the political will was less pronounced. In all three provinces, the main supporters of the process were provincial civil servants. One explanation for this difference was the fact that this project was partly initiated and largely funded by the Dutch Ministry for Housing, Regional Development, and the Environment (VROM). The ministry wanted a general model for assessing sustainable development at the provincial level in the Netherlands (with a 75 percent overlap concerning requirements and indicators).

As explained earlier, the first step in the Telos model is to determine the stocks, requirements, and indicators for each of the three forms of capital. In Brabant this was done in three steps. The first step was desk research, a thorough analysis of existing scientific publications and policy documents. The second step was to discuss the results of this research with experts and adapt them if necessary. The third step was to discuss these outcomes with a selected group of thirty stakeholders, consisting of key informants from all relevant sectors (government, civil society, business) in Brabant. They had to decide what the final stocks, requirements, and indicators should be. The decision was based on consensus.

In the other three provinces, the procedure was somewhat different. The stocks, requirements, and indicators already defined in Brabant served as a starting point. They were presented in each of the provinces to a selected group of stakeholders, mostly provincial civil servants, as well as specialists and generalists. The procedure for doing this was as follows.

After an introduction, where the method was presented along with the stocks, requirements, and indicators already defined in Brabant, subgroups were formed for each type of capital. Those subgroups were made up of officials from the province who were specialized in the policy fields relevant to the capital concerned. Each subgroup was given the assignment to examine the stocks, requirements, and indicators defined for Brabant and, if necessary, adjust them to local circumstances. Once this was done, we verified whether the outcomes formulated by those provincial civil servants were also accepted by the other categories of stakeholders. This was done in interactive sessions with selected representatives. When necessary, adjustments were made. The results of those sessions were presented and discussed again in a workgroup made up of some of the members mentioned earlier. This group made the final decision about what the stocks, requirements, and indicators should be.

OPERATIONALIZATION AND DATA

The next step is the most labor-intensive one. The indicators are put into use and data are collected, processed, and weighted on the basis of these indicators. The first real obstacles appeared at this time. In Brabant and in the other three provinces, little time was available for this part of the process. We soon found out that much of the data needed were available only at the national level. Sometimes it was difficult to break these national data down into subsets usable at the regional level. Sometimes regional data available in theory were in fact not accessible. Even worse, sometimes the available data did not match the indicators, for quantitative or qualitative reasons.

In fact, the process of collecting, interpreting, weighting, and combining regional data proved to be the main obstacle in the whole method and was very time consuming.

EVALUATION OF THE PROCESS

In all four provinces, the learning process was considered as important as, if not more important than, the factual results of the assessments. The main purpose was to develop and test a method for assessing regional sustainable development.

In 1999, when we started developing our method, well-developed models to assess regional sustainable development did not exist, although there already was a growing body of useful literature in principle. For instance, we used the studies of Hardi and Zdan (1997) and Bossel (1999). One of the few existing region-oriented methods was developed in British Columbia, but that approach differed from the method we had in mind. It opted for a strongly structured, statically based model (Pierce and Dale 1999; Robinson 1996).

Our desire to emphasize the learning process was a good reason to call evaluation meetings with experts and stakeholders in all the provinces. A lot of useful comments came out of these sessions. The most relevant are as follow:

Positive.

- The reason why the elements that make up a capital are chosen and brought together is worked out well in the model. It also increases our insight into the ways they are connected.
- The method reinforces and conceptualizes the interaction and communication between scientific disciplines, science and nonscience, and different arenas in society (policy, civil society, and business).
- The method lends itself well to propagating the idea of sustainable development and the assessment of it on a regional scale to a wide audience. This is partly because of the simple and attractive way in which the concept and the results can be presented and visualized.

Negative.

- The connection and interaction between the three forms of capital are not worked out well enough.
- Partly as a result of this, the debate often got bogged down in a discussion about the effectiveness of concrete measures, instruments, or funds. (However, this was also caused by giving stakeholders a strong voice.)
- The method of choosing requirements, indicators, and sometimes also stocks was too subjective and therefore was too sensitive to change. Moreover, the way the norms were selected was not always transparent.
- Although participants in the end tended to appreciate the method, they had some problems getting acquainted with it at first. The degree of abstraction is considered high. Disciplines and topics are linked in new ways.

Differences in Place and Time

As explained earlier, the idea of developing a model to assess regional sustainable development was broadly supported in the province of Brabant, more widely so than in the other three provinces. But there are more differences. In 1999, when the project started in Brabant, the economy in the Netherlands—and even more in Brabant—was booming, and it was a time of political optimism. The assessment in the other provinces was done after the ICT bubble burst and the 9/11 attacks took place. The promotion of sustainable development became a lower priority in the Netherlands, and the perception of it changed. Economic growth became a top priority again, along with another and rather new concern: security. These concerns clearly affected the perceptions and wishes of the stakeholders and influenced the selection in all three provinces. This is one of the possible drawbacks of relying heavily on stakeholders.

However, physical, economic, and sociocultural disparities were more important in explaining the differences between the provinces. Some of these differences are as follow:

• Brabant and Limburg have sandy soils; Zeeland and Flevoland have clay soils.
• In Limburg minerals are extracted; this is not the case in Zeeland and Flevoland.
• Brabant has a lot of manufacturing industries. These are nearly absent from Flevoland, where agriculture and the service industries are important.
• Aging of the population is not an issue in Flevoland but is an important one in the other provinces (e.g., in the countryside in Brabant).

The Telos Method Itself as an Indicator

The Telos method is, first of all, itself an indicator, intended to show what the situation is at a certain moment in a certain political and geographic area with regard to sustainable development. It is a kind of alarm system designed to show what is going well and what needs attention. Indicating trends is considered more important than providing static measurements.

However, the better one can measure, the better this purpose will be served. Measuring presupposes the availability of sufficient, reliable, comparable data. Often these requirements have not been fulfilled, as explained earlier. This does not mean that the method cannot be used. Certainly in the initial phase the process is already on track, and thus is a success, if it frames the opinions of people who make decisions, helps to develop a common language, stimulates discussions, and draws attention to the need to develop adequate norms and data and if it provides some concrete indications on the basis of already existing data.

The Telos method is still in its infancy. Its main goal is to offer politicians, civil servants, nongovernment organizations, firms, and citizens a tool to discuss sustainable development in general and its state in their own region.

The Results Compared

It is worthwhile to discuss some of the concrete outcomes of the assessments in the different provinces. However, we will limit ourselves to comparing the selection procedures of the stocks, requirements, and indicators in the different provinces.

STOCKS

At first, the idea was to keep very strictly to the set of stocks already formulated by Telos on the basis of desk research. This proved not to be completely possible. Debates with local experts and stakeholders about the structure of capital led to adaptations, especially in the sociocultural capital (Telos 2003). Second, stocks already proved to be defined to some extent by temporal and spatial conditions. This became very clear when we applied the method in the different provinces:

• In Limburg, the stocks coincided to a large degree with those formulated for Brabant, although their interpretation differed greatly. The mineral resources stock was added to economic capital. It is very likely that this stock will henceforth be included in every new regional account. But before we can do this, we have to redefine the natural resource stock in ecological capital.
• The discussion in Zeeland made it clear that the consumption stock, before inclusion in sociocultural capital, should be deleted. The fact that the experts and stakeholders in Zeeland saw no use for the natural resource stock or for a new stock dealing with mineral resources was interesting.

REQUIREMENTS

When the Telos method was applied in other provinces, the hope was to identify a large correspondence in the requirements defined and an overlap of at least 75 percent. The underlying reasoning was that this would allow the development of a generic, usable regional model.

The Venn diagram in Figure 19.6 shows the overlap for all provinces. The diagram consists of four rectangles, each representing a province. The area where all provinces overlap is black. Overlap between three (dark gray) or two (light gray) provinces is also represented. The white fields represent indicators without any overlap.

The results are remarkable. Only twenty-three requirements overlap everywhere; fourteen are shared by three provinces, fourteen by two provinces, and twenty-eight requirements proved to be completely province-specific.

Limburg has defined the most requirements (fifty-two) and also has the most province-specific ones (fourteen). This is partially because of the extra stock added in Limburg. This resulted in four extra requirements. Zeeland used the smallest number of requirements (thirty-nine). But in Zeeland two stocks were left out. If we look at the overlap with Brabant, we see that this overlap is the largest for Flevoland and the smallest for Limburg. As mentioned earlier, Limburg has the largest number of province-specific requirements.

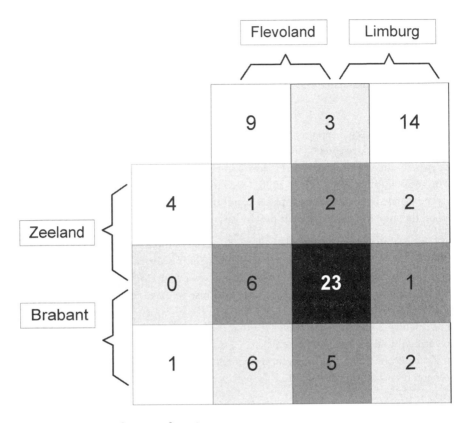

Figure 19.6. Venn diagram of requirements.

INDICATORS

We also assessed the overlap of indicators, as shown in Figure 19.7.

Limburg has the largest number of indicators (123), 24 of which are province-specific. Sixty-three indicators are shared by all four provinces, represented in the area where all rectangles overlap. If we look at the overlap of Brabant with the other provinces, we see that Brabant and Limburg share the greatest number of overlaps (83) but also the greatest number of differences (69). But Flevoland and Zeeland have approximately as many overlaps with Brabant (80 and 81 indicators, respectively).

OVERLAP

To calculate the total overlap between the provinces, we formulated the following model. If a requirement or indicator occurs in all four provinces (f4, or a frequency of 4), there is 100 percent overlap; if a requirement occurs in three provinces, the overlap is 75 percent, and so on.

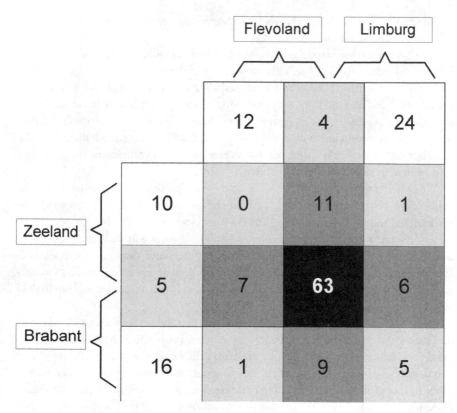

Figure 19.7. Venn diagram of indicators.

$$\text{Overlap} = f_4 \times 100\% + f_3 \times 75\% + f_2 \times 50\% + f_1 \times 25\%$$

Application of this formula led to the following results[3]:

Requirements: 60% overlap
Indicators: 64% overlap

Conclusions

In 1999 Telos started to develop a method for assessing regional sustainable development in the province of Brabant. Since then, a model has gradually been developed and applied five times, twice in Brabant and once in three other provinces. At the moment we are engaged in evaluating the outcomes of these assessments. Out of this evaluation, some preliminary conclusions have emerged.

The first conclusion is that the model withstood our testing, but it has become clear that some major adaptations are needed.

One weakness of our model is in the way in which the interactions between the three forms of capital are worked out. This was also one of the main comments that emerged from the stakeholders' evaluations. One could argue that this criticism is beside the point. The Telos method does not try to model processes such as interactions but to describe a state of affairs. It is enough to specify that the development of one capital may not prevent the development of another, as Telos did, and investigate whether this is the case. But this argument is a sophism. Even these kinds of investigations require a clear view of how the kinds of capital are interconnected.

We are well aware of this. At the same time, we realize that the task of describing the connections between the capitals is perhaps the most difficult of all if you adopt the three-capital model. Resolving this issue is a high priority for Telos but not something we can do and want to do on our own. The more closely you look at the problems involved, the deeper you have to dig. It is no coincidence that this question arises everywhere and consumes much of the energy and time of researchers involved in conceptualizing and modeling sustainable development. It is a matter we want to look at in close cooperation with others.

Another weakness of the Telos model is the fact that the requirements and indicators developed, and to a lesser extent the stocks, are too sensitive to change. Local and temporal considerations have too great an impact on the manner in which they are chosen and defined. This compromises our ability to repeat and compare assessments done at different moments in time or in different places. This was an unforeseen and unwanted consequence of our decision to involve local stakeholders in all the phases of the model's construction. It was good for commitment but not for the robustness of the model.

We want to define more sharply where and how stakeholders should play a role. We think that the assessments done so far provide us with clues on how to proceed. For instance, we think that the stocks and some of the requirements can be made far less susceptible to changes in the perceptions of stakeholders or to the impact of time. At the same time, stakeholders could be more involved in the weighting process.

We are still at the beginning of this process. We want to proceed carefully and take the time to analyze related models. The situation is different from that in 1999. We do not have to start from scratch, and now there are far more methods to assess sustainable development at the local level.

We are especially interested in models developed for totally different regional settings than ours and in models that are built with a greater stakeholder involvement.

Notes

1. At the level of capital, Telos opts for the idea of strong sustainability instead of weak sustainability. For an extended and interesting discussion about this issue, see the CRITINC working papers, available at www.kcele.ac.uk/depts/spire/working_Papers/

CRITINC/CRITINC_Working_Papers.htm.

2. We realize that this statement is far too crude if we look at it from a broader (e.g., global) perspective. However, the first objective of Telos was to develop a method to assess the state of affairs with regard to regional sustainable development in a province of the Netherlands. Therefore, we started with a local focus in defining the issues to address. For the sake of acceptance of the model we decided to take up the problem of flows and far-reaching supraregional interactions and transfers later.

3. The calculations were not corrected for the new stock that was included only in Limburg. If two provinces decided to accept a new indicator, this counted twice. The same indicators measure the same issue.

Literature Cited

Benton, T. 1999. Sustainable development and accumulation of capital: Reconciling the irreconcilable? Pp. 199–229 in *Fairness and futurity: Essays on environmental sustainability and social justice*, edited by A. Dobson. Oxford: Oxford University Press.

Bossel, H. 1999. *Indicators for sustainable development: Theory, method, applications.* Winnipeg: IISD.

Dobson, A. 1996. Environmental sustainabilities: An analysis and a typology. *Environmental Politics* 5:401–428.

Dobson, A. 1999. *Fairness and futurity: Essays on environmental sustainability and social justice.* Oxford: Oxford University Press.

Hardi, P., and T. Zdan. 1997. *Assessing sustainable development: Principles and practise.* Winnipeg: IISD.

Holland, A. 1999. Sustainability: Should we start from here? Pp. 46–68 in *Fairness and futurity: Essays on environmental sustainability and social justice*, edited by A. Dobson. Oxford: Oxford University Press.

Holland, A. 2000. Sustainable development: The contested vision. Pp. 1–10 in *Global sustainable development in the 21st century*, edited by K. Lee, A. Holland, and D. McNeill. Edinburgh: Edinburgh University Press.

Pierce, J. T., and A. Dale. 1999. *Communities, development, and sustainability across Canada.* Vancouver: UBC Press.

Prescott-Allen, R. 2001. *The Wellbeing of Nations: A country-by-country index of quality of life and the environment.* Washington, DC: Island Press.

Robinson, J. B. 1996. *Life in 2030: Exploring a sustainable future for Canada.* Vancouver: UBC Press.

Rotmans, J., J. Grosskurth, M. Van Asselt, and D. Loorbach. 2001. *Sustainable development: From draft to implementation* [in Dutch: *Duurzame ontwikkeling: Van concept naar uitvoering*]. Maastricht: ICIS.

SFSO. 2001. *Swiss Federal Statistical Office MONET project: From the definition to the postulates of sustainable development.* Neufchâtel: SFSO.

Telos. 2002a. *De duurzaamheidsbalans van Noord-Brabant 2001: De verantwoording.* Tilburg: Telos.

Telos. 2002b. *The sustainability balance 2001: Method.* Tilburg: Telos.

Telos. 2003. *The road towards a socially sustainable society* [in Dutch: *Op weg naar een sociaal duurzame samenleving*]. Tilburg: Telos.

UN. 2002. *Report of the World Summit on Sustainable Development.* United Nations A/CONF. 199/20. New York: UN.

WCED. 1987. *Our common future.* Oxford: Oxford University Press.

20

Sustainability Assessment Indicators: Development and Practice in China

Rusong Wang and Juergen Paulussen

China is experiencing rapid urbanization and industrial transition. In China one can find nearly all levels of development, from highly developed cities to poor, underdeveloped rural communities. Ecosustainability can be ensured only with an understanding of the complex interactions between environmental, economic, political, and sociocultural factors and with careful planning and management grounded in ecological principles.

Sustainability assessment indicators (SAIs) in China were initiated in the 1980s and came to the forefront in the 1990s, when China was developing its Agenda 21 (People's Republic of China 1997). China's Agenda 21 is management oriented. It is based on four pillars: a comprehensive strategy and policy of sustainable development, sustainable social development, sustainable economic development, and rational use of resources and environmental protection. A set of indicators has been developed by research institutes and central, regional, and local government organizations. Five kinds of coordination are emphasized by the central government: between regions of different sizes, between urban and rural areas, between social and economic development, between humans and nature, and between self-sufficiency and external symbiosis. Different ministries have initiated various projects and campaigns to promote sustainability at different scales since 1992, such as the Comprehensive Experimental Community for Sustainable Development, initiated by the Ministry of Science and Technology; the National Eco-Agricultural Counties, initiated by the Ministry of Agriculture; and Ecological Demonstration Districts (ecoprovinces, ecocities, ecovillages), initiated by the State Environmental Protection Agency (Figure 20.1).

Based on the experiences of these pilot studies, different ministries have created certain indicators to measure and assess these test units. The State Environmental Protection Agency (SEPA) has promulgated a set of ecopolis assessment indicators to measure ecosustainability by cultivating an ecologically integrative and biologically vivid

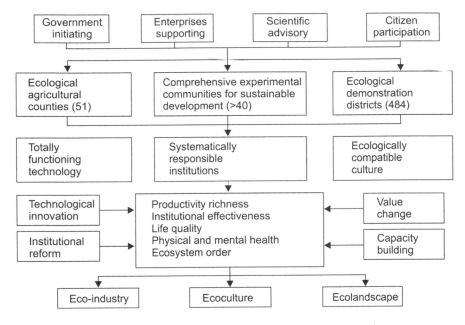

Figure 20.1. Sustainability demonstration of ecopolis development in China.

landscape (ecoscape), economically productive and ecologically efficient production (eco-industry), and large-scale and long-term responsible culture (ecoculture). There are twenty-two indices for ecoprovinces, twenty-eight indices for prefecture-level ecocities, and thirty-six indices for ecocounties (SEPA 2003). The indicators include economic productivity, scientific and technological creativity, ecological integrity, governance coordinating ability, social integrity, and external openness.

Academic Assessment of Sustainable Development Using Sustainability Indicators Based on the SENCE Approach

Grounded in ancient Chinese human ecological philosophy, the Social–Economic–Natural Complex Ecosystem (SENCE) concept was developed by Ma and Wang (1984) to assess the sustainability of human-dominated ecosystems (Figure 20.2). SENCE's natural subsystem consists of the traditional Chinese Five Elements: metal (minerals), wood (living organisms), water, fire (energy), and soil (nutrients and land). Its economic subsystem includes the functional components of production, transportation, consumption, regeneration, and regulation. Its social subsystem includes technology, institutions, and behavior. Whereas the natural subsystem is the basis and framework for all activities (outer pentagon), the social and economic sub-

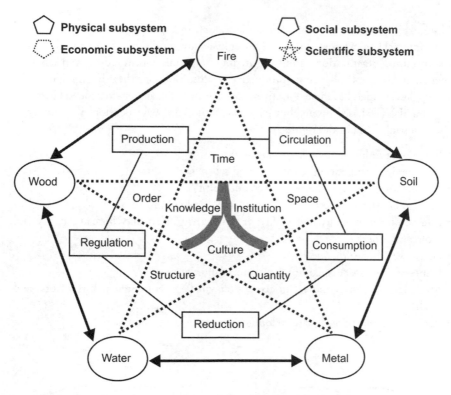

Figure 20.2. Social–Economic–Natural Complex Ecosystem (SENCE) and its sustainability dimension.

systems are the components of the system, where decisions are made, directing the whole system to or from sustainability. Science provides people with an understanding of the system's contexts in time, space, quantity, structure, and order, and it provides decision makers with images, strategies, tools (e.g., software) for systematic planning and management (Wang et al. 1996). The critical issues in assessing sustainability are to frame the complicated interactions, simplify and integrate the diverse relationships, and develop a practical instrument for promoting sustainable development.

A series of SENCE-based combined models for assessing sustainability was developed and implemented in the 1980s. This group model consists of mechanism explanation models (internal mechanisms of competition, symbiosis, circulation, and self-sufficiency; temporal and spatial processes, patterns, and order; the balance between the four driving forces of energy, money, power, and spirit; and the human interference of technology, institutions, and culture), planning models (panobjective ecological programming and conjugate ecological planning), and management models (ecoservice, ecometabolism, and eco-institution monitoring and supervision and capacity building).

Sustainability assessment involves multiple scales (time, space, and administrative ones, both internal and external, upper and lower scale, long and short term, centralized and decentralized organization), multiple attributes (population, resource, and environment), and multiple objectives (productive wealth, functional health, and faith). The use of these models has three objectives: social, economic, and environmental benefits.

The SENCE-based sustainability indicators in China were developed according to the following framework (Figure 20.3):

• Function: producing, living, and sustaining
• Cybernetics: competition, symbiosis, and self-reliance
• Driving forces: energy, money, power, and spirit
• Human interference: technology, institutions, and culture
• Effects: efficiency, equity, and vitality; or productivity, life quality, human capacity, and ecological order
• Goals: wealth, health, and faith
• Context: time, space, quantity, structure, and order
• Awareness: scientists and technicians, policymakers, entrepreneurs, residents, and media
• Evolution level: survival, predevelopment, and postdevelopment

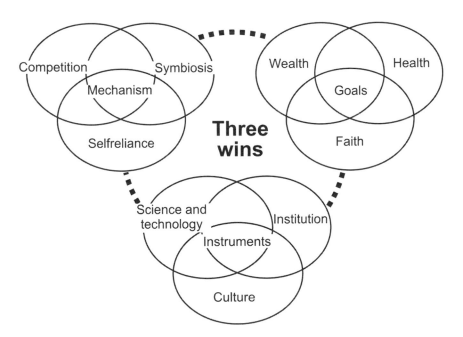

Figure 20.3. Three dimensions of sustainable development.

• Stability: structural stability (diversity and dominance), process stability (growth rate and fluctuation range), and functional stability (self-reliance and openness) (Wang et al. 2004b).

Based on this framework, SAIs in China give primary attention to the harmony between mechanisms, concepts, methods, analysis, aggregation, and implementation and the coupling between natural services and human well-being (Wang et al. 2004a).

Methodological Considerations for the Assessment of Sustainability Indicators in China

There are ten kinds of problems for SAIs in China.

Definition standards. Some indicator sets are targeted to identify ecological units, but they consist of a large number of socioeconomic indicators. The understanding of *ecological* in indicator sets is different from the scientific definition of *ecological.* It merges two aspects without resolving contradictions.

Imbalance in evaluation. Environmental protection projects often consider economic and social effects. On the other hand, the ecological effects of economic and social programs often are evaluated only by environmental agencies and are seldom integrated into an overall decision-making process at the city and province levels, although the central government pays more attention to sustainability. There is a risk of overvaluing economic aspects and favoring prosperous areas with high economic power but excessive consumption levels.

Local or regional and short-term orientation. In many Chinese indicator systems, some key issues of global concern, such as emissions of CO_2 and other greenhouse gases, are not sufficiently expressed by the indicators.

Problematic link of ecological relevant key issues to gross domestic product (GDP). In many indicator systems, the consumption of energy and water, two critical key resources in China, is linked to the economic development level, expressed by GDP. Sometimes there is a threshold (e.g., "energy consumption should not be more than 1.4 tons of coal equivalent per 10,000 yuan GDP," "water consumption should not be more than 150 m^3 per 10,000 yuan GDP"). This definition is not useful because it does not encourage rich cities and counties to reduce their water and energy consumption. In fact, linking key issues to GDP may encourage cities and counties to increase their GDP through higher per capita consumption of critical resources. Furthermore, the link may contribute to disharmony between rich and poor areas, particularly when water and energy consumed by rich cities (e.g., in the three main coastal agglomerations around Beijing, Shanghai, and Guangzhou) are generated in and withdrawn from poorer regions.

Quantitative and qualitative indicators. Quantitative criteria, such as energy consumption and water consumption, should be supplemented by the assessment of how energy is generated and how water is supplied in sustainable ways.

Aggregation of results. Research results, such as the ecological performance of a city, are often highly aggregated when published. Proving results and comparing scores for different cities are often difficult for non-insiders (e.g., the percentage of green areas can be measured easily but is not very useful because of the extreme geographic differences between cities).

Data generation and data availability. The availability of data is a key issue for SAIs in China. Data in China come from public statistics; specific statistics and institutional databases of administrative bodies; internal statistics and databases of enterprises, research institutes, and nongovernment organizations; and geographic information systems and survey-based and remote sensing data, generated from aerial photography and satellite images. But the difficulties are numerous:

- Databases often are not sufficient.
- Databases often are not reliable.
- Data needed to assess sustainability and ecology either do not exist or are not made available to researchers, planners, and the public.
- References of data sources often are not complete, and transparency is low, often because of high competition in research.
- Exchange of key data between scientific institutions and planning institutions could be improved in order to safeguard the quality of research and enable researchers and planners to generate results from a common base.

For these reasons, economic criteria are often given priority over ecological phenomena, and important but difficult-to-measure ecological indicators are neglected in favor of economic ones.

Weighting of indicators. Results gained through indicator systems are often purported to be scientifically based, objective, and free of individual values. However, values and weighting are unavoidable and necessary in the development and use of indicator systems. Even if no explicit weighting has been applied (and variables have been set equal), the indicators have been valued. Evaluating indicators is an important step in the assessment process. Furthermore, the selection of indicators for the assessment itself is an act of evaluating. Therefore, values hidden behind the indicators and the weighting procedure should be made transparent. Levin (1997) states that "all the important decisions are made with tremendous dependence on embedded values" and that "sustainability is about values—valuing other species, and valuing other humans, living now and in the future." In light of the increasing public discussion about ecological and socioeconomic issues, strategies, and values in China, this aspect of evaluation is of great importance.

Substitution of indicators. Some indicators can be substituted for others. A good economy can "buy" some achievements in other fields (e.g., end-of-pipe treatment, precautions, social benefits). An intact environment is a precondition for a high-level, high-tech, high-wellness economy and society. In the externalization of costs and impacts, institutions, enterprises, and individuals tend to relay costs and negative impacts to others (e.g., other communities, third parties).

Preference dilemma. On the basis of scientific results, which problems should be solved first by policy? In China, the budget for environmental measures often is very limited. Government institutions usually give preference to economic goals and indicators ("pollution first, environment later").

Case Study for SAIs: Yangzhou Ecocity Development

The city of Yangzhou is located at the juncture of the Grand Canal and the Yangtze River, with an area of 6,638 km² and a population of 4.47 million people. Yangzhou has more than 2,500 years of history and is surrounded by beautiful landscapes. The pace of urbanization and industrialization in the Yangtze delta region is dramatic, especially on the south bank of the Yangtze River, along the Suzhou–Wuxi–Changzhou corridor. To catch up with its neighbors' development while avoiding the heavy environmental pollution and ecological deterioration, the city of Yangzhou has been engaged in an ecocity development project since 1999. An ecocity plan was made by the Chinese Academy of Sciences and adopted by the city's People's Congress (Wang and Xu 2004).

The Yangzhou ecocity development project has three main goals:

To cultivate an ecological industry through economic transformation from a product- and consumption-oriented economy to a social and ecological service function–oriented economy

To cultivate the ecological landscape through a transition from fossil fuel–driven agriculture and pollution-intensive industry to a clean, green, self-sustaining ecoscape through comprehensive ecosystem management and ecological engineering

To cultivate ecological culture through a transition from traditional living styles and values (production modes) to a culture of harmony between people and nature (via value change and capacity building)

Two systems of indicators are being used simultaneously to assess the city's overall potential for sustainable development. The first system consists of sector-oriented indicators for sustainability assessment, which follow the current statistical system in China, classified into economy, environment, and society, where the historical data are available. There are three, seventeen, and seventy-nine basic indicators in class I, II, and III, respectively. The class I indicator, "Economy," includes five aggregated indicators (economic level, resource efficiency, development potential, ecological industrial, and enterprise behaviors), based on nineteen basic indicators; "Environment" includes four aggregated indicators (environmental quality, pollutant emission, pollution treatment, and ecological conservation and design), based on thirty-two basic indicators; and "Society" includes eight aggregated indicators (social equity, education and medical treatment, living quality, population dynamics, consumption behaviors, cultural landscape, social ethics, and government behaviors), based on thirty-eight basic indicators.

The second system is based on the potentials of sustainability. The indicators measure sustainability development status, dynamics, and strength, based on nine aggregated and twenty-five basic indicators selected from seventy-nine sectoral indicators (Table 20.1).

Development status represents the quality of economic growth, social living conditions, and environmental quality. *Development dynamics* represents the efficiency and speed of economic growth, social stability, environmental improvement, and ecological restoration. *Development potentials* represents the capability of the economic structure and administrative structures, decision-making ability, and ecological services and carrying capacity.

Table 20.2 shows the results of aggregation using the aforementioned indicators in Yangzhou ecocity planning (2005–2020), using a polygon-based approach (Wang and Xu 2004).

Since 2000, a series of institutional, legislative, technical, educational, and financial capacity-building measures has been implemented based on these indicators. Some domestic and international cooperative projects were initiated, including a project called "Ecocity Planning and Management," conducted by the German GTZ Institute. By 2005, some key indicators were exceeded. For example, the goals for GDP per capita and the GDP revenue ratio have been exceeded by 113 percent and 127 percent, respectively, and the comprehensive index of urban environment has been upgraded from number 7 to number 1 in the ranking of all Jiangsu Province cities assessed by the provincial government since 2004.

Conclusions

In the past decade, many Chinese research institutes and universities active in the environmental and ecological field have conducted research on sustainability indicators and set up indicator systems. Common traits and trends include the following:

- *Reduction and simplification:* Many researchers are trying to reduce the number and complexity of indicators.
- *Three pillars:* The division of indicators into three major groups or pillars—ecological, economic, and social—is widely applied in China. Compared with many other countries, in China there is not much discussion about the problems and shortcomings of this approach.
- *Anthropocentric approach and data pragmatism:* Although a large amount of economic and social data has been collected in recent decades, the availability and exchange of reliable environmental data are still a problem in China. Non-anthropocentric data are particularly rare. With a few exceptions, sufficient data are available only for anthropocentric aspects. Therefore, the selection of criteria and indicators is often determined by data availability.

Table 20.1. Indicators for system sustainability assessment in Yangzhou ecocity development.

First Class (D)	Second Class (C)	Third Class (B)	Fourth Class (A)	Target Reference Value	2000	2005	2010	2020
Sustainability	Development status	Economic growth	GDP per capita (10^4 yuan)	7	1.05	1.80	2.80	5.80
			Land productivity (10^4 yuan/km^2)	5,000	711.2	1,100	1,850	4,000
		Living conditions	Average life expectancy (years)	80	72	73	75	78
			Housing index	1.2	0.67	0.75	0.85	1
		Environmental quality	Percentage of water bodies with quality better than NES class III	100%	41.4%	60%	80%	95%
			Annual percentage of days with air quality better than NES class III	100%	83.9%	90%	95%	95%
			Forest cover	40%	13.8%	15%	20%	25%
			Percentage of people satisfied with their environment	100%	69.5%	80%	90%	95%
	Development dynamics	Economic dynamics	Annual GDP growth rate	15%	10.5%	11%	8%	7%
			Energy efficiency (industry GDP 10,000 yuan/ton of standard coal equivalent)	3	0.85	1	1.6	2.8
			Percentage of government revenue in total GDP	22%	7%	10%	15%	20%

(continued)

Table 20.1. Indicators for system sustainability assessment in Yangzhou ecocity development (*continued*).

First Class (D)	Second Class (C)	Third Class (B)	Fourth Class (A)	Target Reference Value	2000	2005	2010	2020
Development potentials		Social dynamics	Reciprocal of Gini index (social equality)	3.300	3.226	2.857	2.632	2.941
		Environmental dynamics	Restoration rate of degraded land	100%	80%	94%	96%	100%
			Ratio of discharged industrial sewage that was treated and met national standard	100%	93.8%	95%	99%	99%
			Recycling and reuse rate for household garbage	100%	40%	60%	80%	100%
			Use rate of domestic animal waste	100%	35%	55%	70%	90%
		Economic potentials	Authentication ratio of ISO 14000 enterprises	100%	10%	30%	50%	90%
			Annual investment in fixed assets as a percentage of GDP	40%	27%	30%	33%	38%
			Percentage of jobs in R&D	20%	2.9%	8.0%	14%	18%

(continued)

Table 20.1. Indicators for system sustainability assessment in Yangzhou ecocity development (*continued*).

First Class (D)	Second Class (C)	Third Class (B)	Fourth Class (A)	Target Reference Value	2000	2005	2010	2020
		Social potentials	Average education of adults (years)	16	8	10	12	14
			Average professional education of civil servants (years)	7	2	4	5.5	6.5
			Percentage of government policies in accordance with ecocity planning	100%	70%	90%	100%	100%
		Ecoservice enhancement potentials	Investment in the environment as a percentage of GDP	8%	1.69%	2.1%	2.5%	4%
			Percentage of preserved areas in the territory	30%	5%	12%	15%	20%
			Percentage of citizens who participate in and are aware of environmental management	95%	90%	35%	50%	75%

NES = national environmental standards in China.

Table 20.2. Sustainability assessment of Yangzhou ecocity planning (2005–2020).

Second Class	Third Class	2000 Index	Grade	2005 Index	Grade	2010 Index	Grade	2020 Index	Grade
Development status	Economic growth	0.10	IV	0.25	III	0.44	III	0.87	I
	Living conditions	0.40	III	0.46	III	0.56	II	0.75	I
	Environmental quality	0.19	IV	0.39	III	0.64	II	0.78	I
	Comprehensive index	0.20	IV	0.38	III	0.60	II	0.82	I
Development dynamics	Economic dynamics	0.28	III	0.42	III	0.58	II	0.72	II
	Social dynamics	0.95	I	0.71	II	0.58	II	0.76	I
	Environmental dynamics	0.30	III	0.53	II	0.76	I	0.94	I
	Comprehensive index	0.35	III	0.48	III	0.63	II	0.81	I
Development potentials	Economic potentials	0.05	IV	0.20	IV	0.42	III	0.80	I
	Social potentials	0.17	IV	0.47	III	0.73	II	0.88	I
	Ecoservice enhancement	0.17	IV	0.34	III	0.52	II	0.75	I
	Comprehensive index	0.12	IV	0.33	III	0.55	II	0.81	I
Overall sustainability index		0.22	IV	0.44	III	0.64	II	0.85	I

The index ranges from 0 to 1, with 0 the worst and 1 the best, and is divided into four grades: grade I (>0.75–1), grade II (>0.50–0.75), grade III (>0.25–0.50), and grade IV (0–0.25).

- *Focus on numeric data:* Many Chinese decision makers favor numeric data.
- *Transparency of results:* In Chinese publications, a high level of data aggregation and standardization is common, and data transparency is low. Thus, verifying final results (e.g., overall ecological performance) and comparing them (e.g., scores of different cities) are difficult for non-insiders.

The SENCE approach to the assessment of sustainability indicators in China is designed to help decision makers, researchers, planners, entrepreneurs, and ordinary people understand how the ecocomplex is functioning, systematically and ecologically, and how their actions are connected with their social, economic, and long-term ecological interests. Experience and evidence on the assessment of sustainability indicators and their application in China show that although there are still some gaps between theory and practice, between more and less powerful decision makers, and even between scientists from different backgrounds, the trial-and-error approach in ecopolis development at different scales in China has promoted urban and regional sustainable development in the past two decades. In China, transforming scientific results on ecological issues into policy action targeting sustainability is still a major challenge.

Acknowledgments

We would like to thank Erich Schienke for his careful revision of this chapter.

Literature Cited

Levin, H. 1997. Internet forum, organized February 19 at greenbuilding-request@crest.org.

Ma, S. J., and R. S. Wang. 1984. Social–Economic–Natural Complex Ecosystem. *Acta Ecologica Sinica* 4(1):1–9.

People's Republic of China. 1997. *China's Agenda 21. National report on sustainable development.* Available at www.iclei.org/la21/map/acca21.htm.

SEPA. 2003. *The constructing indicators for the ecological country, the ecological city and the ecological province.* Beijing: China State Environmental Protection Agency.

Wang, R., D. Hu, X. Wang, and L. Tang. 2004a. *Urban eco-service.* Beijing: Meteorological Press.

Wang, R., S. Lin, and Z. Ouyang. 2004b. *Theory and practice of Hainan eco-province development.* Beijing: Chemical Industry Press.

Wang, R., and H. Xu. 2004. *Methodology of ecopolis planning with a case of Yangzhou.* Beijing: China Science and Technology Press.

Wang, R., J. Zhao, and Z. Ouyang. 1996. *Wealth, health and faith: Sustainability study in China.* Beijing: China Environmental Science Press.

21

Core Set of UNEP GEO Indicators Among Global Environmental Indices, Indicators, and Data

Jaap van Woerden, Ashbindu Singh, and Volodymyr Demkine

Numerous international and regional organizations, government agencies, and scientific bodies have launched a variety of environmental indicator initiatives encompassing different areas of the environment. The UN and other international agencies have developed a number of different sets of environment-related indicators distinguished by certain objectives.

The UN Environment Programme (UNEP) Global Environment Outlook (GEO) project was initiated in response to the environmental reporting requirements of Agenda 21 and to a UNEP Governing Council decision of May 1995 that requested the production of a comprehensive biennial global state of the environment report. UNEP initially produced the GEO report about every two and a half years starting in 1997 (GEO1, GEO2 [GEO 2000], and GEO3) and now plans to publish GEO4 in 2007 because the report cycle has been changed to 5 years. GEO carries out integrated environmental assessments, provides global and regional overviews of the state of environment, develops an outlook using scenarios, and suggests options for action. The assessment report is prepared through a participatory process and complemented with educational and training material, comprehensive databases, and other information systems. Other outputs of the GEO project include regional, subregional, and national integrated environmental assessments, technical and other background reports, a Web site (www.unep.org/geo), products for young people (GEO for Youth), and a core database, the GEO Data Portal (geodata.grid.unep.ch).

GEO is a largely science-driven assessment based on sound facts and figures. Data and indicators have been at the very basis of the GEO assessment and reporting activities. Major data issues are addressed by one of the working groups established at the beginning of the GEO process, the GEO Data Working Group. The use of data for scientific analysis and for illustrations in the GEO reports has been, to a large extent, a process of learning by doing. However, the pragmatic approach has been backed by various attempts to frame the major core data sets and indicators. This has resulted in a living GEO Core Data/Indicator Matrix (UNEP 2005a), which lists the key indicators for the major global environmental issues and the data sets and sources from which they can be obtained or derived.

At the twenty-second session of the UNEP Governing Council/Global Ministerial Environment Forum (GC/GEMF) in 2003, governments asked UNEP to prepare an annual Global Environment Outlook statement to highlight significant environmental events and achievements during the year and raise awareness of emerging environmental issues identified by scientific research and other sources. It aims to bridge the gap between science and policy and to make environmental information easily accessible to policymakers and other readers. The *Year Book* presents, in a clear and timely manner, an analytical overview of issues and developments that, for better or worse, have most influenced the environment during the year and may continue to be major factors in the years ahead.

Keeping abreast of environmental issues as they unfold, the *Year Book* is released early every year between the comprehensive GEO reports. The *GEO Year Book 2003*, the first in the annual series, was launched at the eighth special session of the GC/GMEF on March 29, 2004 (see www.unep.org/geo/yearbook/yb2003). The second edition, the *Year Book 2004/5* (see www.unep.org/geo/yearbook/yb2004), was launched at the twenty-third session of GC/GMEF in February 2005. In addition to giving overviews of major global and regional environmental developments and emerging issues, the *GEO Year Books* present a small number of GEO indicators showing major headline trends for the themes being addressed under the broader GEO assessment process. Although the availability of reliable, up-to-date global data sets still limits the choice, the core set of indicators selected for this report aims to give a consistent, quantitative overview of major environmental changes on an annual basis. The GEO indicators are selected from a wider set of data and indicators that have been used in integrated environmental assessment practice over the years to illustrate major environmental issues, as summarized in the GEO Core Data Matrix, with some seventy indicators, and is backed by the comprehensive online GEO Data Portal with currently more than 450 variables.

The major starting point of the indicators selected for inclusion in the *GEO Year Book* is to show one or two solid quantitative trends—a handful at most—for the major environmental issues addressed through the GEO assessment process. The indicators are presented, where possible, at the global and regional levels and as a time series for up to several decades. The illustrations (graphs, maps, and tables) are extracted from a core database with more detailed and comprehensive data captured in the GEO Data Portal, which, in turn, is based on a harmonized and accepted data strategy and method and supported by the GEO Data Working Group in close cooperation with reliable and authoritative data holders (examples of such indicators are shown in Figures 21.1, 21.2, and 21.3).

Thus, the GEO indicators present a selected set of headline environmental trends, which are not assumed to be comprehensive. The selected indicators are a mix of environmental pressures, states, impacts, and responses, but they do not intend, nor are they able, to capture all aspects of all global and regional environmental problems. The underlying data that are used to compile the indicators come from internationally recognized sources and are readily available for recent years for most parts of the world or at least are expected to become available in the near future.

Perhaps not surprisingly, the GEO indicators overlap with those under Goal #7 of the Millennium Development Goals (MDGs) on the environment. Both sets share the starting point of focusing on a consistent, harmonized set of environmental indicators, with solid statistical data as time series for most regions or countries of the world. Unlike the MDGs and the set of indicators resulting from extensive consultations and agreements, the GEO indicator set is less fixed and can be adapted regularly. The GEO indicators also expand on certain additional or specific environmental issues of relevance to MDG-7.

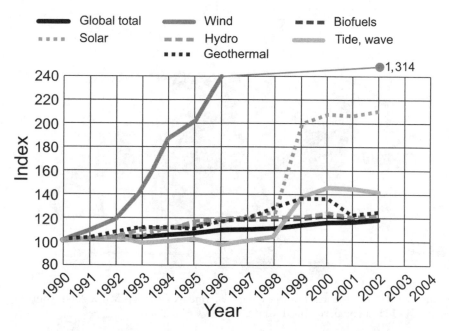

Figure 21.1. Renewable energy supply index by sector and global total, 1990–2002 (1990 = 100) (courtesy of the GEO Data Portal, compiled from International Energy Agency).

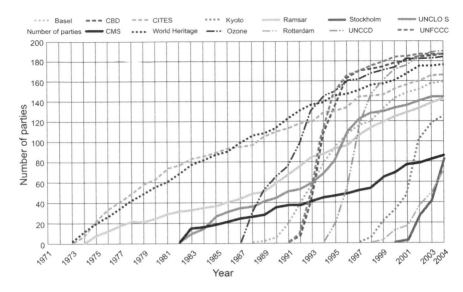

Figure 21.2. Number of parties to multilateral environmental agreements, 1971–2004 (courtesy of the GEO Data Portal, compiled from MEA Secretariats).

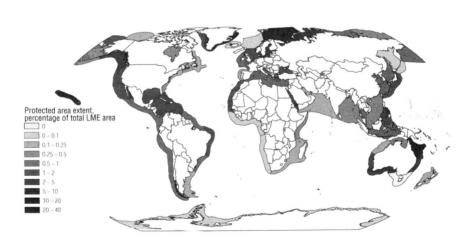

Figure 21.3. Protected area coverage of large marine ecosystems (LMEs), 2004 (courtesy of the GEO Data Portal, compiled from UNEP-WCMC 2004).

However, across-the-board presentation of the key global and regional aggregated trends for many environmental issues is severely limited by the lack of comprehensive and good-quality underlying data, mainly statistical data collected at the country level. Although some progress in data gathering and global observation is being made, data gaps and shortcomings are still profound. In addition to the need to show trends, it is equally important to mention that certain environmental issues that deserve headlines cannot be highlighted adequately by means of indicators, at least not yet. Currently, this holds true for environmental problems such as marine pollution, urban issues, and land degradation. For still other issues, such as freshwater use and forest cover change, at this stage only a snapshot of 1 or 2 years or just a few regions can be presented, not comprehensive yearly trends. Despite great effort to build up a sound set of key environmental indicators, as in many other core sets, the current GEO indicator collection may seem somewhat sketchy and can be said to be unbalanced with respect to the selected themes and issues. Several indicators are merely proxies of the real issues that they reflect. Related to the magnitude of the issue, there are more proxy indicators on climate change as compared to other priority areas. Because of the lack of good underlying data, some important environmental issues are missing at this stage. For example, there are no comprehensive indicators on urban environmental issues, no land indicators besides forest cover, and no direct climate trends; also, water quality cannot yet be illustrated adequately.

The first *GEO Year Book* (2003) contained seventeen indicators structured along themes and issues, as shown in Table 21.1 (UNEP 2004).

All indicators were updated in the second *GEO Year Book* (UNEP 2005b), and several new ones were added: renewable energy, consumption of hydrochlorofluorocarbons and methyl bromide, marine protected areas, water quality, and urban air pollution. At the same time, indicators on glacier retreat, population affected by disasters, and forest cover change were omitted.

The value of the inclusion of the GEO indicator set in the *GEO Year Book* lies predominantly in the illustrated overview of global and regional trends of major environmental issues portrayed by means of a varied but easy-to-understand set of graphics. All indicators are shown in visual form (charts, maps, and tables) combined with a short descriptive text in order to convey the message in a compact and straightforward manner. Thus, the set provides a concise picture of the global environment, complementing the more detailed information provided through the comprehensive GEO assessments and their regular reports. The lack of sufficient and sound data makes it impossible to complete the picture for all environmental issues, but the progress in global observation and monitoring and in scientific methods is encouraging and is expected to result in better sets of GEO indicators each year (Figures 21.4 and 21.5 show examples of GEO indicators).

In the coming years, the GEO indicator set will be revised, updated, and expanded where possible. More and better data will become available, and methods to develop simple indicators from complex data will become more sophisticated.

Notwithstanding the expected progress, a lot remains to be done. There still is a

Table 21.1. *GEO Year Book 2003* indicators.

Theme	Issue	Indicators
Atmosphere	Climate change	Energy use per US$1,000 GDP Total carbon dioxide emissions Total carbon dioxide emissions per capita Glacier mass balance
	Stratospheric ozone depletion	Consumption of chlorofluorocarbons
Natural disasters	Human vulnerability to extreme natural events	Number of people killed Number of people affected
Forests	Deforestation	Proportion of land area covered by forest
Biodiversity	Species loss	Number of threatened animal species Number of threatened plant species
	Habitat loss	Ratio of area protected to maintain biological diversity to surface area
Coastal and marine areas	Unsustainable use of living marine resources	Catch of living marine resources
Freshwater	Sustainable water use	Per capita water use Water use as a percentage of quantity of annual renewable water resources
	Access to improved water supply and sanitation	Population with access to improved water supply Population with access to improved sanitation
Global environmental issues	International environmental governance	Number of parties to multilateral environmental agreements

substantial lack of environmental data and information, which limits our ability to monitor developments and adequately show trends. The data gap will persist for some time and must be addressed through improved environmental monitoring, data collection, and compilation of indicators and indices at different geographic scales.

Thus, the use of indicators such as the GEO core set should be seen as a process of continuous development, not just a one-time exercise. A regular assessment of the indicators used, their relevance for and impact on policymakers and the wider public, and their continuous adjustment and updating to reflect emerging issues, better data, and new challenges and insights must be implemented in parallel.

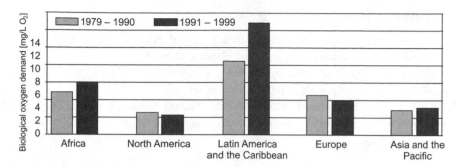

Figure 21.4. Mean biological oxygen demand in surface waters by selected region, 1979–1990 and 1991–1999. Note: Data for West Asia not available (courtesy of the GEO Data Portal, compiled from UNEP/GEMS-Water 2004).

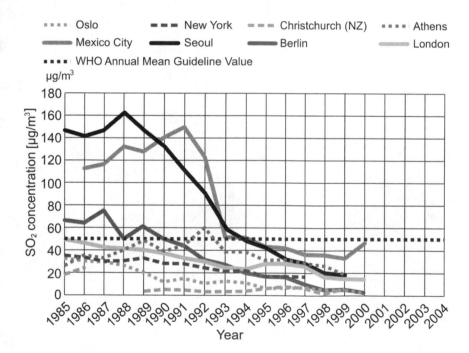

Figure 21.5. Concentrations of SO$_2$ in the air in selected cities, 1985–2000 (courtesy of the GEO Data Portal, compiled from OECD Environmental Data Compendium 2002).

Literature Cited

UNEP. 2004. *GEO year book 2003*. Nairobi: United Nations Environment Programme. Available at www.unep.org/geo/yearbook/yb2003/.

UNEP. 2005a. *GEO Data Working Group meeting report*, Geneva, June 16–17, 2003. Available at www.unep.org/geo/pdfs/dwgm_report.pdf.

UNEP. 2005b. *GEO year book 2004/5*. Nairobi: United Nations Environment Programme. Available at www.unep.org/geo/yearbook/yb2004/.

22

Further Work Needed to Develop Sustainable Development Indicators

Edgar E. Gutiérrez-Espeleta

By the late 1980s and early 1990s, a new paradigm started to rise among those who recognized that important things for development had not been developed. Redevelopment was then proposed (to develop what was not developed in the past) and later evolved into what is now known as sustainable development.

Sustainable development was conceived as a strategy to sustain development within certain limits, where both technology and social organization can be ordered and improved in such a manner that a path can be opened for a new era of economic growth (WCED 1987).

In light of the holism and synergism that the concept of sustainable development brought about and the need to improve societal diversity and access to opportunities for all, sectoral policies as they are understood and practiced today are not all useful. Economic, social, or environmental sectors, dimensions, or attributes of development are not sufficient to characterize what is needed to sustain the quality of life for all.

As we understand them today, economic problems do not have economic solutions, nor do environmental problems have environmental solutions. If one thing is clear today, it is that solutions to societal problems must address the society and its relationship with nature as a whole.

A New Approach Is Needed

Just as gross national product (GNP) has been overused for policy orientation, so have the traditional dimensions of development.

When human beings decided to settle down as societies (i.e., to live together and use natural resources), the main conflict between the natural system and the societal system was established. How societies deal with that interaction will lead us to sustain or maintain the human species on planet Earth or to shape a different planet that cannot sustain human life anymore.

Development is the result of interactions with new properties or attributes emerging from its own processes. These attributes, not yet identified, relate and synthesize a broad and complex range of topics that concern society. Furthermore, these attributes should clearly describe the fundamental interactions between societal systems and natural systems.

Toward Synthetic Indices for Sustainable Development

Indicators and indices, as aggregated figures, are an intrinsic part of the decision-making process. They belong to the stream of information we use to make decisions and plan our actions.

Indicators are tools used to provide solid bases for decision making at all levels and to contribute to the self-regulating sustainability of integrated environment and development systems (United Nations 1993). Reliable environmental information therefore is needed to frame policy, set priorities, and assess results (UNEP 1994).

The United Nations Environment Programme (UNEP) *GEO 3* (2002) establishes that "high quality, comprehensive and timely information on the environment remains a scarce resource, and finding the 'right' information can pose problems: data are more difficult and expensive to obtain. It is also difficult to find indicators that capture and reflect the complexity of the environment and human vulnerability to environmental change. Environmental data acquisition remains a basic need in all countries."

This lack of data limits our ability to develop and use higher levels of information in response to the call to integrate environment and development in policy formulation issued by the UN Conference on Environment and Development (UNCED), held in Rio de Janeiro in June 1992. This highlights the need for dependable environmental data for use in policy construction, priority setting, and result assessment on the part of the government and civil society. The improvement and use of available environmental and socioeconomic information is a major prerequisite for the development of national policy and international understanding.

The World Bank (1997) stated, "The development of useful environmental indicators requires not only an understanding of concepts and definitions, but also a thorough knowledge of policy needs. In fact, the key determinant of a good indicator is the link from measurement of environmental conditions to practical policy options." Practical policy options imply a relationship between environmental and societal affairs, but any decision has a price, whether it is environmental, social, or economic, so a policy's impact ultimately depends on the priorities of the decision maker. Thus, the integration of both areas must provide a solid platform for supporting the path toward sustainable development indicators and synthetic indices.

Proposal for a New Typology of Sustainable Development Indicators

Different approaches could be taken to develop indicators under the sustainable development paradigm. However, they must be able to meet the challenge of fully integrating the social, economic, environmental, and institutional aspects of development, in accordance with the main conclusions of UNCSD in 1997. The national decision-making process requires indicators sensitive to change, supported by reliable, readily available data, relevant to the issue, and understood and accepted by intended users.

Starting from these premises, indicators and indices can be classified according to the way they are generated; that is, a classification can be established according to the sort of attributes they try to describe and integrate synthetically.

In the main categories "Environment" and "Society," an issue is defined as a topic of interest to decision makers. The issue encompasses one or more characteristics (properties or components), from which one or several are selected for measurement. These characteristics can be either quantitative or qualitative. Depending on the origin and number of characteristics selected for use in developing an index or indicator (i.e., the extent of complexity related to the synthetic information linked to the selected issue or issues), indicators can be classified as a first, second, third, fourth, or fifth generation. First-generation indicators measure a single characteristic of an environmental (or societal) issue; most state-of-the-environment (or society or economy) indicators fall in this category. Second-generation indicators measure various characteristics of various environmental (or societal) issues, and third-generation ones are an integrative measure of a single characteristic of a selected environmental issue combined with a single characteristic of a societal issue. Fourth-generation indicators are integrative measures that combine one or more characteristics of an environmental issue and one or more characteristics of various societal issues. Fifth-generation indicators are integrative, composite indicators that measure multiple characteristics of more than one environmental issue, integrated, in a thoughtful fashion, with multiple characteristics of various societal issues. Fifth-generation indicators provide the initial tools for scenario building, keeping the "bricks" explicit in order to reveal relationships at lower levels.

Graphically this typology can be seen in Figure 22.1.

Topfer (1998) points out that "the pace at which the world is moving towards a sustainable future is far too slow." In order to reverse this global trend and maintain the spirit of Agenda 21, we need to produce fourth- and fifth-generation indicators without

Class generation	System		Example
	Environmental	Social	
I	For just one of the systems One issue → one or several measurements		Air pollution quantified as CO_2 ambient concentration.
II	For just one of the systems Several issues → several measurements		Gross primary school enrollment ratio encompasses population and educational issues, and total primary enrolment and population in primary school-age bracket are used as their respective characteristics.
III	One issue → one measurement	One issue → one measurement	A relation between air pollution and consumption patterns could be established by measuring CO_2.
IV	One issue → one or several measurements	Several issues → several measurements	An indicator for ambient pollution might be able to relate CO_2 concentration and emissions of CO_2 from (a) liquid fuels consumption; (b) land use change, and (c) industrial processes.
V	Several issues → several measurements	Several issues → several measurements	For air pollution, CO_2 and CH_4 concentration can be taken as environmental characteristics. Several societal characteristics could be used to call attention to this problem, such as CO_2 from fossil fuels; CO_2 from land use change; CH_4 from agriculture; CH_4 from waste, etc.

Figure 22.1. Typology of sustainable development indicators.

neglecting the less complex ones, also very much needed. This is a daunting challenge for national sustainable development programs.

Example: The Latin American and Caribbean Initiative for Sustainable Development

If an inventory of existing indicators were performed, many of them undoubtedly would fall into the class of first- and second-generation indicators. In other words, we have been providing specifics while failing to show national decision makers, clearly and explicitly, the relationship between the environment and societal consumption patterns. Could this be one of the reasons we have failed to move politicians to a more proactive approach to sustainable development, as pointed out by Topfer?

The example in this section illustrates how much we have accomplished in trying to show integration and main patterns in the whole, using the set of indicators proposed by the Forum of Ministers of Environment of Latin America and Caribbean to monitor compliance with the Latin American and Caribbean Initiative for Sustainable Development (ILAC).

ILAC is a political response from the Ministers of Environment of Latin America and Caribbean to the need to bring practical meaning to the processes of the World Summit on Sustainable Development held in Johannesburg in 2002.[1] It is an opportunity to assess progress achieved in compliance with the commitments adopted at UNCED in Rio de Janeiro in 1992 and to adopt effective actions in search of solutions for the new sustainable development challenges. In this way, ILAC defines goals for biological diversity; water resource management; vulnerability, human settlements, and sustainable cities; social issues such as health, inequity, and poverty; economic aspects including competitiveness, trade, and production and consumption patterns; and institutional aspects.

Beyond public management, this political platform recognizes the active participation of the entrepreneurial sector and civil society to promote actions that can generate sustainable productive activities and, at the same time, allow the conservation and sustainable use of environmental goods and services essential to life.

It is a priority for ILAC, as an ethical principle, to set up competitive sustainable development models built on public policies formulated to develop science and technology, financing, human resource education, institutional frameworks, environmental good and service appraisals, and indicators of sustainability suitable to social, economic, environmental, and policy conditions of each country or to the subregional needs. For this purpose, the fourteenth meeting of the Forum of Ministers of Environment of Latin America and Caribbean (November 2003) approved an initial indicator matrix proposed by Brazil, Colombia, Costa Rica, Cuba, Mexico, Peru, and Saint Lucia and supported by the World Bank, the Inter-American Development Bank, the Pan-American Health Organization, the UN Statistical Division, and the UNEP Regional Office for LAC, which acted as facilitator. This matrix contains a set of forty-three indicators to monitor thirty-eight operational proposals for twenty-five goals.

Once this matrix was approved by the ministers, a network of national focal points (see www.vulnerabilityindex.net/EVI_2005.htm) was created to discuss methodological sheets (metadata) for each selected indicator. This process is still under way. The UNEP Regional Office for Latin America and the Caribbean is inviting countries to produce national reports using ILAC indicators. Costa Rica was the first country to prepare such a report, followed by Mexico, to be published soon, and others to come.

Because the indicators proposed for Objective 7 of the Millennium Development Goals (MDGs) fell short in describing LAC region monitoring needs, it has been pointed out that ILAC indicators can be used to complement them. For example, MDG indicator 28, "carbon dioxide emissions per capita," is of little use in LAC to monitor environmental sustainability because the contribution of this region to global emissions is less than 6 percent, and the per capita figure for North America is almost eight times that for LAC (www.yale.edu/esi/). Using ILAC indicators to complement those for MDGs will provide a better chance to monitor Objective 7, "ensure environmental sustainability."

Using this set of forty-eight indicators to test the proposed typology, 77 percent of them fall in the first class and the rest in the second (Table 22.1). One might argue that because there is a need to monitor specific goals, there is a justification to use first- and second-generation indicators, but it also helps identify how much work is still needed in order to provide a better integration of aspects in development. The way of trying to develop a picture of the forest by using single trees is very thorny and uncertain. Integrative tools are needed to provide better appraisals of the path toward a broad set of development goals.

This problem is also shared by the UNCSD set of fifty-seven indicators proposed in 2001. This set concentrates on first- and second-generation indicators. Possible explanations for this situation include the following: We still have a very fragmented way of analyzing development; the limited availability of national environmental statistics leads decision makers to pick just first- and second-generation indicators, leaning toward the social side; the mainstream way of thinking leads experts to identify or elaborate on pieces and not on the whole of development; data coverage is insufficient; up-to-date data are not available; or countries are inconsistent in the concepts and methods they use for data gathering and processing and in their presentation of the corresponding reports (Gutiérrez-Espeleta 2003).

Conclusions and Recommendations

The situation just described seems to be typical in data-related matters on development. Gutiérrez-Espeleta (2003) presents the state of development of social indicators and concludes, "Although [it] is true and needs to be recognized that social indicators are lagging behind [economic ones] and that there lacks a conceptual framework to integrate them, it is also important to recognize that society, its organizations and windows for individual participation, are those that should define the desired orientation; the goals at five, ten or fifteen years time, and consequently support those indicators that allow for tracking down of the social ensemble in that journey. That is important and fundamental, if social indicators are truly to serve for an effective assessment of the social parameters considered relevant towards a society for all." Something very similar can be said on the environmental side but under worse conditions.

Whereas social indicators have been in use for some time—since it was discovered that economic development alone does not alleviate poverty or provide opportunity—

Table 22.1. ILAC Indicators according to proposed classification.

Themes	Guiding Goals	Operational Proposal	Number of Indicators	Indicator Class I	II	III	IV	V
Biological diversity	4	4	4	4				
Water resource management	4	7	7	5	2			
Vulnerability, human settlements, and sustainable cities	7	10	13	11	2			
Social issues, including health, inequity, and poverty	3	7	8	4	4			
Economic issues, including competitiveness, trade, and production and consumption patterns	3	4	5	4	1			
Institutional arrangements	4	6	6	5	1			
Total	25	38	43	33	10			

and information systems have been put in place as part of UN initiatives such as Household Surveys, environmental statistics remain to be developed in most countries of the world. This problem has been recognized since UNCED 92 (Chapter 40 in Agenda 21), but little progress has been achieved, helping to promote the fragmented vision of development.

The mismatch between the worldwide discourse on sustainable development and funding allocations is obscuring the importance of the balanced use of our environment. Decisions and funds are needed to develop an integrative view of development and define the parameters or dimensions that need to be monitored, to bring in new experts with a different view who pay more attention to synergisms, to increase the dialogue between decision makers and the scientific community, and to develop new ways of presenting information to decision makers and the public. However, the most relevant task for the near future is to overcome the fragmented way of seeing development.

Despite the importance of first- and second-generation indicators and their usefulness for decision making, it seems that the task for the years to come is to develop more high-level indicators to assist the national decision-making process. This is already on the international agenda, but unfortunately only a few attempts have been made despite the impetus created by the publication of the Human Development Index in the early 1990s as a synthetic index for the social side and as a second-generation index.

Some recent attempts are promising in that they show the interest of some scientific communities in developing higher-order indices. Two examples of these efforts are the Environmental Vulnerability Index (EVI) and the Environmental Sustainability Index (ESI), both developed in response to political decisions in different fora.

The EVI "is among the first of tools now being developed to focus environmental management at the same scales that environmentally significant decisions are made, and focus them on planned outcomes. The scale of entire countries is appropriate because it is the one at which major decisions affecting the environment in terms of policies, economics and social and cultural behaviours are made. If environmental conditions are monitored at the same time as those concerning human systems, there is better opportunity for feedback between them. Without exception, the environment is the life-support system for all human systems and therefore an integral part of the developmental success of countries" (www.vulnerabilityindex.net/EVI_2005.htm). Developed by the South Pacific Applied Geoscience Commission, the UNEP, and their partners, the EVI tries to respond to Section C5 of the Barbados Programme of Action.

The EVI uses fifty indicators to estimate the environmental vulnerability of a country to future shocks. Combined by simple averaging to provide a single index, it offers synthetic measures of aspects of vulnerability (hazards, resistance, and damage), subindices relevant to policy (climate change, exposure to natural disasters, biodiversity, desertification, water, agriculture and fisheries, human health aspects), and an overall national score. Although the soundness of the metric being used is beyond the scope of this chapter, the EVI is a good example of an index developed for a purpose that uses a conceptual framework to integrate several relevant components.

Another example of the efforts being made to provide better tools for decision making is the ESI (www.yale.edu/esi). In this case, a scientific team has made a proposal that overcomes the fragmented vision of development. Using twenty-one indicators derived from seventy-six statistical variables, a composite index is proposed to show national environmental stewardship. These twenty-one indicators permit comparisons across a range of issues that fall into the five broad categories of "environmental systems," "reducing environmental stresses," "reducing human vulnerability to environmental stresses," "societal and institutional capacity to respond to environmental challenges," and "global stewardship." "By facilitating comparative analysis across national jurisdictions, these metrics provide a mechanism for making environmental management more quantitative, empirically grounded, and systematic."

These composite indices (EVI and ESI) try to convey visions toward a better understanding of national development dynamics. Other difficulties (e.g., mathematical algorithms for integrating variables, and model validation and acceptance) have arisen, but at least an integrative framework has been developed and new ways of improving development are proposed. With these indices, the concept of sustainable development goes beyond the traditional three-pillar approach and aims at developing a metric to measure synergies between the natural and social systems. More of these efforts are needed to advance understanding of the new sustainability paradigm and to begin writing a new story on development.

With higher-order indicators, policymakers would have better tools to understand the main patterns of how society is using its natural endowment. Perhaps with these indicators and an understanding of their components, the holistic approach to policymaking can be achieved.

Note

1. There is no Latin American or Caribbean (LAC) country included in Annex I of United Nations Framework Convention on Climate Change (UNFCCC).

Literature Cited

Gutiérrez-Espeleta, E. E. 2003. Social indicators: A brief interpretation of their state of development. In *Social development in Latin America: Issues for public policy*, edited by C. Sojo. New York: World Bank.

Topfer, K. 1998. Foreword. In *Where we stand: A state of the environment overview for the Global Environment Facility*. A report from the Global Environment Outlook (GEO) Program. UNEP.

UNEP. 1994. *Environmental data report 1993–94*. Oxford: United Nations Environmental Programme.

UNEP. 2002. *Global environment outlook 3*. London: Earthscan.

United Nations. 1993. *Agenda 21: Report of the United Nations Conference on Environ-*

ment and Development. Chapter 40. Rio de Janeiro, June 3–14, 1992. New York: United Nations.

WCED (World Commission on Environment and Development). 1987. *Our common future.* Oxford: Oxford University Press.

World Bank. 1997. *Expanding the measure of wealth: Indicators of environmentally sustainable development.* Environmentally Sustainable Development Studies and Monographs Series no. 17. Washington DC: World Bank.

23

The Yale and Columbia Universities' Environmental Sustainability Index 2005

Compiled by Tomáš Hák

The Environmental Sustainability Index (ESI) was introduced for the first time at the Economic Forum's Annual Meeting in Davos, Switzerland, in 2000. The report "Pilot Environmental Sustainability Index" was part of an exploratory effort to measure the ability of economies to achieve environmentally sustainable development. Since then, the ESI has been further developed and published in 2001, 2002, and 2005.

The latest ESI was formally released, again in Davos, in January 2005 by the Yale Center for Environmental Law and Policy (YCELP), Yale University, and the Center for International Earth Science Information Network (CIESIN) at Columbia University, in collaboration with the World Economic Forum Geneva, Switzerland, and the Joint Research Centre of the European Commission in Ispra, Italy.

ESI is part of the effort to shift environmental decision making to firmer analytic foundations using environmental indicators and statistics. The most important function of the ESI is as a policy tool for identifying issues that deserve greater attention within national environmental protection programs and across societies more generally. It also provides a way of identifying governments that are leading the way (as well as the laggards) with regard to any particular issue included in the ESI. In this regard, the heart of the ESI is not the rankings but rather the underlying indicators and variables used for a comparative analysis across national jurisdictions.

Approach and Method

The basic concepts are sustainability as a characteristic of dynamic systems that maintain themselves over time (not a fixed endpoint that can be defined) and environmental sustainability as a long-term maintenance of valued environmental resources in an evolving human context. The overall picture created by the index does not define sustainability but instead provides a gauge of a country's present environmental quality and capacity to maintain or improve conditions in the future. The ESI is still under development in both methodological improvements and data collection methods.

ESI must cope with several challenges commonly encountered in the computation of composite indices: variable selection, missing data treatment, aggregation, and weighting methods.

At the top level of aggregation, the ESI loosely uses the driving force, pressure, state, impact, response (DPSIR) framework. It centers on the state of environmental systems, both natural and managed. Then it measures pressures on those systems, including natural resource depletion and pollution rates. The ESI also includes impacts as human vulnerability to environmental change and responses as a society's capacity to cope with environmental stresses and each country's contribution to global stewardship. Thus, the broad categories "environmental systems," "reducing environmental stresses," "reducing human vulnerability," "social and institutional capacity," and "global stewardship" form the core components of the ESI. Below this level of aggregation each of these five components encompasses three to six indicators of environmental sustainability. These twenty-one indicators, in total, are considered the building blocks of environmental sustainability. Each indicator builds on two to twelve data sets, for a total of seventy-six underlying variables (Figure 23.1, Table 23.1).

The issues reflected in the indicators and the underlying variables were chosen through an extensive review of the environmental literature, assessment and analysis of available data, and broad-based consultation with policymakers, scientists, and indicator experts. The seventy-six variables cover issues that are local in scope as well as those

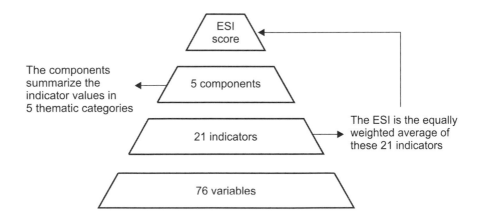

Figure 23.1. ESI architecture: aggregation scheme (Esty et al. 2005, modified).

Table 23.1. ESI components and indicators.

5 Components	21 Indicators
Environmental systems	Air quality
	Biodiversity
	Land
	Water quality
	Water quantity
Reducing stresses	Reducing air pollution
	Reducing ecosystem stresses
	Reducing population growth
	Reducing waste and consumption pressures
	Reducing water stress
	Natural resource management
Reducing human vulnerability	Environmental health
	Basic human sustenance
	Reducing environment-related natural disaster vulnerability
Social and institutional capacity	Environmental governance
	Eco-efficiency
	Private sector responsiveness
	Science and technology
Global stewardship	Participation in international collaborative efforts
	Greenhouse gas emissions
	Reducing transboundary environmental pressures

that are global in scale. Although countries at different levels of development and with diverse national priorities may choose to focus on different aspects of environmental sustainability, all of the issues included in the ESI are of some relevance to all countries. Despite the great diversity of national priorities and circumstances, a uniform weighting of the twenty-one indicators was chosen in order to keep the aggregation easy to understand.

At the lowest level of aggregation, each indicator is itself an equally weighted sum of the two to twelve underlying variables. The variables are standardized by the means of z scores. The z scores for each variable are constructed by subtracting the mean from the observation and dividing the result by the standard deviation of the variable. They preserve the relative position of each country for each variable while providing a neutral way to aggregate the variable into indicators. The ESI score is then calculated as an equally weighted average of the twenty-one indicator scores.

It is obvious that because of the structure of the ESI (unevenly distributed seventy-six variables into twenty-one indicators and five elements), the individual variable weights vary in their contribution to the overall ESI score in proportion to the number of variables in a given indicator. This hidden weighting implies that the relative contribution of variables to the total ESI score ranges from 2 percent for an indicator with only two variables (e.g., greenhouse gas emissions) to 0.3 percent for an indicator with twelve variables (e.g., environmental governance). By giving each variable within an indicator the same weight and weighting each of the twenty-one indicators equally, ESI provides an imperfect but clear starting point for analysis. An interactive version of the ESI, which will allow the user to adjust the indicator or component weights and calculate a new score, is planned in order to improve the policy utility of the ESI.

ESI Results

The ESI ranking provides a relative gauge of environmental sustainability in 146 countries (these countries met the criteria for inclusion in the 2005 ESI, such as country size, variable, and indicator coverage) (Table 23.2). The ESI results cannot be compared between editions; there are too many refinements in the methods and improvements in variables for such comparisons at present.

The higher a country's ESI score, the better positioned it is according to selected variables. However, as is often the case with composite indices, it is difficult to interpret results. Besides ranking, the ESI can be useful in the search for the best practices in environmental decision making. Because of difficulties with reaching consensus on the index or even component level, that search might be best conducted at the indicator or variable level. It shows that countries at different levels of development face distinct environmental challenges, such as the pollution pressures of industrialization on one hand and the stresses and impacts of poverty and incapacity on the other. ESI also demonstrates that economic success contributes to the potential of environmental success (high environmental performance) but does not guarantee it because it is affected by many various factors.

ESI has been both positively accepted and subject to criticism, like most other attempts to measure such complex issues (Parris and Kates 2003, Wackernagel 2001). The main criticisms may be summarized as follow:

- The ESI has an inherently Northern bias; it favors developed countries by including too many measures of capacity and favoring technological innovations.
- The equal weighting of the ESI is arbitrary or inappropriate; it underemphasizes certain critical aspects of environmental sustainability.
- The index architecture is inappropriate; environmental sustainability cannot be summarized in a single index that combines too many disparate elements (even in the terms of causality) in one, thus rendering it meaningless.
- Many countries that score high on the ESI, such as the Nordic countries, have levels

le 23.2. ESI ranking and scores.

Rank	Country	ESI Score	ESI Rank	Country	ESI Score
	Finland	75.1	41	Netherlands	53.7
	Norway	73.4	42	Chile	53.6
	Uruguay	71.8	43	Bhutan	53.5
	Sweden	71.7	44	Armenia	53.2
	Iceland	70.8	45	United States	52.9
	Canada	64.4	46	Myanmar	52.8
	Switzerland	63.7	47	Belarus	52.8
	Guyana	62.9	48	Slovakia	52.8
	Argentina	62.7	49	Ghana	52.8
	Austria	62.7	50	Cameroon	52.5
	Brazil	62.2	51	Ecuador	52.4
	Gabon	61.7	52	Laos	52.4
	Australia	61.0	53	Cuba	52.3
	New Zealand	60.9	54	Hungary	52.0
	Latvia	60.4	55	Tunisia	51.8
	Peru	60.4	56	Georgia	51.5
	Paraguay	59.7	57	Uganda	51.3
	Costa Rica	59.6	58	Moldova	51.2
	Croatia	59.5	59	Senegal	51.1
	Bolivia	59.5	60	Zambia	51.1
	Ireland	59.2	61	Bosnia & Herzegovina	51.0
	Lithuania	58.9	62	Israel	50.9
	Colombia	58.9	63	Tanzania	50.3
	Albania	58.8	64	Madagascar	50.2
	Central African Republic	58.7	65	Nicaragua	50.2
	Denmark	58.2	66	United Kingdom	50.2
	Estonia	58.2	67	Greece	50.1
	Panama	57.7	68	Cambodia	50.1
	Slovenia	57.5	69	Italy	50.1
	Japan	57.3	70	Bulgaria	50.0
	Germany	56.9	71	Mongolia	50.0
	Namibia	56.7	72	Gambia	50.0
	Russia	56.1	73	Thailand	49.7
	Botswana	55.9	74	Malawi	49.3
	Papua New Guinea	55.2	75	Indonesia	48.8
	France	55.2	76	Spain	48.8
	Portugal	54.2	77	Guinea–Bissau	48.6
	Malaysia	54.0	78	Kazakhstan	48.6
	Congo	53.8	79	Sri Lanka	48.5
	Mali	53.7	80	Kyrgyzstan	48.4

Table 23.2. ESI ranking and scores (continued).

ESI Rank	Country	ESI Score	ESI Rank	Country	ESI Score
81	Guinea	48.1	120	Sierra Leone	43.4
82	Venezuela	48.1	121	Liberia	43.4
83	Oman	47.9	122	South Korea	43.0
84	Jordan	47.8	123	Angola	42.9
85	Nepal	47.7	124	Mauritania	42.6
86	Benin	47.5	125	Libya	42.3
87	Honduras	47.4	126	Philippines	42.3
88	Côte d'Ivoire	47.3	127	Viet Nam	42.3
89	Serbia & Montenegro	47.3	128	Zimbabwe	41.2
90	Macedonia	47.2	129	Lebanon	40.5
91	Turkey	46.6	130	Burundi	40.0
92	Czech Republic	46.6	131	Pakistan	39.9
93	South Africa	46.2	132	Iran	39.8
94	Romania	46.2	133	China	38.6
95	Mexico	46.2	134	Tajikistan	38.6
96	Algeria	46.0	135	Ethiopia	37.9
97	Burkina Faso	45.7	136	Saudi Arabia	37.8
98	Nigeria	45.4	137	Yemen	37.3
99	Azerbaijan	45.4	138	Kuwait	36.6
100	Kenya	45.3	139	Trinidad & Tobago	36.3
101	India	45.2	140	Sudan	35.9
102	Poland	45.0	141	Haiti	34.8
103	Niger	45.0	142	Uzbekistan	34.4
104	Chad	45.0	143	Iraq	33.6
105	Morocco	44.8	144	Turkmenistan	33.1
106	Rwanda	44.8	145	Taiwan	32.7
107	Mozambique	44.8	146	North Korea	29.2
108	Ukraine	44.7			
109	Jamaica	44.7			
110	United Arab Emirates	44.6			
111	Togo	44.5			
112	Belgium	44.4			
113	Democratic Republic of the Congo	44.1			
114	Bangladesh	44.1			
11	Egypt	44.0			
116	Guatemala	44.0			
117	Syria	43.8			
118	El Salvador	43.8			
119	Dominican Republic	43.7			

of natural resource use per capita beyond those that the biosphere can indefinitely sustain (other indicators such as the Ecological Footprint do a better job of measuring it).

Despite the fact that measuring trends with respect to environmental sustainability is conceptually difficult, ESI provides a way to benchmark performance and facilitates efforts to identify the best practices. The statistical foundation of the 2005 ESI represents a significant improvement from earlier versions. It was used both for the ESI construction (e.g., imputation of missing data, sensitivity analysis) and its interpretation. The approaches and methods for combining data sets into a single index continue to be refined (e.g., the authors are already thinking about the "ideal ESI" that would incorporate issues such as environmental impacts of trade, investment, and consumption flows; transboundary environmental pressures; solid and hazardous waste generation; and stresses on ecosystem functioning). Thus the problem of persistent data gaps seems to be the most serious impediment to obtaining a full and unbiased picture of environmental sustainability.

Literature Cited

Esty, D. C., M. A. Levy, T. Srebotnjak, and A. de Sherbinin. 2005. *2005 Environmental Sustainability Index: Benchmarking national environmental stewardship*. New Haven, CT: Yale Center for Environmental Law & Policy.

Parris, T. M., and R. W. Kates. 2003. Characterizing and measuring sustainable development. *Annual Review Environmental Resources* 28(13):1–28.

Wackernagel, M. 2001. *Shortcomings of the Environmental Sustainability Index*. Notes by Mathis Wackernagel, "Redefining Progress," February 10, 2001. Available at www.anti-lomborg.com/ESI%20critique.rtf.

Annex: Menu of Selected Sustainable Development Indicators

Sets of Indicators

Sets for International Level

CBD 2010 TARGET INDICATORS

This list of indicators was agreed upon in Convention on Biodiversity/Conference of the Parties 7 (CBD/COP7) to evaluate chosen targets (UNEP 2003c). This set is used also for reporting on indicators and monitoring at national level (UNEP 2003b).

EUROPEAN ENVIRONMENT AGENCY (EEA) CORE SET OF INDICATORS

The proposed EEA set contains 354 indicators (main indicators and subindicators); 206 of these are from more developed areas (issues of climate change, air pollution, ozone depletion, and water, excluding ecological quality, waste and material flows, energy, transport, and agriculture), and 148 are from less developed areas (biodiversity, terrestrial environmental, water ecological quality, tourism and fisheries). See themes.eea.eu.int/indicators/.

ENVIRONMENTAL PRESSURE INDICATORS

Environmental Pressure Indicators has been published twice by Eurostat (Eurostat 1999, 2001c) as a result of the Commission Communication on Environmental Indicators and Green National Accounts in 1994 (COM(94)670). The most recent edition

---

This annex was inspired by Dr. Peter L. Daniels from the Faculty of Environmental Sciences at Griffith University, Nathan, Australia.

contains forty-eight indicators, covering nine environmental policy fields, including a breakdown by sector where possible and relevant. See europa.eu.int/comm/eurostat/.

Eurostat Set of Sustainability Indicators

The report *Measuring Progress Towards a More Sustainable Europe* (Eurostat 2001b) contains sixty-three indicators, of which twenty-two are mainly social, twenty-one are mainly economical, and sixteen are mainly environmental. The publication draws on and extends the UN Commission on Sustainable Development (UNCSD) list of fifty-nine core sustainable development indicators; this list is structured along a more policy-oriented classification than the previous one, according to the relevant sustainability dimensions (four), themes (fifteen), and subthemes (thirty-eight). As a result, more than 66 percent of the indicators presented are comparable with those in the UNCSD core list. See www.eu-datashop.de.

Global Environment Assessment and Reporting Under the Global Environmental Outlook (GEO) Program

The *Global Environment Outlook* (UNEP 2003a) is an analysis of environmental conditions around the world on the basis of environmental indicators. It is a comprehensive and authoritative review undertaken by approximately thirty-five regional and global Collaborating Centers. GEO presents a regionally differentiated analysis of the state of the world's environment and scenario-based outlooks into the future. It highlights global as well as region-specific concerns and makes recommendations for policy action. See www.unep.org/geo/geo3/.

Health System Achievement Index

The World Health Organization (WHO) created an index for comparing health system performance in 191 countries, in terms of both the overall level of goal achievement and the distribution of that achievement (Murray et al. 2001). Five indicators are included (level of health, health inequality, responsiveness, responsiveness inequality, and fairness of financial contribution). See www.who.int/health-systems-performance/docs/overall -framework_docs.htm.

Indicators on Transport and Environment Integration in the EU (TERM)

The Transport and Environment Reporting Mechanism (TERM) (EEA 2002) was initially set up to develop a comprehensive set of indicators of the sustainability of transport in conjunction with the EEA. Annual publications are produced by the EEA (synthetic report) and Eurostat (statistics and indicators). The 2002 edition is the first to include the thirteen accessing countries. See reports.eea.europa.eu/environmental _ issue_report_2002_24/en

INDICATORS TO MEASURE DECOUPLING OF ENVIRONMENTAL PRESSURE FROM ECONOMIC GROWTH

The report *Indicators to Measure Decoupling of Environmental Pressure from Economic Growth* (OECD 2002a) explores a set of thirty-one decoupling indicators covering a broad spectrum of environmental issues. Sixteen indicators relate to the decoupling of environmental pressures from total economic activity under the headings of climate change, air pollution, water quality, waste disposal, material use, and natural resources. The remaining fifteen indicators focus on production and use on four specific sectors: energy, transport, agriculture, and manufacturing. The term *decoupling* refers to breaking the link between environmental "bads" and economic goods. Decoupling can be measured by decoupling indicators that have an environmental pressure variable as the numerator and an economic variable as the denominator. See www.oecd.org/dataoecd/0/52/1933638.pdf.

ORGANISATION FOR ECONOMIC CO-OPERATION AND DEVELOPMENT (OECD) CORE SET OF INDICATORS

The OECD Core Set (OECD 2001) helps track environmental performance and progress toward sustainable development. Key indicators drawn from the Core Set inform the public about key issues of common concern to OECD countries. Sectoral indicators help integrate environmental concerns into sectoral decisions (e.g., transport, agriculture). When developing environmental indicators, OECD countries have agreed to use the pressure, state, response (PSR) model as a common harmonized framework. They have identified indicators based on their policy relevance, analytical soundness, and measurability and have developed guidance on how to use and interpret the indicators. See www.oecd.org/home/.

OECD SUSTAINABLE CONSUMPTION INDICATORS

The framework that was adopted to structure the work on sustainable consumption indicators resembles that of other OECD work on sectoral indicators. It is based on an adjusted pressure, state, response (PSR) model and distinguishes three themes: environmentally significant consumption trends and patterns (i.e., major driving forces and indirect pressures), interactions between consumption patterns and the environment (i.e., direct pressures on the environment and on natural resources and related impacts), and economic and policy aspects covering key policy and other societal responses (regulatory instruments, economic instruments, information and social instruments) (OECD 2002b). See www.oecd.org.

Integrated Environmental and Economic Accounting (SEEA) INDICATORS

The last version of SEEA has been undertaken under the joint responsibility of the United Nations, the European Commission, the International Monetary Fund, the

OECD, and the World Bank (World Bank et al. 2003). Much of the work was done by the London Group on Environmental and Natural Resource Accounting, through a review process that started in 1998. SEEA 2003 is an accounting framework that comprises four categories of accounts with relevant indicators: accounts of material and energy flows, accounts that are relevant to the good management of the environment (e.g., expenditures made by businesses, governments, and households to protect environment), accounts for environmental assets measured in physical and monetary terms (e.g., timber stock accounts showing opening and closing timber balances and the related changes over the course of an accounting period), and accounts that consider how the existing System of National Accounts might be adjusted to account for the impact of the economy on the environment. See unstats.un.org/unsd/env Accounting/seea.htm

SET OF COMPETITIVENESS INDICATORS

In addition to the competitiveness indices, the Global Competitiveness Report published by the World Economic Forum comprises sets of indicators that include country performance indicators (e.g., gross domestic product [GDP] per capita and real growth in GDP), government and fiscal policy indicators (e.g., composition of public spending and government subsidies), institutional indicators (e.g., time with government bureaucracy and use of courts), infrastructure indicators (e.g., roads and cellular telephones), and human resource indicators (e.g., publicly funded schools and quality of health care) (World Economic Forum 2003).

SET OF UNITED NATIONS DEVELOPMENT PROGRAMME (UNDP) INDICATORS

Together with the highly aggregated Human Development Index, the regularly published *Human Development Report* (e.g., UNDP 2003) also features a set of predominantly social indicators, such as population with access to improved sanitation, undernourished people, public expenditures on education and health care, Internet users, and imports and exports of goods and services. All these indicators are arranged according to the level of human development as quantified by the Human Development Index. See http://hdr.undp.org/reports/.

STRUCTURAL INDICATORS

At the Lisbon Special European Council in March 2000, the European Union set itself a strategic goal for the next decade: "to become the most competitive and dynamic knowledge-based economy in the world capable of sustainable economic growth with more and better jobs and greater social cohesion." The European Commission was asked to draw up an annual progress report (the so-called *Synthesis Report*) based on an agreed set of structural indicators. These are by definition macro-level and performance-oriented indicators, focused on short-term development. The 2003 *Synthesis Report* presented forty-two indicators organized along five policy domains (employment, innovation, economic reform, social cohesion, and environment) and some general economic background indicators. See epp .eurostat.ec.europa.eu/portal/page?_pageid=1133,47800773,1133_47802558&_dad =portal&_schema=PORTAL.

THE BALATON GROUP INDICATORS

The list of indicators selected by the Balaton Group is different from most indicators lists. Of the thirty-three indicators, only one third are related to the UNCSD list. The indicator list uses the "Daily Triangle" as an integrating framework, creating a hierarchy from ultimate means (natural capital) to ultimate ends (well-being) and to relate nature health to human activity (technology, economy, politics, and ethics). The indicators are organized into four groups: indicators for natural capital, indicators for built capital, indicators for human and social capital, and indicators for ultimate ends (Meadows 1998). See www.nssd.net/pdf/Donella.pdf.

UNCSD THEME INDICATOR FRAMEWORK AND SPECIFIED INDICATORS

As part of the implementation of the Work Programme on Indicators of Sustainable Development adopted by the CSD at its Third Session in April 1995, a working list of 134 indicators and related methodology sheets were developed, improved, and tested at the national level by countries. Based on voluntary national testing and expert group consultation, a revised set of fifty-eight indicators and methodology sheets are now available for all countries to use (United Nations 2001).

WORLD BANK DEVELOPMENT INDICATORS

World Development Indicators (WDI) is the World Bank's premier annual compilation of data about development. WDI 2003 includes approximately 800 indicators in eighty-seven tables, organized in six sections: "World View," "People," "Environment," "Economy," "States and Markets," and "Global Links." The tables cover 152 economies and fourteen country groups with basic indicators for a further fifty-five economies. See web.worldbank.org/WBSITE/EXTERNAL/DATASTATISTICS/ 0,,contentMDK:20523710~hlPK:1365919~menuPK:64133159~pagePK:64133150 ~piPK:64133175~theSitePK:239419,00.html

WRI WORLD RESOURCES

WRI's *World Resources 2002–2004* comprises the latest core country data from more than 150 countries and new information on poverty, inequality, and food security. It includes indicators of potential risks to human health from environmental threats, social indicators of development, and basic economic indicators. *World Resources* is published every other year, and each publication has its own subfocus. The latest publication is subtitled *Decision for the Earth: Balance, Voice, Power* and focuses on good governance issues. See pubs.wri.org/pubs_description.cfm?PubID=3764.

SUSTAINABILITY DASHBOARD

The Joint Research Center in Ispra, Italy, developed the Dashboard of Sustainability as a free, noncommercial software application that allows one to present complex relationships between economic, social, and environmental issues in a highly communicative format aimed at decision makers and citizens interested in sustainable development. It is also particularly recommended to students, university lecturers, researchers, and indicator experts. For the WSSD, the Consultative Group on Sustainable Development Indicators (CGSDI) published the "From Rio to Jo'burg" Dashboard, with more than sixty indicators for more than 200 countries, an excellent tool for doing one's own 10-year assessment since the Rio Summit. See esl.jrc.it/dc/.

ECONOMY-WIDE MATERIAL FLOW INDICATORS

The Eurostat's material flow indicators are based on economy-wide material flow analysis, which quantifies physical exchange between the national economy, the environment, and foreign economies on the basis of total material mass flowing across the boundaries of the national economy. Material inputs into the economy consist primarily of extracted raw materials and produced biomass that has entered the economic system (e.g., biomass composed of harvested crops and wood). Material outputs consist primarily of emissions to air and water, landfilled wastes, and dissipative uses of materials (e.g., fertilizers, pesticides, and solvents). The most commonly used material flow indicators are usually divided into several groups: input, output, and consumption indicators (Eurostat 2001a).

Sets for National Level

HEADLINE INDICATORS OF SUSTAINABLE DEVELOPMENT FOR THE UK

The set developed by the Department of Environment, Food and Rural Affairs (DEFRA) comprises fifteen indicators that cover the three pillars of sustainable development: economic growth, social progress, and environmental protection. Assessments are made for each of the fifteen headline indicators on the basis of "Change Since 1970," "Change Since 1990," and "Change Since the Strategy." The "Change Since the Strategy" assessment highlights progress since the baseline assessment of indicators in "Quality of Life Counts" (DEFRA 1999), following the government's sustainable development strategy in 1999. The last assessment of headline indicators was published in the 2003 edition of *Achieving a Better Quality of Life* (DEFRA 2003).

NATIONAL SUSTAINABLE DEVELOPMENT INDICATORS FOR FINLAND

The main responsible body for national sustainable development indicators for Finland is the Ministry of the Environment. The indicators are arranged according to three dimensions of sustainable development: ecological, economic, and sociocultural. For these dimensions a set of issues was identified (e.g., climate change, acidification, natural resources, the workforce, lifestyles, and illnesses) and indicators for each issue developed. The whole set contains eighty-three indicators, with links between the indicators. There is no aggregation or weighting of indicators. See www.environment.fi/default.asp?node=12282&lan=en.

SUSTAINABLE DEVELOPMENT INDICATORS FOR SWEDEN

This set was developed with the participation of the Ministry of the Environment, Statistics Sweden, and the Swedish Environmental Protection Agency. The indicators are arranged according to four major themes: efficiency, contribution and equality, adaptability, and values and resources for coming generations. Within these themes, the indicators encompass economic, environmental, and social dimensions. Links between indicators are indicated, and a cross-reference matrix has been developed. No aggregation or weighting of indicators is performed. See www.scb.se/templates/Product_21323.asp.

INDICATORS OF SUSTAINABLE DEVELOPMENT FOR THE NETHERLANDS

This set was developed by the government and government-related institutions (planning agencies). The indicators are organized along two axes: sociocultural (financial–economic and ecological–environmental) and time and geography (here and now, here and later, elsewhere, now and later). It focuses on themes important for future generations (later) and on the influence of exports, imports, and financial flows on other (especially developing) countries (elsewhere), here and now, here and later, now and later. See international.vrom.nl/pagina.html?id=7388.

INDICATORS OF SUSTAINABLE DEVELOPMENT OF THE CZECH REPUBLIC

This set was developed during the testing phase of the UNPD project "Towards Sustainable Development in the Czech Republic: Building National Capacities" (Kovanda et al. 2002). It comprises sixty-three indicators divided into three categories (environmental, social, and economic). This set is used by the Czech Ministry of Environment as an underlying basis for the Internet portal on the indicators of sustainable development. See indikatory.env.cz/index.php?lang=en.

ENVIRONMENTAL PERFORMANCE INDICATORS FOR NEW ZEALAND

The Ministry for the Environment of New Zealand is developing a set of environmental indicators. The indicators in the Environmental Performance Indicators program measure and report the pressures being put on the environment, the current and historical state of the environment, and the effectiveness of any responses made to protect or repair the environment. It includes fourteen categories, such as air, climate, energy, and waste. See www.mfe.govt.nz/state/monitoring/epi/index.html.

CANADA'S NATIONAL ENVIRONMENTAL INDICATOR SERIES

This publication was prepared by the National Indicators and Reporting Office of Environment Canada. It is based on indicators presented in the National Environmental Indicator Series and is a follow-up to *Tracking Key Environmental Issues*, released in 2001. It is divided into four categories: ecological life support systems, human health and well-being, natural resources sustainability, and human activities. The Web portal also contains headline indicators related to the full set. See www.ec.gc.ca/soer-ree/English/default.cfm.

Sets for Regional and Local Level

EUROPEAN COMMON INDICATORS

The European Common Indicators is a monitoring initiative focused on sustainability at the local level. Ten common local sustainability indicators were identified through a bottom-up process; these are now being tested. Used in combination with other indicators and other evaluation methods, the European Common Indicators can contribute to a comprehensive local or regional monitoring strategy.

REGIONAL VERSIONS OF THE UK NATIONAL HEADLINE INDICATORS OF SUSTAINABLE DEVELOPMENT

These are published regularly in the publication *Regional Quality of Life Counts* (e.g., DEFRA 2002), which contains regional information for the nine English Government Office Regions, where available, for the fifteen headline issues. The 2002 issue is the third edition of *Regional Quality of Life Counts*, the first was published in December 2000, and the second was published in June 2002. In some cases it has not been possible to reproduce the national indicator at a regional level, so proxy information has been included. It has not been possible to produce regional information for the headline indicator on housing conditions. However, for the first time, some regional information on carbon dioxide emissions for the climate change headline indicator is available.

SET OF URBAN INDICATORS, HABITAT

Urban indicators are regularly collected in a sample of cities worldwide in order to report on progress in the twenty key areas of the Habitat Agenda at the city level. The Global Urban Indicators Database 2 contains policy-oriented indicators for more than 200 cities worldwide. Two different types of data are included in the minimum set. Key indicators, comprising indicators that are both important for policy and easy to collect, are either numbers, percentages, or ratios. Qualitative data or checklists, which assess areas that cannot easily be measured quantitatively, are audit questions generally accompanied by checkboxes for "yes" or "no" answers. See www.unhabitat.org/pmss/getPage.asp?page-bookView&book=1535.

Individual Indicators

CITY DEVELOPMENT INDEX, HABITAT

The City Development Index (CDI) is defined at the city level and could also be taken as a measure of average well-being and access to urban facilities by individuals. The high statistical significance and usefulness of the index indicate that it is actually measuring something real. CDI is a measure of depreciated total expenditure over time on human and physical urban services and infrastructure, and it is a proxy for the human and physical capital assets of the city. The CDI was developed as a prototype for Habitat II to rank cities according to their level of development. It is used in this report as a benchmark for comparative display of several of the key indicators from the United Nations Center for Human Settlements (UNCHS) (Habitat) Global Urban Indicators Database.

CORRUPTION PERCEPTION INDEX, GLOBAL CORRUPTION BAROMETER, AND BRIBE PAYERS INDEX

This group of indicators presents a joint initiative of the University of Passau and Transparency International. The Corruption Perception Index (CPI) is a poll of polls, reflecting the perceptions of businesspeople, academics, and risk analysts, both resident and nonresident. First launched in 1995, the 2003 CPI draws on seventeen surveys from thirteen independent institutions. A rolling survey of polls provided to Transparency International between 2001 and 2003, the CPI 2003 includes only countries that feature in at least three surveys. Whereas the CPI aims at assessing levels of corruption across countries, the Global Corruption Barometer (GCB) is concerned with attitudes that the general public forms about these levels of corruption. One question in the GCB asks respondents how significantly corruption affects their personal and family life. The resulting attitudes can vary widely and do not necessarily correlate with levels of corruption. Respondents in some countries may be capable of living with high levels of corruption, whereas for others even low levels of corruption provoke serious concerns. The CPI and GCB are complemented by Transparency International's Bribe Payers Index (BPI), which addresses the propensity of companies from top exporting countries to bribe in emerging markets. See www.transparency.org/tools/measurement.

DOW JONES SUSTAINABILITY INDEX

Launched in 1999, the Dow Jones Sustainability Indices (DJSIs) are the first global indices tracking the financial performance of the leading sustainability-driven companies worldwide. The Dow Jones STOXX sustainability indices consist of a pan-European and a Eurozone index: the Dow Jones STOXX sustainability index (DJSI STOXX) and the Dow Jones EURO STOXX sustainability index (DJSI EURO STOXX). For both of these indices a composite and a specialized index are available, with the latter excluding companies that generate revenue from alcohol, tobacco, gambling, armaments, or firearms. See www.sustainability-indexes.com/.

ECOLOGICAL FOOTPRINT

The Ecological Footprint (EF) was published for the first time in 1996 by Wackernagel and Rees. The EF of a specified population can be defined as the area of ecologically productive land needed to maintain its current consumption patterns and absorb its wastes with the prevailing technology. People consume resources from all over the world, so their footprint can be thought of as the sum of these areas, wherever on the planet they are located (Wackernagel and Rees 1996).

ENVIRONMENTAL SPACE

The concept of environmental space has been promoted by Friends of the Earth in Europe as a way of measuring sustainability and quantifying the inequity in environmental impact between the North and the South. In practical terms, environmental space is the total amount of energy, nonrenewable resources, agricultural land, and forests that each person in a given population can use without causing irreversible environmental damage or depriving future generations of the resources they will need. The total amount of environmental space therefore is limited by the carrying capacity of the earth. The concept of a fair share in environmental space is based on the premise that all people have a right to an equitable share in the earth's resources and therefore is used to highlight the discrepancy in consumption patterns between different countries, communities, and lifestyle choices (Spangenberg 1995).

ENVIRONMENTAL SUSTAINABILITY INDEX

The Environmental Sustainability Index (ESI) is a measure of overall progress toward environmental sustainability, developed for 142 countries. The ESI scores are based on a set of twenty core indicators, each of which combines two to eight variables, for a total of sixty-eight underlying variables. The ESI permits cross-national comparisons of environmental progress in a systematic and quantitative fashion. It represents a first step toward a more analytically driven approach to environmental decision making. See www.ciesin.org/indicators/ESI/.

FREEDOM COUNTRY SCORES

Since 1972, Freedom House has published an annual assessment of the state of freedom in all countries (and selected territories), now known as Freedom in the World. Individual countries are evaluated based on a checklist of questions on political rights and civil liberties that are derived in large measure from the Universal Declaration of Human Rights. Each country is assigned a rating for political rights and a rating for civil liberties based on a scale of 1 to 7, with 1 representing the highest degree of freedom and 7 the lowest level of freedom. The combined average of each country's political rights and civil liberties ratings determines an overall status of "Free," "Partly Free," or "Not Free." See www.freedomhouse.org/template.cfm?page=15&year=2006.

Genuine Progress Indicator (GPI)

This indicator uses a similar method as the Index of Sustainable Economic Welfare (ISEW) and makes twenty-seven adjustments to GDP. Its purpose is to provide a better indicator for well-being (Hamilton and Deniss 2000).

Genuine Savings

Genuine savings has been estimated and published in the World Bank's World Development Indicators. The rationale of the genuine savings approach is that persistently negative rates of genuine savings must lead to declining well-being. Genuine savings is calculated by subtracting natural resource depletion and pollution damages from net saving (net saving is gross saving minus the value of depreciation of produced assets). Resource depletion is measured as the total rents on resource extraction (bauxite, copper, gold, iron ore, lead, nickel, silver, tin, coal, crude oil, natural gas, and phosphate rock) and harvest (forests). So far, pollution damages are calculated only for carbon dioxide (Hamilton 2001).

Gross Domestic Product

GDP represents the total value of the goods and services produced by an economy over some unit of time (e.g., a month, a season, a year). The "domestic" part of the name comes from the fact that, unlike gross national product (GNP), it does not consider imports or exports in the calculation.

Growth Competitiveness Index and Business Competitiveness Index

The Growth Competitiveness Index (GCI) was developed by Jeffrey D. Sachs of Columbia University and John W. McArthur of the Earth Institute and was presented in *Global Competitiveness Report 2001–2002*. The Business Competitiveness Index (BCI) was developed by Michael Porter of Harvard University and was first introduced in *Global Competitiveness Report 2000*. The GCI uses both hard (publicly available) data and data from the World Economic Forum's Survey to estimate three component indices: the technology index, the public institutions index, and the macroeconomic environment index. The three components are then combined to calculate the overall GCI. To derive the overall BCI, two subindices are computed. The subindices measure the sophistication of company operations and strategy and the quality of the national business environment, respectively (World Economic Forum 2003).

Human Development Index

The Human Development Index is a summary composite index that measures a country's average achievements in three basic aspects of human development: longevity, knowledge, and a decent standard of living. Longevity is measured by life expectancy at birth; knowledge is measured by a combination of the adult literacy rate and the combined primary, secondary, and tertiary gross enrollment ratio; and standard of living is measured by GDP per capita. The index can take values between 0 and 1. Countries with an index over 0.800 are part of the High Human Development group. Between 0.500 and 0.800, countries are part of the Medium Human Development group, and below 0.500 they are part of the Low Human Development group (UNDP 2003).

INDEX OF ENVIRONMENTAL FRIENDLINESS

Statistics Finland developed the model for the Index of Environmental Friendliness. It is a general model for aggregating direct and indirect pressure data to problem indices and to an overall index. The scope of the model is designed to cover the key environmental problems related to the greenhouse effect, ozone depletion, acidification, eutrophication, ecotoxicological effects, resource depletion, photo-oxidation, biodiversity, radiation, and noise.

Because of shortcomings in either the aggregation methods or data availability, the practical testing of the model takes place with respect to the greenhouse effect, ozone depletion, acidification, eutrophication, ecotoxicological effects, and resource depletion. Also, the most important indirect emissions of electricity and heat consumption, waste, and wastewater treatment were attributed to the data evaluation in proportion to their purchases. See www.stat.fi/tk/yr/ye22_en.html.

INDEX OF SUSTAINABLE ECONOMIC WELFARE

The ISEW is proposed by Friends of the Earth. It is an indicator of economic welfare and represents an attempt to measure the underlying economic, social, and environmental factors that create real progress. The index has personal consumption spending as its base. A series of adjustments are made to consumption to arrive at the index value for a given year. The ISEW represents an important index of underlying long-term trends in real welfare. With careful use among a basket of other indicators, the ISEW informs policymakers and the general public of the factors that add to and subtract from welfare (Daly and Cobb 1989).

LIVING PLANET INDEX

The Living Planet Index (LPI) is an indicator promoted by the World Wildlife Fund (WWF). It tries to assess the overall state of the earth's natural ecosystems, which includes national and global data on human pressures on natural ecosystems arising from the consumption of natural resources and the effects of pollution. The 1999 LPI measures primarily abundance and is derived from an aggregate of three different indicators of the state of natural ecosystems: the area of the world's natural forest cover, populations of freshwater species around the world, and populations of marine species around the world.

Each of these individual component indices is set at 100 in 1970, and they are given an equal weighting. The overall LPI has declined by 30 percent between 1970 and 1995, implying that the world has lost 30 percent of its natural wealth in the space of a generation (WWF 2002).

Natural Capital Index

The Natural Capital Index (NCI) was developed as an assessment tool for the Convention on Biological Diversity (UNEP 1997). It defines natural capital as the product of ecosystem quantity and quality. Ecosystem quality is calculated as a function of ecosystem quality variables such as abundance of various species, ecosystem structures, and species richness and expressed as the ratio between the current and a baseline state. The index potentially ranges from 0 to 100. An NCI of 100 for agricultural areas means that the total area is converted into agricultural land with a quality of 100 percent, signifying the pre-industrial or extensive agricultural state. The components of the NCI, trends in habitat area (quantity) and in the abundance of a selected set of species (quality), are part of the set indicators for evaluating the 2010 target, agreed upon in the CBD/COP7 in Kuala Lumpur, 2004. A pressure-based NCI has been applied in the United Nations Environment Programme (UNEP) Global Environment Outlook.

Well-Being Index (Barometer of Sustainability)

This index was developed by the World Conservation Union (IUCN). It combines thirty-six indicators of health, population, wealth, education, communication, freedom, peace, crime, and equity into a Human Well-Being Index and fifty-one indicators of land health, protected areas, water quality, water supply, global atmosphere, air quality, species diversity, energy use, and resource pressures into an Ecosystem Well-Being Index. The two indices are then combined into a Well-Being/Stress Index that measures how much human well-being each country obtains for the amount of stress it places on the environment (Prescott-Allen 2001).

Methodology Overview for Indicators

UN Division for Sustainable Development *Indicators of Sustainable Development: Guidelines and Methodologies*, UN, 2001.

This publication represents the outcome of a work program on indicators of sustainable development approved by the Commission on Sustainable Development at its Third Session in 1995. The successful completion of the work program is the result of an intensive effort of collaboration between governments, international organizations, academic institutions, nongovernmental organizations, and individual experts aimed at developing a set of indicators for sustainable development for use at the national level. The thematic framework, guidelines, methodology sheets, and indicators set out in this publication have benefited from this extensive network of cooperation and consensus building.

GUINOMET, I. *The Relations Between Indicators of Sustainable Development. An Overview of Selected Studies.* FIFTH EXPERT GROUP MEETING ON INDICATORS OF SUSTAINABLE DEVELOPMENT, NEW YORK, APRIL 7–8, 1999.

This paper includes a brief overview of twenty-five studies concerned in sustainable development indices methodology. The study is divided into four parts: studies on linkages, studies on aggregation, studies on geographical integration, and other work on linkages and aggregation of Indicators of Sustainable Development (ISD).

SMEETS, E., AND R. WETERINGS. *Environmental Indicators: Typology and Overview,* EEA, 10/22/1999.

This is an introduction to the EEA typology of indicators and the driving forces, pressure, state, impact, response (DPSIR) framework used by the EEA in its reporting activities. This report should help policymakers understand the meaning of the information in indicator reports. The paper should also help define common standards for future indicator reports from the EEA and its member states.

SAISANA, M., AND S. TARANTOLA. *State-of-the Art Report on Current Methodologies and Practices for Composite Indicator Development,* JRC, ISPRA, 2002, EUR 20408/EN (2002).

This report examines a number of methods with a view to clarifying how they relate to the development of composite indicators. Several methods are investigated, such as aggregation systems, multiple linear regression models, principal component analysis and factor analysis, Cronbach's alpha, neutralization of correlation effect, efficiency frontier, distance to targets, experts opinion (budget allocation), public opinion, and analytic hierarchy process. The report also examines twenty-four published studies on this topic in a number of fields such as environment, economy, research, technology, and health, including practices from the Directorates General of the European Commission.

HASS, J. L., F. BRUNVOLL, AND H. HØIE. *Overview of Sustainable Development Indicators Used by National and International Agencies,* OECD STATISTIC, 2002, JT00130884.

This paper presents a general overview of recent work on sustainable development indicators in OECD countries. It provides an overview of ongoing work for developing agreed-upon indicators that measure progress across the three dimensions of sustainable development (economic, social, and environmental). The paper then takes a more specific look at the approaches to sustainable development indicators adopted by different countries and highlights the challenges of having one set of standard international indicators across the various countries.

OECD. *Aggregated Environmental Indices: Review of Aggregation Methodologies in Use*, OECD WORKING GROUP ON ENVIRONMENTAL INFORMATION AND OUTLOOKS, 2001, JT00125240.

This report responds to the increasing interest in and reservations about aggregated environmental indices that are provided to the public and to high-level decision makers and was prepared as part of the OECD program on environmental indicators, steered by the Working Group on Environmental Information and Outlooks. It complements the work carried out since 1990 that resulted in the adoption, at OECD level, of a common framework for environmental indicators.

OECD. *Indicators to Measure Decoupling of Environmental Pressure from Economic Growth*, GENERAL SECRETARIAT, OECD 2002, JT00126227.

This report was prepared by the OECD Secretariat in response to the request issued by the OECD Council at ministerial level (May 2001) that the OECD assist its member countries in realizing their sustainable development objectives. The council suggested that the OECD undertake the specific task of developing agreed indicators to measure progress across all three dimensions of sustainable development. This includes indicators that can measure the decoupling of economic growth from environmental degradation and that might be used in conjunction with other indicators in OECD's economic, social, and environmental peer review processes.

GLAUNER, C., AND T. WIEDMANN. *Comparative Analysis of Indicator Sets for Sustainable Development*, VDI TECHNOLOGY CENTER, FUTURE TECHNOLOGIES DIVISION, DÜSSELDORF, 2000.

This report is part of the European Science and Technology Observatory (ESTO) Study on National and Regional Programs and Strategies for Sustainable Development. It is a comparative analysis of seventeen indicator sets for sustainable development. Information about the indicator sets was delivered by the partner institutions within this ESTO project, and Verein Deustcher Ingenieure Technologiezentrum (VDI-TZ) performed its own investigations on several indicator systems. The table on page 6 of the report shows the institutions that made the main investigations on the respective indicator systems.

UNITED NATIONS DEPARTMENT OF ECONOMIC AND SOCIAL AFFAIRS (UNDESA). *Report on Aggregation of Indicators for Sustainable Development*, BACKGROUND PAPER NO. 2, CSD 9TH SESSION, NEW YORK, 2001.

The primary objective of this study by the UNDESA is to outline and recommend possible approaches and methods currently available to derive aggregated indicators of sustainable development, based on the final themes, subthemes, and a core set of indicators of the CSD framework. All initiatives analyzed in the report of Eurostat, "The Relationship Between Indicators of Sustainable Development," have been considered, although only those that were relevant to aggregation have been described. Other relevant initiatives have also been considered.

UNDESA. *Indicators of Sustainable Development: Framework and Methodologies,* BACKGROUND PAPER NO. 3, CSD 9TH SESSION, NEW YORK, 2001.

This report has been prepared as the culmination of the CSD Work Program on Indicators of Sustainable Development (1995–2000). It provides a detailed description of key sustainable development themes and subthemes and the CSD approach to the development of indicators of sustainable development for use in decision-making processes at the national level.

UNDESA. *Information and Institutions for Decision-Making: Report of the Secretary-General,* CSD 9TH SESSION, NEW YORK, 2001.

This report was prepared by the UNDESA Secretariat as task manager for chapters 8, 38, 39, and 40 of Agenda 21, in cooperation with the UNEP and the UNDP as task manager of chapter 37 of Agenda 21, with the contributions of other UN agencies and international organizations. The report is a brief factual overview intended to inform the CSD on key developments in the subject area.

JOINT ECE/EUROSTAT WORK SESSION ON METHODOLOGICAL ISSUES OF ENVIRONMENT STATISTICS *Report of the Work Session on Methodological Issues of Environment Statistics,* OTTAWA, CANADA, OCTOBER 1–4, 2001.

Selected Papers

WORKING PAPER NO. 2: *Eco-Efficiency Indicators in German Environmental Economic Accounting.*

WORKING PAPER NO. 10: *Eco-Efficiency Indicators as a Step to Indicators of Sustainable Development.*

The papers presented at this conference and subsequent discussions highlighted the fact that sustainable development means economic and social development as well as environmental issues and that development work must involve economic, social, and environmental statisticians as well as policymakers. The importance of communicating the results and getting feedback from the audience was also emphasized.

FREUDENBERG, M. *Composite Indicators of Country Performance: A Critical Assessment,* DIRECTORATE FOR SCIENCE, TECHNOLOGY AND INDUSTRY, OECD 2003, JT00153477.

This paper reviews the steps in constructing composite indicators and their inherent weaknesses. A detailed statistical example is given in a case study. The paper also offers suggestions on how to improve the transparency and use of composite indicators for analytical and policy purposes.

JACKSON, L. E., J. C. KURTZ, AND W. S. FISHER, EDS. *Evaluation Guidelines for Ecological Indicators,* U.S. ENVIRONMENTAL PROTECTION AGENCY, OFFICE OF RESEARCH AND DEVELOPMENT, RESEARCH TRIANGLE PARK, NC, 2000, EPA/620/R-99/005.

This document presents fifteen technical guidelines for evaluating the suitability of an ecological indicator for a particular monitoring program. The guidelines are organized within four evaluation phases: conceptual relevance, feasibility of implementation, response variability, and interpretation and utility.

Metainformation on Indicators

COMPENDIUM OF SUSTAINABLE DEVELOPMENT INDICATOR INITIATIVES

The International Institute for Sustainable Development created a large database of indicator initiatives. The current version, which houses information on about 600 initiatives, shows in-depth information on each initiative, including the type of initiative, the nature of public involvement, geographic scope, complete contact information, and project goals. See www.iisd.org/measure/compendium/.

COMPOSITE INDICATORS: AN INFORMATION SERVER ON COMPOSITE INDICATORS

The Joint Research Center runs a site developed to present methodology, case studies, articles, books, software, workshops, and any news related to composite indicators in a concise way. Composite indicators are organized in five broad thematic categories: environment; society; economy; innovation, technology, and information; and globalization. Detailed descriptions of each indicator include the scope of the index, the normalization method applied to the indicators, the weighting method, whether correlation and sensitivity analysis during the construction of the composite indicator were considered, and discussions (papers, reports, workshops) related to the composite indicator in question. See farmweb.jrc.cec.eu.int/ci/.

Literature Cited

Daly, H. E., and J. B. Cobb. 1989. *For the common good: Redirecting the economy toward community, the environment, and a sustainable future.* Boston: Beacon.

DEFRA. 1999. *Quality of life counts.* London: DEFRA.

DEFRA. 2002. *Regional quality of life counts.* London: DEFRA.

DEFRA. 2003. *Achieving a better quality of life.* London: DEFRA.

EEA. 2002. *Paving the way for EU enlargement: Indicators of transport and environment integration (TERM).* Copenhagen: EEA.

Eurostat. 1999. *Towards environmental pressure indicators for the EU.* Luxembourg: Eurostat.

Eurostat. 2001a. *Economy-wide material flow accounts and derived indicators. A methodological guide.* Luxembourg: Eurostat.

Eurostat. 2001b. *Measuring progress towards a more sustainable Europe: Proposed sustainable development indicators.* Luxembourg: Eurostat.

Eurostat. 2001c. *Towards environmental pressure indicators for the EU,* 2nd ed. Luxembourg: Eurostat.

Hamilton, C., and R. Deniss. 2000. *Tracking well-being in Australia: The Genuine Progress Indicator 2000,* Discussion Paper no. 35. Canberra: The Australia Institute.

Hamilton, K. 2001. *Indicators of sustainable development: Genuine savings. Note for technical discussion on sustainable development indicators.* Paris: OECD.

Kovanda, J., M. Scasny, and T. Hák. 2002. Core set of sustainability indicators of the Czech Republic. Pp. 355–391 in *Education, information and indicators,* edited by B. Moldan, T. Hák, and K. Kolářová. Volume IV of the Final Compendium of the UNDP Project "Towards Sustainable Development of the Czech Republic: Building National Capacities." Prague: Charles University Environment Center.

Meadows, D. 1998. *Indicators and information systems for sustainable development.* A report to the Balaton Group. Hartland Four Corners, Vermont: The Sustainability Institute.

Murray, C. J. L., J. Lauer, A. Tandon, and J. Frenk. 2001. *Overall health system achievement for 191 countries,* EIP/GPE, Geneva: World Health Organization.

OECD. 2001. *OECD environmental indicators. Towards sustainable development.* Paris: OECD.

OECD. 2002a. *Indicators to measure decoupling of environmental pressures from economic growth.* SG/SD(2002)1. Paris: OECD.

OECD. 2002b. *Towards sustainable household consumption? Trends and policies in OECD countries.* Paris: OECD.

Prescott-Allen, R. 2001. *The well-being of nations. A country-by-country index of quality of life and the environment.* Washington, DC: Island Press.

Spangenberg, J. H. (ed.). 1995. *Towards sustainable Europe. A study from the Wuppertal Institute for Friends of the Earth, Europe.* London: Friends of the Earth Publications.

UNDP. 2003. *Human development report.* Oxford: Oxford University Press.

UNEP. 1997. *Recommendation for a core set of indicators of biological diversity. Convention of Biological Diversity.* Convention on Biological Diversity, Third Meeting, Montreal, September 1–5, Item 7.3 of the provisional agenda. UNEP/CBD/SBSTTA/3/9 and inf. 13, inf. 14.

UNEP. 2003a. *GEO: Global environment outlook 3.* London: Earthscan.

UNEP. 2003b. *Monitoring and indicators: Designing national-level monitoring programs and indicators.* Convention on Biological Diversity, Ninth Meeting, Montreal, November 10–14, Item 5.3 of the provisional agenda. UNEP/CBD/SBSTTA/9/10, July.

UNEP. 2003c. *Proposed biodiversity indicators relevant to the 2010 target.* Convention on Biological Diversity, Ninth Meeting, Montreal, November 10–14, Items 5.3 and 7.2 of the provisional agenda, UNEP/CBD/SBSTTA/9/INF/26.

United Nations. 2001. *Indicators of sustainable development: Guidelines and methodologies.* New York: United Nations.

Wackernagel, M., and W. Rees. 1996. *Our ecological footprint. Reducing human impact on the earth.* Gabriola Island, BC: New Society Publishers.

World Bank, United Nations, the European Commission, the International Monetary Fund, and the OECD. 2003. *Integrated environmental and economic accounting 2003.* Final draft circulated for information before official editing.

World Economic Forum. 2003. *Global competitiveness report 2003–2004.* Davos: World Economic Forum.

WWF. 2002. *Living planet report 2002.* Gland, Switzerland: WWF.

Contributors

Christof Amann
IFF Department of Social Ecology
Schottenfeldgasse 29
A-1070 Vienna, Austria

Alain Ayong Le Kama
Commissariat Général du Plan
Université de Grenoble
18 rue de Martignac
F-75018 Paris, France

Tom Bauler
Free University of Brussels
Université Libre de Bruxelles–IL
EAT
50 avenue Roosevelt
B-1050 Brussels, Belgium

Theo Beckers
Telos (Brabant Centre for
Sustainable Development)
Warandelaan 2, P.O. Box 90153
5000 LE Tilburg, The Netherlands

Reinette (Oonsie) Biggs
Natural Resources and Environment
Council for Scientific and Industrial
Research
P.O. Box 395
Pretoria 0001, South Africa

Peter Bosch
European Environment Agency
Kongens Nytorv 6
DK-1050 Copenhagen K, Denmark

Ben J. E. ten Brink
Netherlands Environmental
Assessment Agency (RIVM-MNP)
P.O. Box 1
3720 BA Bilthoven,
The Netherlands

Marion Cheatle
Officer-in-Charge
Division of Early Warning Assessment
P.O. Box 30552
Nairobi, 00100 Kenya

Slavoj Czesany
Czech Statistical Office
Na Padesatem 81
Prague 10, Czech Republic

John Dagevos
Telos (Brabant Centre for
   Sustainable Development)
Warandelaan 2, P.O. Box 90153
5000 LE Tilburg, The Netherlands

Arthur Lyon Dahl
International Environment House
11–13 Chemin des Anémones
CH-1219 Chatelaine, Geneva,
   Switzerland

Peter Daniels
Australian School of Environmental
   Studies
Urban Research Programme
Griffith University
Brisbane 4111, Australia

Volodymyr Demkine
Environmental Affairs Officer
Division of Early Warning and
   Assessment UNEP United
   Nations Environment Programme
P.O. Box 30552
Nairobi, 00100 Kenya

Ann Dom
European Environment Agency
Kongens Nytorv 6
DK-1050 Copenhagen K,
   Denmark

Ian Douglas
School of Geography
University of Manchester
Oxford Road
Manchester M13 9PL,
   United Kingdom

Nina Eisenmenger
IFF Department of Social Ecology
Schottenfeldgasse 29
A-1070 Vienna, Austria

Karl-Heinz Erb
IFF Department of Social Ecology
Schottenfeldgasse 29
A-1070 Vienna, Austria

Marina Fischer-Kowalski
IFF Department of Social Ecology
Schottenfeldgasse 29
A-1070 Vienna, Austria

Peder Gabrielsen
European Environment Agency
Kongens Nytorv 6
DK-1050 Copenhagen K, Denmark

David Gee
European Environment Agency
Kongens Nytorv 6
DK-1050 Copenhagen K, Denmark

Gisbert Glaser
International Council for Science
   (ICSU)
51 boulevard de Montmorency
75016 Paris, France

Jasper Grosskurth
Pantopicon Bvba
Lange Winkelhaakstraat 26
B-2060 Antwerp, Belgium

Edgar E. Gutiérrez-Espeleta
Director, Development
  Observatory
University of Costa Rica
San José, Costa Rica 2060

Wim Haarmann
Telos (Brabant Centre for
Sustainable Development)
Warandelaan 2, P.O. Box 90153
5000 LE Tilburg,
The Netherlands

Helmut Haberl
IFF Department of Social Ecology
Schottenfeldgasse 29
A-1070 Vienna, Austria

Tomáš Hák
Charles University Environment
  Center
U Krize 8
158 00 Prague 5, Czech Republic

Stephen Hall
Department for Environment,
  Food and Rural Affairs
5/F15, Ashdown House
123, Victoria Street
London SW1E 6DE, United
  Kingdom

Mohd Nordin Hj. Hasan
ICSU Regional Office for Asia and
  the Pacific

902-4 Jalan Tun Ismail
50480 Kuala Lumpur,
Malaysia

Miroslav Havránek
Charles University Environment
  Center
U Krize 8
158 00 Prague 5, Czech Republic

Frans Hermans
Telos (Brabant Centre for
  Sustainable Development)
Warandelaan 2, P.O. Box 90153
5000 LE Tilburg, The Netherlands

Jochen Jesinghaus
Joint Research Centre
Institute for Systems, Informatics
  and Safety Via Fermi 1
I-21020 Ispra (Varese), Italy

Sylvia Karlsson
Finland Futures Research Centre
Turku School of Economics and
  Business Administration
Hämeenkatu 7
FIN-33100 Tampere, Finland

Luuk Knippenberg
Telos (Brabant Centre for
  Sustainable Development)
Warandelaan 2, P.O. Box 90153
5000 LE Tilburg, The Netherlands

Jan Kovanda
Charles University Environment
  Center
U Krize 8

158 00 Prague 5,
Czech Republic

Fridolin Krausmann
IFF Department of Social Ecology
Schottenfeldgasse 29
A-1070 Vienna, Austria

Petra Kusková
Charles University Environment
   Center
U Krize 8
158 00 Prague 5, Czech Republic

Gregor Laumann
German Aerospace Centre (DLR)
Project Management Agency
Environmental Research and
   Technology
Heinrich-Kohen-Str 1
D-53227 Bonn, Germany

Erich Lippert
Czech Ministry of Environment
Vrsovicka 65
Prague 10, Czech Republic

Jock Martin
European Environment Agency
Kongens Nytorv 6
DK-1050 Copenhagen K,
   Denmark

Jacqueline McGlade
European Environment Agency
Kongens Nytorv 6
DK-1050 Copenhagen K, Denmark

Peter Mederly
RegioPlan
Mostná 13
94901 Nitra, Slovakia

Jan Melichar
Charles University Environment
   Center
U Krize 8
158 00 Prague 5, Czech Republic

Bedřich Moldan
Charles University Environment
   Center
U Krize 8
158 00 Prague 5, Czech Republic

Imre Overeem
Telos (Brabant Centre for
   Sustainable Development)
Warandelaan 2, P.O. Box 90153
5000 LE Tilburg, The Netherlands

Juergen Paulussen
CAS-Chinese Academy of Sciences
RCEES-Research Center for Eco-
   Environmental Sciences
18 Shuangging Road
Beijing 100085, China

Véronique Plocq-Fichelet
SCOPE
51 boulevard de Montmorency
F-75016 Paris, France

Christoph Plutzar
Vienna Institute for Nature
   Conservation and Analysis
   (VINCA)

Giessergasse 6/7
A-1090 Vienna, Austria

Robert Prescott-Allen
PADATA
627 Aquarius Road
Victoria, BC V9C 4G5, Canada

Teresa Ribeiro
European Environment Agency
Kongens Nytorv 6
DK-1050 Copenhagen K,
  Denmark

Louise Rickard
European Environment Agency
Kongens Nytorv 6
DK-1050 Copenhagen K,
  Denmark

Jan Rotmans
DRIFT/KSI
Postbus 1738
3000 DR Rotterdam, The
  Netherlands

Kenneth G. Ruffing
Coordinator of the African
Economic Outlook
OECD Development Centre
2 rue André Pascal
75775 Paris Cedex 16, France

Yasmin von Schirnding
Healthy Environments for Children
  Alliance, SDE-HECA
World Health Organization
20 Avenue Appia
CH-1211 Geneva 27, Switzerland

Robert J. Scholes
Natural Resources and Environment
Council for Scientific and Industrial
  Research
P.O. Box 395
Pretoria 0001, South Africa

Ashbindu Singh
Division of Early Warning and
  Assessment
UNEP United Nations
  Environment Programme
1707 H. Street N.W., Suite 300
Washington, DC 20006, USA

Joachim Spangenberg
Sustainable Europe Research
  Institute
Grosse Telegraphenstrasse 1
D-50676 Köln, Germany

David Stanners
European Environment Agency
Kongens Nytorv 6
DK-1050 Copenhagen K, Denmark

David Vařkáč
Agency for Nature Conservation
  and Landscape Protection of the
  Czech Republic
Kališnická 4–6
130 23 Prague 3, Žižkov, Czech
  Republic

Rusong Wang
CAS-Chinese Academy of Sciences
RCEES-Research Center for Eco-
  Environmental Sciences

18 Shuangging Road
P.O. Box 2871
Beijing 100085, China

Jean-Louis Weber
European Environment Agency
Kongens Nytorv 6
DK-1050 Copenhagen K, Denmark

Helga Weisz
IFF Department of Social Ecology
Schottenfeldgasse 29
A-1070 Vienna, Austria

Jaap van Woerden
International Environment House
11 Chemin des Anémones
1219 Chatelaine, Geneva,
  Switzerland

Edwin Zaccai
Université Libre de Bruxelles,
  Centre d'Etudes du
  Développement Durable
CP 130/02
50 avenue F. Roosevelt
B-1050 Brussels, Belgium

Tomasz Zylicz
Warsaw University
Department of Economics
44–50 Dluga Street
00-241 Warsaw, Poland

SCOPE Series List

SCOPE 1–59 are now out of print. Selected titles from this series can be downloaded free of charge from the SCOPE Web site (http://www.icsu-scope.org).

SCOPE 1: *Global Environment Monitoring*, 1971, 68 pp

SCOPE 2: *Man-made Lakes as Modified Ecosystems*, 1972, 76 pp

SCOPE 3: *Global Environmental Monitoring Systems (GEMS): Action Plan for Phase I*, 1973, 132 pp

SCOPE 4: *Environmental Sciences in Developing Countries*, 1974, 72 pp

SCOPE 5: *Environmental Impact Assessment: Principles and Procedures*, Second Edition, 1979, 208 pp

SCOPE 6: *Environmental Pollutants: Selected Analytical Methods*, 1975, 277 pp

SCOPE 7: *Nitrogen, Phosphorus and Sulphur: Global Cycles*, 1975, 129 pp

SCOPE 8: *Risk Assessment of Environmental Hazard*, 1978, 132 pp

SCOPE 9: *Simulation Modelling of Environmental Problems*, 1978, 128 pp

SCOPE 10: *Environmental Issues*, 1977, 242 pp

SCOPE 11: *Shelter Provision in Developing Countries*, 1978, 112 pp

SCOPE 12: *Principles of Ecotoxicology*, 1978, 372 pp

SCOPE 13: *The Global Carbon Cycle*, 1979, 491 pp

SCOPE 14: *Saharan Dust: Mobilization, Transport, Deposition*, 1979, 320 pp

SCOPE 15: *Environmental Risk Assessment*, 1980, 176 pp

SCOPE 16: *Carbon Cycle Modelling*, 1981, 404 pp

SCOPE 17: *Some Perspectives of the Major Biogeochemical Cycles*, 1981, 175 pp

SCOPE 18: *The Role of Fire in Northern Circumpolar Ecosystems*, 1983, 344 pp

SCOPE 19: *The Global Biogeochemical Sulphur Cycle*, 1983, 495 pp

SCOPE 20: *Methods for Assessing the Effects of Chemicals on Reproductive Functions, SGOMSEC 1*, 1983, 568 pp

SCOPE 21: *The Major Biogeochemical Cycles and their Interactions*, 1983, 554 pp

SCOPE 22: *Effects of Pollutants at the Ecosystem Level*, 1984, 460 pp

SCOPE 23: *The Role of Terrestrial Vegetation in the Global Carbon Cycle: Measurement by Remote Sensing*, 1984, 272 pp

SCOPE 48: *Sulphur Cycling on the Continents: Wetlands, Terrestrial Ecosystems and Associated Water Bodies,* 1992, 345 pp

SCOPE 49: *Methods to Assess Adverse Effects of Pesticides on Non-target Organisms, SGOMSEC 7,* 1992, 264 pp

SCOPE 50: *Radioecology after Chernobyl,* 1993, 367 pp

SCOPE 51: *Biogeochemistry of Small Catchments: a Tool for Environmental Research,* 1993, 432 pp

SCOPE 52: *Methods to Assess DNA Damage and Repair: Interspecies Comparisons, SGOMSEC 8,* 1994, 257 pp

SCOPE 53: *Methods to Assess the Effects of Chemicals on Ecosystems, SGOMSEC 10,* 1995, 440 pp

SCOPE 54: *Phosphorus in the Global Environment: Transfers, Cycles and Management,* 1995, 480 pp

SCOPE 55: *Functional Roles of Biodiversity: a Global Perspective,* 1996, 496 pp

SCOPE 56: *Global Change, Effects on Coniferous Forests and Grasslands,* 1996, 480 pp

SCOPE 57: *Particle Flux in the Ocean,* 1996, 396 pp

SCOPE 58: *Sustainability Indicators: a Report on the Project on Indicators of Sustainable Development,* 1997, 440 pp

SCOPE 59: *Nuclear Test Explosions: Environmental and Human Impacts,* 1999, 304 pp

SCOPE 60: *Resilience and the Behavior of Large-Scale Systems,* 2002, 287 pp

SCOPE 61: *Interactions of the Major Biogeochemical Cycles: Global Change and Human Impacts,* 2003, 384 pp

SCOPE 62: *The Global Carbon Cycle: Integrating Humans, Climate, and the Natural World,* 2004, 526 pp

SCOPE 63: *Alien Invasive Species: A New Synthesis,* 2004, 352 pp.

SCOPE 64: *Sustaining Biodiversity and Ecosystem Services in Soils and Sediments,* 2003, 308 pp

SCOPE 65: *Agriculture and the Nitrogen Cycle,* 2004, 320 pp

SCOPE 66: *The Silicon Cycle: Human Perturbations and Impacts on Aquatic Systems,* 2006, 296 pp

SCOPE Executive Committee 2005–2008

President
Prof. O. E. Sala (Argentina)

Vice-President
Prof. Wang Rusong (China-CAST)

Past-President
Dr. J. M. Melillo (USA)

Treasurer
Prof. I. Douglas (UK)

Secretary-General
Prof. M. C. Scholes (South Africa)

Members
Prof. W. Ogana (Kenya-IGBP)
Prof. Annelies Pierrot-Bults (The Netherlands-IUBS)
Prof. V. P. Sharma (India)
Prof. H. Tiessen (Germany)
Prof. R. Victoria (Brazil)

Index